McGraw-Hill Ryerson

Mathematics 12

Preparing for College & Apprenticeship

Authors

Jeff Brosseau, B.A., B.Ed.
Greater Essex County District School Board

Patricia Byers, B.Sc.N., B.A.
Georgian College

Wayne Erdman, B.Math., B.Ed.
Toronto District School Board

Brian McCudden, M.A, M.Ed., Ph.D.
Toronto, Ontario

Krysta Mehler, B.Sc., H.B.Comm, B.Ed.
Greater Essex County District School Board

Tess Miller, B.Sc., B.Ed.
Durham District School Board

Lillian Kovachis Perivolaris, B.A., B.Ed., M.A.
Toronto District School Board

Peter Saarimaki, A.T.C.L., B.Sc., M.Ed.
Toronto, Ontario

Assessment/Pedagogy Consultants

Brian McCudden
Peter Saarimaki

College Education Consultant

Patricia Byers

Occupations/Careers Consultants

Patricia Byers
Tess Miller

Technology Consultant

Lillian Kovachis Perivolaris

Special Contributers

Roland W. Meisel, Port Colborne
Dan Phillips, Georgian College

McGraw-Hill Ryerson

Toronto   Montréal   Boston   Burr Ridge, IL   Dubuque, IA   Madison, WI   New York
San Francisco   St. Louis   Bangkok   Bogotá   Caracas   Kuala Lumpur   Lisbon   London
Madrid   Mexico City   Milan   New Delhi   Santiago   Seoul   Singapore   Sydney   Taipei

McGraw-Hill
Ryerson Limited

A Subsidiary of The **McGraw·Hill** Companies

McGraw-Hill Ryerson
Mathematics: Preparing for College & Apprenticeship 12

ISBN 0-07-086486-1/0-07-091712-4

http://www.mcgrawhill.ca

2 3 4 5 6 7 8 9 0 TRI 0 9 8 7 6 5 4 3 2

Printed and bound in Canada

Care has been taken to trace ownership of copyright material contained in this text. The publishers will gladly take any information that will enable them to rectify any reference or credit in subsequent printings.

CorelDRAW® and Corel® Quattro® Pro 8 are trademarks or registered trademarks of Corel Corporation or Corel Corporation Limited in Canada, the United States and/or other countries.

Microsoft® Excel are either registered trademarks or trademarks of Microsoft Corporation in the United States and/or other countries.

Fathom Dynamic Statistics™ Software and *The Geometer's Sketchpad*®, Key Curriculum Press, 1150 65th Street, Emeryville, CA 94608, 1-800-995-MATH.

AutoCAD LT® are either registered trademarks or trademarks of Autodesk, Inc., in the USA and/or other countries. All other brand names, product names, or trademarks belong to their respective holders.

National Library of Canada Cataloguing in Publication Data

Main entry under title:
　　　McGraw-Hill Ryerson mathematics : preparing for college and apprenticeship 12
Includes index.
ISBN 0-07-086486-1

　　1. Mathematics. I. Brosseau, Jeff II. Title: Mathematics : preparing for college and apprenticeship 12.
QA39.2.M227 2002　　510　　　　C2001-904136-5

PUBLISHER: Diane Wyman
DEVELOPMENTAL EDITORS: Maggie Cheverie, Tom Gamblin
SENIOR SUPERVISING EDITOR: Carol Altilia
COPY EDITOR: Debbie Wright
PERMISSIONS EDITORS: Maria DeCambra, Krista Alexander
EDITORIAL ASSISTANT: Erin Parton
JUNIOR EDITORS: Christopher Cappadocia, Cheryl Stallabrass
ASSISTANT PROJECT COORDINATORS: Melissa Nippard, Janie Reeson
PRODUCTION SUPERVISOR: Yolanda Pigden
PRODUCTION COORDINATOR: Paula Brown
COVER DESIGN: Greg Devitt
INTERIOR DESIGN: Tom Dart/First Folio Resource Group, Inc.
ART DIRECTION: Tom Dart/First Folio Resource Group, Inc.
ELECTRONIC PAGE MAKE-UP: Tom Dart, Greg Duhaney, Alana Lai, Claire Milne/First Folio Resource Group, Inc.
COVER IMAGE: Alan Marsh/Firstlight.ca; Don Farrall/PhotoDisc

COPIES OF THIS BOOK
MAY BE OBTAINED BY
CONTACTING:

McGraw-Hill Ryerson Ltd.

WEB SITE:
http://www.mcgrawhill.ca

E-MAIL:
orders@mcgrawhill.ca

TOLL-FREE FAX:
1-800-463-5885

TOLL-FREE CALL:
1-800-565-5758

OR BY MAILING YOUR ORDER TO:
McGraw-Hill Ryerson
Order Department
300 Water Street
Whitby, ON L1N 9B6

Please quote the ISBN and title when placing your order.

Student text ISBN:
0-07-086486-1

Student e-book ISBN:
0-07-090961-X

Acknowledgements

Reviewers of *McGraw-Hill Ryerson Mathematics: Preparing for College & Apprenticeship 12*

The authors and editors of McGraw-Hill Ryerson Mathematics: Preparing for College & Apprenticeship 12 wish to thank the reviewers listed below for their thoughtful comments and suggestions. Their input has been invaluable in ensuring that this text meets the needs of the students and teachers of Ontario.

Larry Ash
Durham Board of Education

Yolanda Baldasaro
Hamilton-Wentworth Catholic District School Board

Mary Card
Toronto District School Board

Mike Doyle
School District of Greater Niagara

Steve Etienne
District School Board of Niagara

Mary-Beth Fortune
Peel District School Board

Susan Harvey
Peel District School Board

Marilyn Hurrell
Thunder Bay

Jeff Irvine
Peel District School Board

Ann Kajander
Lakehead University and Lakehead Public Schools

Roger Knight
Toronto District School Board

Louis Lim
York Region District School Board

Frank Naccarato
Toronto

Terry Paradellis
Toronto District School Board

Susan Rozario
Peel District School Board

Martin Schubert
School District of Greater Niagara

Silvana F. Simone
Toronto District School Board

Al Smith
Kawartha Pine Ridge District School Board

le Sueur van der Riet
York Region District School Board

Accuracy Reviewer

Tom Gamblin, M.A. (Mathematics), Ph.D. (Combinatorics)
Toronto

$\mathcal{C}$ontents

Chapter 3

GEOMETRY IN DESIGN

Chapter 4

SINGLE-VARIABLE STATISTICS

Chapter 5

TWO-VARIABLE STATISTICS

Chapter 6

Interpreting and Analysing Data

Chapter 7

Exploring Quadratic Models

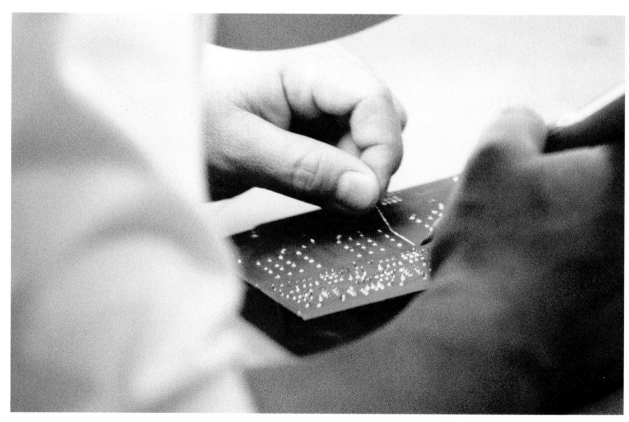

Chapter 8

INTERPRETING GRAPHICAL AND ALGEBRAIC MODELS

Introducing Your Textbook

Whether you are planning a career in manufacturing, health care, technology, or graphic design, *McGraw-Hill Ryerson Mathematics: Preparing for College & Apprenticeship 12* is written to help you develop the mathematical skills you will need to pursue your career choice. Three curriculum strands—applications of geometry, applications of statistics, and analysis of mathematical models—are presented in real-life, career-based contexts. Technologies, such as drawing or statistical software, or web-based resources, are integrated into these strands.

Chapter Organization

Each chapter begins with **Get Ready**, a review of mathematical skills you will need for the chapter. **Purple descriptors** refer to topics in **Appendix A: Prerequisite Skills,** where you can find examples and questions for practice in that particular skill.

The Get Ready section concludes with an outline of the **Chapter**—a specific, real-world problem—to which you will return as you work through the chapter.

Most sections begin with a **Discover**, where you will have an opportunity to explore the concepts of the section. The **Examples** develop the skills and processes you will need to understand the mathematical concepts. Most examples are set in a realistic context, similar to situations you might encounter in the workplace. Full, structured **Solutions** are given.

Key Concepts presents a summary of each section, and is followed by discussion-based questions in **Discuss the Concepts**. These will ensure you have grasped the key ideas of the section before proceeding with the exercises.

Practise questions allow you to practise the knowledge skill of the section. **Apply the Concepts** questions apply concepts to real-life problems that often have a career-based context. **Blue descriptors** identify questions related to particular careers or workplace activities. **Red descriptors** identify the achievement chart categories that the questions address. **Extend the Concepts** questions provide opportunities to enrich your learning.

Technology

Throughout the book, technology tools appropriate to the material are used: graphing calculators, spreadsheet software, dynamic statistics software, the statistical database E-STAT, and geometry and drawing packages. Step-by-step instructions are provided to help you learn new applications. **Appendix B: Technology** gives further help with specific tools.

The **Student e-book** that accompanies this textbook is designed to support your use of technology in the course. As well as the complete textbook and answers to questions with full graphics, you will find large data sets, animations that allow you to explore concepts interactively, source files for the technology-based example solutions given in the textbook, and much more. Watch for the circular e-book content symbol; you will find something in the student e-book to enhance that part of your learning.

The **Web Link** symbol appears whenever you are directed to *www.mcgrawhill.ca/links/MPCA12,* where you will find direct links to recommended web sites that relate to content in the textbook.

Σ-STAT E-STAT is a resource designed by Statistics Canada and made available to schools via the Internet. The database provides a large amount of Canadian census data. E-STAT also includes graphing and mapping tools. You can connect to E-STAT directly from our web site .

Assessment

Achievement Check questions assess your learning of the specific outcomes and achievement chart categories that the section addresses.

The **Chapter Problem Wrap-Up** pulls together your work on the real-world problem presented at the beginning of the chapter, and gives you an open-ended opportunity to develop a solution.

The **Practice Test** at the end of each chapter provides an opportunity to assess your learning of the chapter and is related to the achievement chart categories. The final **Achievement Check** question provides a more open-ended opportunity to demonstrate knowledge learned and to apply critical-thinking skills.

TRIGONOMETRY

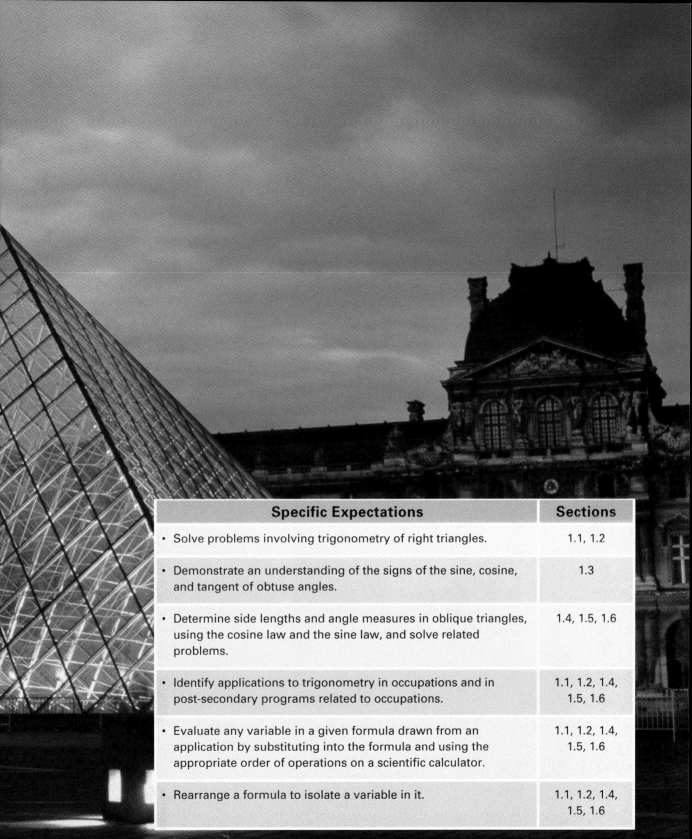

Specific Expectations	Sections
• Solve problems involving trigonometry of right triangles.	1.1, 1.2
• Demonstrate an understanding of the signs of the sine, cosine, and tangent of obtuse angles.	1.3
• Determine side lengths and angle measures in oblique triangles, using the cosine law and the sine law, and solve related problems.	1.4, 1.5, 1.6
• Identify applications to trigonometry in occupations and in post-secondary programs related to occupations.	1.1, 1.2, 1.4, 1.5, 1.6
• Evaluate any variable in a given formula drawn from an application by substituting into the formula and using the appropriate order of operations on a scientific calculator.	1.1, 1.2, 1.4, 1.5, 1.6
• Rearrange a formula to isolate a variable in it.	1.1, 1.2, 1.4, 1.5, 1.6

*G*et Ready

1. **Number Skills** Use a scientific calculator to evaluate each expression, to two decimal places.

   a) $2.3^2 + 5.1^2$

   b) $\dfrac{0.4}{1.6}$

   c) $\dfrac{0.2187}{3.7774}$

   d) $\sqrt{45}$

   e) $180 - (45 + 79)$

   f) $12 \times \dfrac{0.9823}{0.1234}$

   g) $\sqrt{(2.6^2 - 1.9^2)}$

2. **Angles of a triangle** Find the measure of the third angle in each triangle.

   a)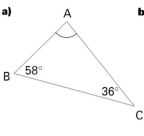

   b)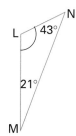

   c)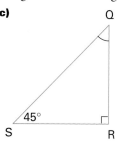

3. **Pythagorean theorem** Use the Pythagorean theorem to find the length of the unknown side in each triangle. Round your answers to the nearest tenth.

   a)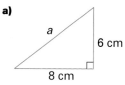

   b)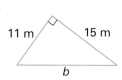

   c)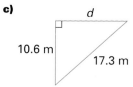

4. **Reducing ratios** Express each ratio in its lowest terms.

   a) $\dfrac{90}{30}$

   b) $14:21$

   c) $\dfrac{24}{18}$

   d) $49:14:21$

   e) $32:36:74$

   f) $30:12:42:150$

5. **Proportions** Solve each proportion.

   a) $\dfrac{x}{2} = \dfrac{5}{8}$

   b) $\dfrac{3}{y} = \dfrac{7}{4}$

   c) $\dfrac{d}{5.3} = \dfrac{2.1}{9.4}$

6. **Graphing Skills**

   a) Plot these points on a coordinate grid.

   A $(2, 5)$ B $(3, -2)$ C $(-4, 4)$ D $(-8, 10)$
   E $(-1, -2)$ F $(-9, 3)$ G $(13, -2)$ H $(3, 7)$

   b) Which points lie in the second quadrant?

For help with drawing software, see page 442 (*The Geometer's Sketchpad*®), page 446 (CorelDRAW®), or page 450 (*AutoCAD LT*®).

7. Drawing diagrams By hand, or using geometry or drawing software, draw an accurate diagram to represent each of the following situations. Label the diagrams.

a) A 4-m ladder is leaning against a building. The foot of the ladder rests 1.2 m from the base of the building.

b) A tree 5 m tall is growing vertically on a hill inclined at an angle of 15°.

c) An airplane is flying at an altitude of 1000 m. The angle of descent to the airport is 10°.

8. Trigonometry Determine each value, to four decimal places.

a) sin 30°   **b)** sin 23°   **c)** tan 15°

d) cos 60°   **e)** sin 9°   **f)** sin 84°

g) cos 84°   **h)** sin 45°   **i)** cos 45°

j) tan 45°

9. Trigonometry Find the measure of each angle, to the nearest tenth of a degree.

a) cos A = 0.4167   **b)** cos B = 0.8750   **c)** sin C = 0.5

d) sin D = 0.9127   **e)** cos E = 0.8660   **f)** tan F = 1

g) sin G = 0.8660   **h)** tan H = 2.5619   **i)** cos I = 0.5482

*C*HAPTER PROBLEM

Bridges are designed using many different structural forms. One design makes use of trusses, or diagonal beams, to add strength and stability.

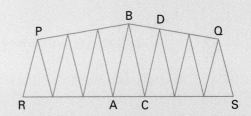

A pedestrian truss bridge is being designed to cross a creek that is 10 m wide. It will be built to these specifications:

- The bridge should have a straight bottom chord (RS) and two top chords (PB and BQ).
- The bridge should have diagonal support beams between the top and bottom chords.
- The support beams angled to the left should all be parallel, and so should the support beams angled to the right.
- The height at the centre should be 4 m.
- The length of the bridge must be between 10.5 m and 12.0 m.

Your assignment is to choose measures for ∠ABC and ∠DBC, and to determine, based on these, if the bridge will be an appropriate length. If not, you should describe what could be done to change the length of the bridge.

As you work through this chapter, look for the questions marked **Chapter Problem**. These will help you complete your assignment.

Using Trigonometry to Find Lengths

The standard approach for an airplane to an airport runway is at an angle of 3° at a height of 150 m. How far from the runway should the pilot begin the descent? You will revisit this problem in Example 3.

Trigonometry is a branch of mathematics that studies the relationships between sides and angles in triangles. The word "trigonometry" comes from the Greek words *trigon*, meaning triangle, and *metron*, meaning measure. Many trades and professions, including surveying, aviation, and ergonomic design, need a good working knowledge of trigonometry.

DISCOVER

Explore Trigonometric Ratios

In right triangles, the ratios of pairs of sides have special names: sine, cosine, and tangent, with short forms sin, cos, and tan. These ratios can be set up as follows.

1. Make a careful drawing of a right triangle. Position the corners so that all three sides have different lengths.
2. Identify the longest side. Label this side the hypotenuse. Where is this side in relation to the right angle?
3. Label one of the acute angles as ∠C. Draw a straight arrow from ∠C to the side opposite ∠C. Label this side **opposite**.
4. Draw a curved arrow from ∠C to the side adjacent (next) to ∠C. Label this side **adjacent**.
5. Measure each side length and ∠C, using a ruler and a protractor.
6. Calculate these values.
    a) sin C, cos C, tan C
    b) $\dfrac{\text{opposite}}{\text{adjacent}}, \dfrac{\text{adjacent}}{\text{hypotenuse}}, \dfrac{\text{opposite}}{\text{hypotenuse}}$
    c) Match up your values from parts a) and b). What do you notice?

You have just explored the trigonometric ratios of a right triangle. As you saw, there is a match between the ratios of side lengths and the values a calculator gives for the trigonometric functions sin, cos, and tan. This relationship is as follows:

$$\sin C = \frac{\text{opposite}}{\text{hypotenuse}}$$

$$\cos C = \frac{\text{adjacent}}{\text{hypotenuse}}$$

$$\tan C = \frac{\text{opposite}}{\text{adjacent}}$$

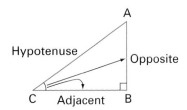

EXAMPLE 1

Using the Sine Ratio

Find the length of side *a*, to one decimal place.

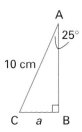

Solution

You are given the measure of one angle and the length of one side. Set up a trigonometric ratio to solve for the unknown side, *a*. The trigonometric ratio must involve ∠A, *a*, and 10.

Using your opposite, adjacent, hypotenuse labelling, you can see that side *a* is *opposite* ∠A, and that the *hypotenuse* is 10 cm. Therefore, use the sine ratio to solve for *a*.

$$\sin A = \frac{\text{opposite}}{\text{hypotenuse}}$$

$$\sin 25° = \frac{a}{10}$$

$$0.422\ 618\ 261\ 7 = \frac{a}{10}$$

$$a = 0.422\ 618\ 261\ 7 \times 10$$

$$a = 4.226\ 182\ 617$$

$$a \doteq 4.2$$

The length of side *a* is 4.2 cm, to one decimal place.

EXAMPLE 2

Using the Cosine Ratio: Roofing

A townhouse roof is to be re-shingled. To calculate the amount of shingles needed, the construction manager needs to know the length of the roof's sloping side. Find the length of side q.

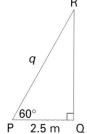

Solution

For $\angle P$, the adjacent side is 2.5 m and the hypotenuse is q. Use the cosine ratio.

$$\cos P = \frac{\text{adjacent}}{\text{hypotenuse}}$$

$$\cos 60° = \frac{2.5}{q}$$

$$0.5 = \frac{2.5}{q}$$

$$0.5q = 2.5$$

$$q = \frac{2.5}{0.5}$$

$$q = 5$$

The length of side q is 5 m.

EXAMPLE 3

Using the Tangent Ratio: Aviation

The standard approach for an airplane to an airport runway is at an angle of 3° at a height of 150 m. How far from the runway, to the nearest ten metres, should the pilot begin the descent?

Solution

Start by drawing and labelling a diagram. Include units of measure.

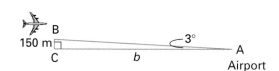

$\angle A$ is given. Let b represent the horizontal distance from the airport when the airplane begins its descent. Opposite = 150 m, adjacent = b. You can now complete the diagram.

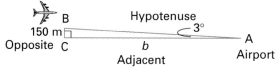

Use the tangent ratio to solve the problem, since it involves the opposite and adjacent sides.

$$\tan A = \frac{\text{opposite}}{\text{adjacent}}$$

$$\tan 3° = \frac{150}{b}$$

$$b = \frac{150}{\tan 3°}$$

$$b = 2862.170\ 503$$

$$b \doteq 2860$$

The descent should begin when the airplane is about 2860 m from the runway, to the nearest ten metres.

KEY CONCEPTS

- The trigonometric ratios of angle C in a right triangle are

$$\sin C = \frac{\text{opposite}}{\text{hypotenuse}}$$

$$\cos C = \frac{\text{adjacent}}{\text{hypotenuse}}$$

$$\tan C = \frac{\text{opposite}}{\text{adjacent}}$$

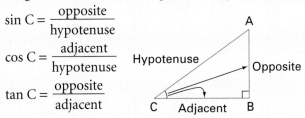

- Trigonometric ratios can be rearranged to find an unknown length, given another side length and an angle. The choice of ratio is based on whether the known and unknown lengths are opposite, adjacent, and/or hypotenuse.

Language Link

How can you remember the sides that are related by each trigonometric ratio? Try this memory device:

SOH CAH TOA

What do the letters represent?

DISCUSS THE CONCEPTS

1. With a calculator, you can see that sin 30° is equal to cos 60°.
   a) Explain why this is true.
   b) Is it true for other pairs of angles? Explain.

2. Describe a work or recreational situation where you could use trigonometry to find a length or distance. Explain what information you would need to make your calculation, and describe how you could gather this information. Also, provide a method for your calculation.

A **Round answers to one decimal place, where necessary.**

1. Copy the triangles, and label each side as "Opposite," "Adjacent," or "Hypotenuse" for the given angle. Then, decide whether to use the sine, cosine, or tangent ratio. Use this ratio to find the length of the labelled side.

   a) 12, w, 75°

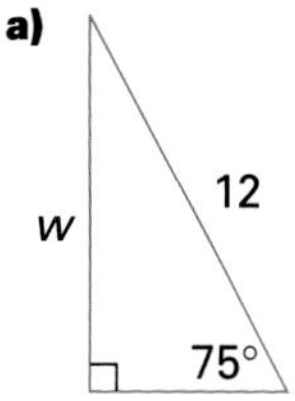

   b) 7, 46°, x

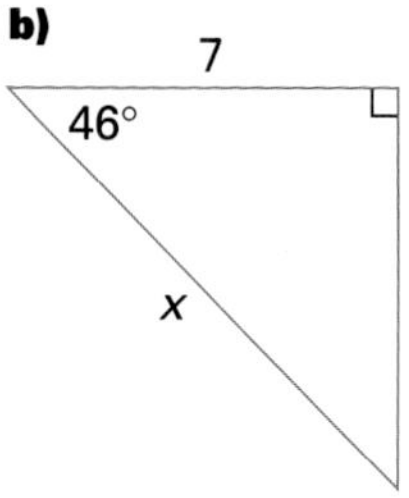

   c) 4, 30°, y

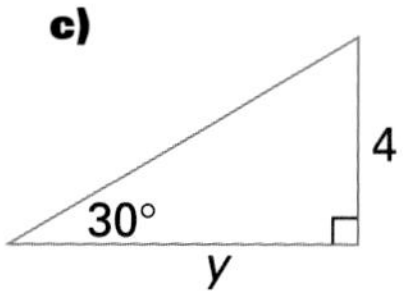

   d) z, 28°, 20

2. Calculate the length of side x in each triangle.

   a) x, 55°, 8

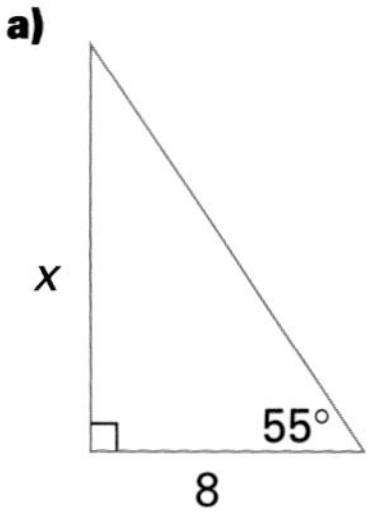

   b) 12, 45°, x

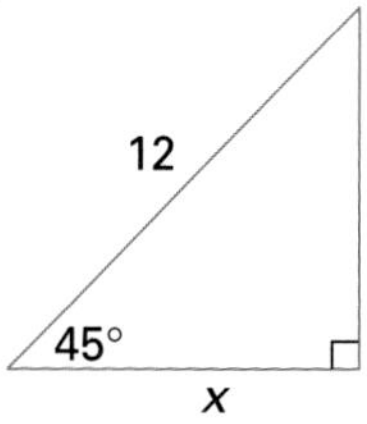

   c) 62°, 10, x

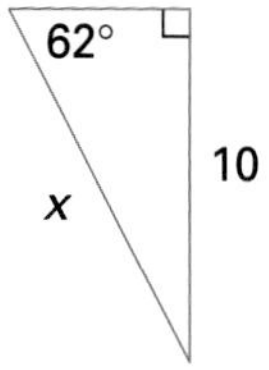

3. **Construction** A guy wire 5 m long is holding up a pole. The guy wire makes an angle of 72° with the ground. How far up the pole is the guy wire attached?

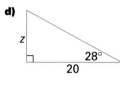

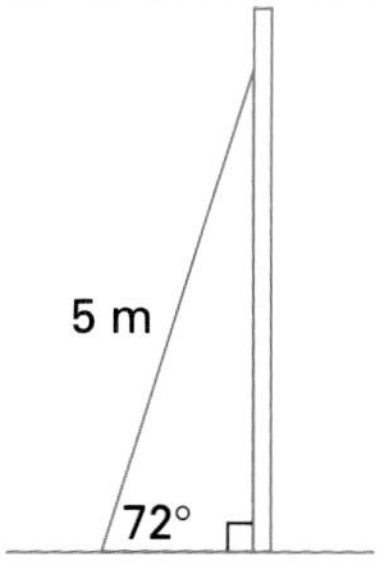

4. **Safety** A ladder is leaning against a wall. The foot of the ladder is 2.3 m from the base of the wall. The ladder makes an angle of 78° with the ground. How long is the ladder?

5. **Firefighting** From the top of a fire tower 50 m tall, the line of sight to a cottage is 18.7° below the horizontal. What is the horizontal distance from the fire tower to the cottage?

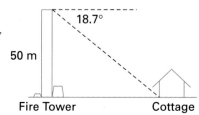

Fire Tower   Cottage

6. Communication One end of a 10-m wire is to be attached to a pole 5 m above the ground. The other end of the wire is to be secured to the ground. Describe how to use trigonometry to determine where the wire should be attached.

EXTEND THE CONCEPTS

C 7. Surveying A land surveyor has mapped out a property as shown in the diagram.

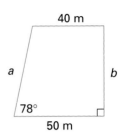

a) How long is side *a*?
b) How long is side *b*?

8. Astronomy To find the distance between the sun and a nearby planet or star, you can use the angles of sight (A and B in the diagram) from Earth to the star exactly six months apart. (You are sighting from opposite sides of Earth's orbit.) One half of the angle formed by these two angles of sight, C, is called the stellar parallax. The distance from Earth to the sun is called one astronomical unit, or 1 AU, and is known to be about 150 000 000 km.

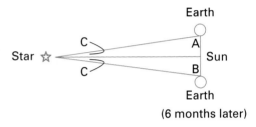

a) Given the parallax C for each star, find its distance from the sun, to the nearest thousand astronomical units.

    i) Alpha Centauri (closest star to the sun), parallax 0.000 21°
    ii) Barnard's Star, parallax 0.000 153 6°
    iii) Vega, parallax 0.000 034 19°
    iv) Polaris (North Star), parallax 0.000 002 102°

b) Why do you think this method is used only for nearby stars?

Science Link

One degree of angle measure can be divided into 60 min or 3600 s, just as an hour is divided. Astronomers use this system to describe small parallax angles. Another distance unit in astronomy is the *parsec*, short for parallax second. One parsec is the distance a star would be from the sun if its parallax were exactly $\left(\dfrac{1}{3600}\right)^{\circ}$ (one second of angle measure). One parsec is about 3.26 light years, a distance light would take 3.26 years to travel.

Using Trigonometry to Find Angles

In many situations that involve trigonometry, you may need to find angle measures based on known lengths or distances. This reverses the process you applied in section 1.1, but you can use the same trigonometric ratios.

EXAMPLE 1

Angle of View

A driver's handbook advises you to adjust the driver's side mirror as follows. The driver should be able to see an object as far as 7 m to the left, when the object is 10 m from the driving position. The angle of view that the adjusted mirror should provide is labelled ∠A in the diagram. Find the measure of ∠A, to one decimal place.

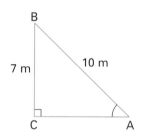

Solution

For ∠A,
opposite = 7
hypotenuse = 10
Therefore, use the sine ratio.

$$\sin A = \frac{\text{opposite}}{\text{hypotenuse}}$$

$$\sin A = \frac{7}{10}$$

$$\sin A = 0.7$$

$$\angle A = \sin^{-1}(0.7) \qquad \text{Read as "inverse sine of A."}$$

$$\angle A = 44.427\,004°$$

$$\angle A \doteq 44.4°$$

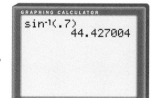

The driver's angle of view is 44.4°, to one decimal place.

EXAMPLE 2

Safety

The national fire safety code states that a ladder used by a firefighter must be placed so that the distance from the foot of the ladder to the wall is one quarter of the length of the ladder. What angle must a 12-m ladder make with the ground? Give your answer to the nearest degree.

Solution

Let $\angle X$ represent the angle the ladder must make with the ground.
You can see that for $\angle X$,

hypotenuse $= 12$

adjacent $= \dfrac{1}{4} \times 12$

adjacent $= 3$

Therefore, use the cosine ratio.

12 m

3 m

$$\cos X = \dfrac{\text{adjacent}}{\text{hypotenuse}}$$

$$\cos X = \dfrac{3}{12}$$

$$\cos X = 0.25$$

$$\angle X = \cos^{-1}(0.25) \qquad \text{Read as "inverse cosine of X."}$$

$$\angle X \doteq 75.52°$$

The ladder must make a 75.6° angle with the ground, to the nearest degree.

Often, surveyors use the **angle of elevation** or the **angle of depression** to describe positions of objects. The diagrams show how these angles are measured.

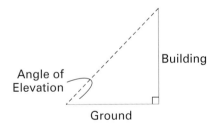

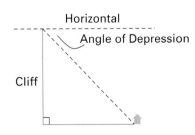

EXAMPLE 3

Surveying

A cabin is 125 m from the base of a 100-m cliff. What is the angle of depression from the top of the cliff to the cabin, to the nearest tenth of a degree?

Solution

Let y be the angle of depression.

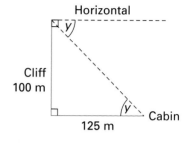

Using the parallel lines in this diagram, the angle made by the top of the cliff, the cabin, and the ground is also equal to y. For this angle,
opposite $= 100$ m
adjacent $= 125$ m
Use the tangent ratio.

$$\tan y = \frac{100}{125}$$

$$y = \tan^{-1}\left(\frac{100}{125}\right)$$

$$y \doteq 38.7°$$

The angle of depression is 38.7°, to the nearest tenth of a degree.

KEY CONCEPTS

- Angles in right triangles can be found using inverse trigonometric ratios. First, set up the trigonometric ratio for the angle, based on the problem or diagram. Then, substitute the known lengths. Finally, use the trigonometric inverse ($\sin^{-1}$, $\cos^{-1}$, or $\tan^{-1}$) to calculate the angle.

- Two angles commonly used in surveying are the angle of elevation and the angle of depression. These angles are always measured relative to the horizontal.

DISCUSS THE CONCEPTS

1. What information do you need to use the inverse sine, cosine, or tangent ratio to find an angle of elevation? Illustrate your answer with a diagram.

2. Describe the difference between an angle of elevation and an angle of depression. Suggest contexts in which each type of angle might occur.

A **Where necessary, round answers to one decimal place.**

1. Which trigonometric ratio would you use to find each of the marked angles? Solve for the marked angle.

a)

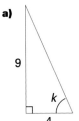

9

4

k

b)

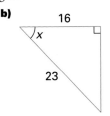

16

x

23

c)

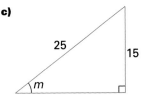

25

15

m

2. Find the length of the unknown side and the measures of all the angles.

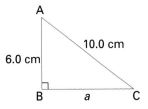

A

6.0 cm

10.0 cm

B   *a*   C

3. **a)** Draw △PQR, where ∠Q = 90°, PQ = 10 cm, and QR = 15 cm.
   b) Find the length of PR and the measures of ∠P and ∠R.

B 4. **Civil Engineering** A road rises 12 m over a horizontal distance of 150 m. What is its angle of elevation?

12 m

150 m

5. **Safety** A 4.5-m ladder rests against a wall. The foot of the ladder is 1.2 m from the wall. What angle does the ladder make with the ground?

Career Link

Public works such as roads and bridges are designed by civil engineers. Find out more about the civil engineering profession.

6. **Aviation** An airplane is beginning final descent to an airstrip from an altitude of 800 m. The airplane is a horizontal distance of 10 km from the airstrip. With at least what angle of depression should the pilot begin the airplane's descent?

7. **Meteorology** A weather balloon is tethered to a rope that is 600 m long.
   a) What is the angle of elevation of the balloon, when it is 400 m high?
   b) What do you have to assume about the rope to answer this question?

8. **Medicine** A paramedic is resetting a fractured bone. The fracture is 1.5 cm from the end of the bone and must be reset by moving the bone down 0.2 cm. Through what angle must the bone be reset?

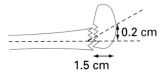

0.2 cm

1.5 cm

9. **Navigation** A lighthouse is 75 m tall. The deck of a trawler is 4 m above the water. The captain, who is almost 2 m tall, stands at the bow of the trawler. At what angle of elevation will the captain sight the top of the lighthouse, when the trawler is 150 m from the lighthouse?

10. **Wheelchair Access** A wheelchair ramp must have an angle of elevation of no more than 12°. A wooden ramp is available that is 8.0 m long and 1.5 m high. Will this ramp meet the requirements? Explain.

11. **Surveying** A surveyor measures the angle of elevation to the top of a building to be 57.8°. The surveyor's measuring instrument is 1.6 m above the ground, 18.5 m from the base of the building. How tall is the building?

12. A 1.3-m pendulum swings through a total horizontal distance of 1.0 m. Through what angle does it swing? Use a diagram to illustrate your answer.

ACHIEVEMENT CHECK   Knowledge/Understanding   Thinking/Inquiry/Problem Solving   Communication   Application

13. **Construction** A plan for a new roofing frame is shown. Find the total length of wood needed to construct it. Also find the angles at the corners.

12 m

5 m

40°

EXTEND THE CONCEPTS

C 14. **Navigation** The top of a 100-m lighthouse can be seen at an angle of elevation of 12° ahead from a ship. Fifteen minutes later, the angle of elevation is 19°.
a) How far did the ship travel in 15 min?
b) What was the speed of the ship?

15. **Medicine** A patient has scoliosis (a curvature of the spine). The diagram shows the measurements made by the radiologist, from the second vertebra to the seventh vertebra. Find the angle x through which the spine needs to be straightened at the fulcrum.

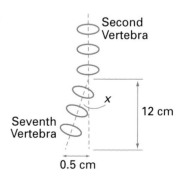

Second Vertebra

Seventh Vertebra

x

12 cm

0.5 cm

Trigonometric Ratios of Obtuse Angles

So far, all of the trigonometry you have studied has involved right triangles and trigonometric ratios of acute angles. In some problems, you will need to use trigonometric ratios of obtuse angles (between 90° and 180°).

DISCOVER

Beyond the Right Angle

1. Find each of the following trigonometric ratios, to four decimal places.

   a) sin 60°   sin 120°   sin 45°   sin 135°

   b) cos 60°   cos 120°   cos 45°   cos 135°

   c) tan 60°   tan 120°   tan 45°   tan 135°

2. **a)** Compare the *signs* of the sine values of the acute angles with those of the obtuse angles.

   b) Compare the signs of the cosine values of the acute and obtuse angles.

   c) Compare the signs of the tangent values of the acute and obtuse angles.

3. **a)** Which ratios in question 2 do not change sign whether the angle is acute or obtuse?

   b) Suggest a reason why the signs of some ratios change when the angle is obtuse, while other ratios do not change sign. State this reason in this form: "If the ratio involves the ... side, then the ... changes when ..."

   c) Verify the conjecture in part b), giving an example for each kind of ratio.

4. Comparing the values in each part of question 1, what else do you notice?

EXAMPLE 1

Robot Elbow Joint

An industrial robot arm is designed with an "elbow" joint that can open up to 180°. To program the robot correctly, an engineer uses the sine and cosine of the elbow angle to predict the position of the robot's "hand." The horizontal distance of the hand is calculated using the sine ratio, and the height above the elbow is found using the cosine ratio. The length of the arm from elbow to hand is 30 cm. What is the position of the hand when the elbow joint is opened to an angle of

a) 65°?   **b)** 125°?

What is effective about determining the hand position in this way?

Solution

a) Let a represent the horizontal distance of the hand from the elbow, and let b represent its height above the elbow. Using the sine ratio,

$$\sin 65° = \frac{a}{30}$$
$$a = 30 \sin 65°$$
$$a \doteq 27.2$$

Using the cosine ratio,

$$\cos 65° = \frac{b}{30}$$
$$b = 30 \cos 65°$$
$$b \doteq 12.7$$

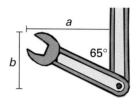

So, the hand is 27.2 cm in front of the elbow, at a height of 12.7 cm above the elbow.

b) In this case, the angle of 125° is obtuse. However, you can still let a represent the horizontal distance of the hand in front of the elbow, and let b represent its height. Using the sine ratio,

$$\sin 125° = \frac{a}{30}$$
$$a = 30 \sin 125°$$
$$a \doteq 24.6$$

Using the cosine ratio,

$$\cos 125° = \frac{b}{30}$$
$$b = 30 \cos 125°$$
$$b \doteq -17.2$$

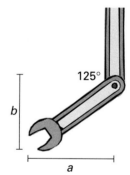

This time, the value of b is negative, which tells you that the robot's hand is *below* the level of the elbow. So, the hand is 24.6 cm in front of the elbow, and it is 17.2 cm below the elbow.

Using this method has two advantages. First, a sensor in the elbow joint can easily tell the software the elbow angle. Second, because the cosine ratio changes sign when the elbow angle is obtuse, a hand position that is below elbow height is automatically reported to the software as a "negative" height.

To extend trigonometry definitions to obtuse angles, you can relate the ratio definitions to angles between the x-axis, the origin, and a point on a coordinate grid. Imagine turning the diagrams in Example 1 on their sides, like this:

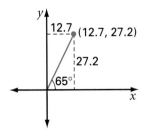 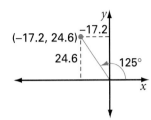

EXAMPLE 2

Obtuse Angle Ratios

Point P(−2, 3) has been plotted on a grid. A line segment joins (−2, 3) to the origin, forming ∠A.

a) Find the sin, cos, and tan ratios of ∠A.

b) Find the measure of ∠A, to the nearest tenth of a degree.

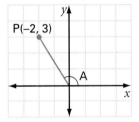

Solution

a) Draw a right triangle by dropping a perpendicular from (−2, 3) to the x-axis.

Now, for ∠A, the opposite side has length 3 and the adjacent side has length −2.

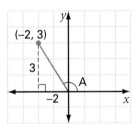

Use the Pythagorean theorem to find the length of the hypotenuse.

$$h^2 = x^2 + y^2$$
$$h^2 = (-2)^2 + 3^2$$
$$h^2 = 4 + 9$$
$$h^2 = 13$$
$$h = \sqrt{13}$$

The ratios are

$$\sin A = \frac{\text{opposite}}{\text{hypotenuse}}$$

$$\sin A = \frac{y}{h}$$

$$\sin A = \frac{3}{\sqrt{13}}$$

$$\sin A \doteq 0.832$$

$$\cos A = \frac{\text{adjacent}}{\text{hypotenuse}}$$

$$\cos A = \frac{x}{h}$$

$$\cos A = \frac{-2}{\sqrt{13}}$$

$$\cos A \doteq -0.555$$

$$\tan A = \frac{\text{opposite}}{\text{adjacent}}$$

$$\tan A = \frac{y}{x}$$

$$\tan A = \frac{3}{-2}$$

$$\tan A = -1.5$$

Notice that the length of the adjacent side is negative. This is because the value of x is negative for point P. So, the cosine and tangent ratios are negative for this obtuse angle, as you saw in the Discover section.

b) By entering each value into a calculator, you can see that you get conflicting results:

$$\sin^{-1}(0.832) = 56.3°$$
$$\cos^{-1}(-0.555) = 123.7°$$
$$\tan^{-1}(-1.500) = -56.3°$$

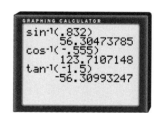

Why do these conflicting results occur? Scientific calculators are programmed to assume that the angle measure is between 0° and 180° when using the $\cos^{-1}$ function, but between −90° and +90° when applying the $\sin^{-1}$ and $\tan^{-1}$ functions. In this case, you know that $\angle A$ is obtuse, so the correct measure is 123.7°, to the nearest tenth of a degree.

The easiest way to be sure you have the correct answer is to determine which quadrant the angle is in. Recall that the four quadrants of the coordinate axes can be labelled as first, second, third, and fourth quadrant, as in the diagram.

As you saw in Example 2, a point in the second quadrant defines an angle between 90° and 180° — an obtuse angle.

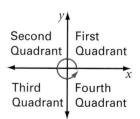

KEY CONCEPTS

- The signs of trigonometric ratios for acute and obtuse angles follow this pattern:

	sin A	cos A	tan A
Acute (0° < A < 90°)	+	+	+
Obtuse (90° < A < 180°)	+	−	−

- The sine, cosine, and tangent ratios, for acute or obtuse angles, can be related to a coordinate graph with

  x = adjacent y = opposite h = hypotenuse (where $h = \sqrt{x^2 + y^2}$)

  and

  $$\sin A = \frac{y}{h} \qquad \cos A = \frac{x}{h} \qquad \tan A = \frac{y}{x}$$

- For obtuse angles, y and h are positive, but x is negative.

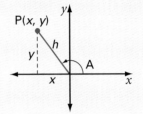

DISCUSS THE CONCEPTS

1. Explain why, for an obtuse angle, the sine ratio is positive, but the cosine ratio is negative.

2. Given that sin A = 0.4067 and cos A = −0.9135, describe how to determine the measure of ∠A.

PRACTISE

Ⓐ 1. Find each value, to four decimal places.

a) sin 150°   **b)** cos 95°

c) tan 170°   **d)** sin 100°

e) cos 163°   **f)** tan 125°

Ⓑ 2. Find the measure of each of these obtuse angles, to the nearest tenth of a degree.

a) sin A = 0.0135   **b)** cos B = −0.1352

c) tan C = −1.4693   **d)** sin D = 0.8923

e) cos E = −0.7619   **f)** tan F = −0.8102

3. Write the three trigonometric ratios for ∠A in each case. Then, use one of these trigonometric ratios to find the measure of ∠A.

a)

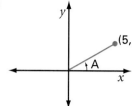

b)

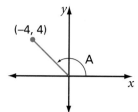

c)

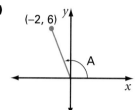

d)

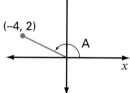

4. A 0.7 m-long windshield wiper is used to clean the rear windshield of a van. It begins in a horizontal position at the base of the windshield and rotates through 120°.
   a) How far does the tip of the wiper stop from the base of the windshield?
   b) What is the width of the coverage of the windshield wiper?

5. A dockside crane is mounted so that the 30-m arm can rotate to either side of the vertical position. Find
   a) the height of the top of the crane
   b) the horizontal distance of the top of the crane from the base mount
   for each of these angles of elevation:
   i) 85°   ii) 115°

C 6. **Thinking/Inquiry/Problem Solving** Extend your understanding of trigonometric ratios to angles greater than 180°.
   a) What are the signs of the sine, cosine, and tangent ratios for angles in the third quadrant? Explain your answer.
   b) What are the signs of the sine, cosine, and tangent ratios for angles in the fourth quadrant? Explain your answer.

The Sine Law

People often need to calculate distances between objects that are not arranged as right triangles. Suppose, in land surveying, a triangular plot of land is marked out by three stakes in the ground. A surveyor will measure one easily accessible distance and all the angles. Even if none of the angles are 90°, the other two distances can be calculated, using the sine law.

DISCOVER

Compare the Ratios

Using *The Geometer's Sketchpad®*, investigate trigonometric ratios in an acute-angled triangle.

1. In the **Display** menu, select **Preferences** and check that the chosen units are **cm**.
2. In the **Graph** menu, select **Snap To Grid**.
3. Draw a vertical line segment CD.
4. Draw a horizontal line segment EF that passes through point D.
5. Complete the construction of acute-angled triangle CFE.

Tech Link

Why does *The Geometer's Sketchpad®* label the first point that you draw as point C? Find out what this program uses points A and B for.

See Appendix B for more information on calculating ratios (page 444) and angles (page 445) with *The Geometer's Sketchpad*®.

6. Using the ratios of sides, determine the sine of ∠CEF. Use the **Calculator** feature in the **Measure** menu, if you wish.

7. Select side CF (the side opposite ∠CEF in triangle CFE) using the pointer. In the **Measure** menu, use the **Length** feature to find the length of CF.

8. Using the **Calculator** feature, calculate the ratio $\dfrac{\sin(\angle CEF)}{CF}$.

9. Repeat steps 6 to 8 for ∠CFE and side CE.

10. Compare the results of steps 8 and 9.

11. What conclusion can be made about the sine ratios and side lengths in a non-right triangle?

You now have evidence to suggest a powerful fact about sine ratios in a non-right triangle. You can explore this fact further by following the reasoning below. Use a standard labelling convention for triangles. Label the side opposite ∠A as side *a*. Likewise, label the side opposite ∠B as side *b*, and the side opposite ∠C as side *c*.

In acute or obtuse triangle ABC, draw AD, the height of the triangle. Label the diagram as shown.

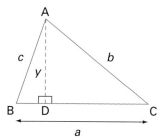

In △ABD,

$$\sin B = \frac{y}{c}$$
$$y = c \sin B \qquad ❶$$

In △ACD,

$$\sin C = \frac{y}{b}$$
$$y = b \sin C \qquad ❷$$

So, comparing ❶ and ❷,

$$c \sin B = b \sin C$$

Divide both sides by *b* and by *c*.

$$\frac{\sin B}{b} = \frac{\sin C}{c}$$

This equation can be extended to state the **sine law** for all three angles and sides.

$$\frac{\sin A}{a} = \frac{\sin B}{b} = \frac{\sin C}{c}$$

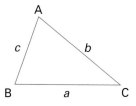

Taking the reciprocal of each ratio, the sine law can also be stated as

$$\frac{a}{\sin A} = \frac{b}{\sin B} = \frac{c}{\sin C}$$

The sine law is true for all triangles. It is a useful tool when working with angles and lengths in non-right triangles.

EXAMPLE 1

Using the Sine Law to Find Lengths

Determine the length of side b, to the nearest tenth of a centimetre.

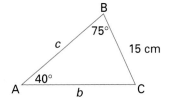

Solution

In this case, you know side a, $\angle A$, and $\angle B$. Use the sine law to find the length of side b.

$$\frac{b}{\sin B} = \frac{a}{\sin A}$$

$$\frac{b}{\sin 75°} = \frac{15}{\sin 40°}$$

$$\sin 75° \times \frac{b}{\sin 75°} = \frac{15}{\sin 40°} \times \sin 75°$$

$$b = \frac{15 \sin 75°}{\sin 40°}$$

$$b \doteq 22.5$$

The length of side b is 22.5 cm, to the nearest tenth.

Notice that in Example 1, you needed two angles and one opposite side length to determine the other opposite side length. In the next example, you have different information.

EXAMPLE 2

Using the Sine Law to Find Angles

Find the measure of $\angle Q$, to the nearest tenth of a degree.

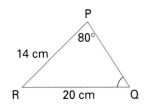

Solution

You know two side lengths, p and q, and one opposite angle, $\angle P$.

This is enough information to calculate the other opposite angle, $\angle Q$.

$$\frac{\sin Q}{q} = \frac{\sin P}{p}$$

$$\frac{\sin Q}{14} = \frac{\sin 80°}{20}$$

$$\sin Q = \frac{14 \sin 80°}{20}$$

$$\sin Q = 0.689\ 365\ 427\ 1$$

$$\angle Q \doteq 43.6°$$

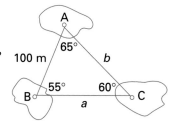

```
GRAPHING CALCULATOR
14*sin(80)/20
         .6893654271
sin⁻¹(Ans)
         43.5798979
```

The measure of $\angle Q$ is 43.6°, to the nearest tenth.

EXAMPLE 3

Urban Planning

The designer of a new city park wants to include three ponds, with a fountain in the middle of each pond. The angles formed by the fountains are shown, along with the distance between fountains A and B. Find the distances between the remaining pairs of fountains, to the nearest metre.

Solution

You are given the length of side c and all the angles. Use the sine law to calculate the other two side lengths.

$$\frac{b}{\sin B} = \frac{c}{\sin C}$$

$$\frac{b}{\sin 55°} = \frac{100}{\sin 60°}$$

$$b = \frac{100 \sin 55°}{\sin 60°}$$

$$b \doteq 95$$

$$\frac{a}{\sin A} = \frac{c}{\sin C}$$

$$\frac{a}{\sin 65°} = \frac{100}{\sin 60°}$$

$$a = \frac{100 \sin 65°}{\sin 60°}$$

$$a \doteq 105$$

Fountains A and C are 95 m apart, to the nearest metre. Fountains B and C are 105 m apart, to the nearest metre.

KEY CONCEPTS

- The sine law relates the angles and sides of any triangle.

$$\frac{\sin A}{a} = \frac{\sin B}{b} = \frac{\sin C}{c}$$

or

$$\frac{a}{\sin A} = \frac{b}{\sin B} = \frac{c}{\sin C}$$

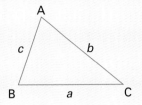

DISCUSS THE CONCEPTS

1. What information do you need to know to use the sine law for finding a side length?

2. What information do you need to know to use the sine law for finding an angle measure?

 1. Calculate the lengths of the indicated sides in each triangle, to one decimal place.

a)

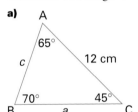

b)

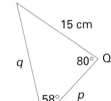

c)

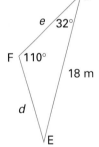

2. Determine the measures of all unknown angles in each triangle, to one decimal place.

a)

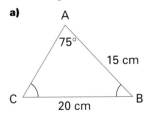

b)

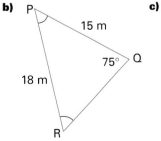

c)

3. Draw each triangle carefully to show the given information. Determine the indicated quantity, to the nearest tenth.
   a) In △ABC, ∠A = 75°, ∠B = 42°, and a = 12 cm. Find b.
   b) In △DEF, ∠D = 120°, ∠F = 24°, and f = 10 cm. Find e.
   c) In △GHJ, ∠G = 65°, g = 8 cm, and h = 7 cm. Find ∠H.
   d) In △KLM, ∠K = 64.3°, k = 11 cm, and l = 9.5 cm. Find ∠L.

B **4.** For each triangle, find all unknown angle measures and side lengths.

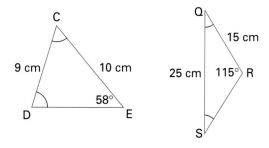

---

5. **Concerts** Three laser beams are installed at points J, K, and N, for a lightshow at a concert. The measurement specifications are shown in the diagram.
   a) How far apart are lasers J and N, to the nearest metre?
   b) How far apart are lasers K and N, to the nearest metre?

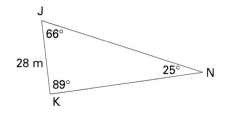

6. **Cartography** Cartographers use trigonometry to create maps. Calculate the angle measurements for the Bermuda Triangle, to the nearest degree.

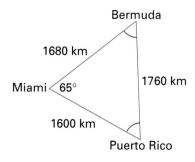

7. **Thinking/Inquiry/Problem Solving** A roof-top solar panel needs to be inclined at an angle between 25° and 30° to the horizontal. Are the measurements shown for the rafters and crossbeams appropriate? Explain.

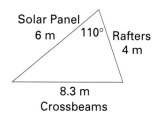

8. A pedestrian truss bridge is being designed to cross a creek that is 10 m wide. Refer to page 3 for the design specifications.

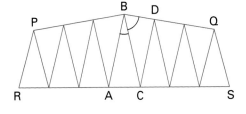

  a) Assign measures to ∠ABC and ∠DBC. Justify your choice of angle measures.

  b) Find the measure of ∠BDC. Which design specification do you need to use for this calculation, and why?

  c) Use an appropriate trigonometric ratio to find the length of diagonal beam BC. Use the specification for the height of the bridge structure.

  d) Calculate the length of diagonal beam CD.

9. A temporary support cable for a radio mast is 150 m long and has an angle of elevation of 30°. Several shorter support cables are attached to steady the mast permanently, with angles of elevation of 70°. How long are these shorter support cables?

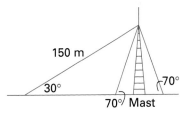

10. **Watersports Safety** A safety committee is reviewing plans for new waterslides at a swimming complex. Two slides are planned, starting from the same point above the pool: Depth Charge, which is a straight run of 50 m, and Helter Skelter, which is a curving slide of 75 m. If the angle of elevation of Depth Charge is 40°, what is the angle at which you would descend on Helter Skelter?

11. **Problem Solving** In a non-right triangle, $a = 3$ cm, $b = 4$ cm, and ∠A = 40°.

  a) Use the sine law to predict the measure of ∠B.

  b) Draw an accurate diagram of the triangle, with AB as the baseline and C as the apex. Add a point B′ on the line AB, between A and B, with the length B′C = 3 cm. Note that ∠AB′C is obtuse.

  c) Apply the sine law to side AC, side B′C, ∠A, and ∠AB′C. What do you notice?

  d) Explain why the data you were given could apply either to △ABC or to △AB′C. This situation is known as the **ambiguous case** of the sine law.

The Cosine Law

On the windiest day, the Skypod in the CN Tower is designed to sway safely through a distance of up to 1 m at a height of 447 m. Through what angle can the Skypod sway?

In a problem like this, you do not have the right information to use the sine law. Fortunately, another relationship, the cosine law, exists for non-right triangles.

DISCOVER

The Right Information?

In non-right triangle ABC, you are given two sides and one angle. Is this enough information to find the remaining side or angles?

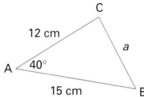

1. Write down the sine law for △ABC.
2. Substitute the known values and the unknowns from the triangle.
3. List all the equations you can form from the sine law with your substitutions.
4. Can you form an equation from the sine law that allows you to solve completely for the unknown quantity? Give reasons why or why not.

In this Discover, the angle you were given is a **contained angle**: ∠A is between sides b and c, which are the known sides of the triangle. However, you do not know the length of the side, a, which is *opposite* ∠A. This is why you could not use the sine law. You need to find an expression for side a in terms of ∠A. Begin by drawing △ABC, with height BD. Label the diagram as shown.

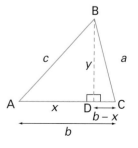

In $\triangle ABD$,

$$\cos A = \frac{\text{adjacent}}{\text{hypotenuse}}$$

$$\cos A = \frac{x}{c}$$

$$x = c \cos A \qquad ❶$$

Equation ❶ relates $\angle A$ to sides c and x, not side a. However, you can use the Pythagorean theorem in $\triangle BCD$ to get another equation in terms of x and a.

$$a^2 = y^2 + (b - x)^2 \qquad ❷$$

Equation ❷ also has y^2. Use the Pythagorean theorem in $\triangle ABD$ to eliminate it.

$$x^2 + y^2 = c^2$$
$$y^2 = c^2 - x^2 \qquad ❸$$

Substitute ❸ into ❷:

$$a^2 = c^2 - x^2 + (b - x)^2$$
$$a^2 = c^2 - x^2 + b^2 - 2bx + x^2$$
$$a^2 = c^2 + b^2 - 2bx$$

Substitute ❶ into this equation:

$$a^2 = c^2 + b^2 - 2b(c \cos A)$$
$$a^2 = b^2 + c^2 - 2bc \cos A$$

This is the **cosine law** for any angle in $\triangle ABC$. It can be written in three ways:

$$a^2 = b^2 + c^2 - 2bc \cos A$$
$$b^2 = a^2 + c^2 - 2ac \cos B$$
$$c^2 = a^2 + b^2 - 2ab \cos C$$

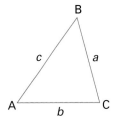

Suppose you know two sides of a triangle and the angle contained between them. You can use the cosine law to find the length of the third side.

EXAMPLE 1

Side Lengths Using the Cosine Law

Determine the length of side b, to the nearest tenth of a metre.

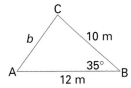

Solution

$$b^2 = a^2 + c^2 - 2ac \cos B$$
$$b^2 = 10^2 + 12^2 - 2(10)(12) \cos 35°$$
$$b^2 = 100 + 144 - 240(0.819\,152\,044\,3)$$
$$b^2 = 47.403\,509\,37$$
$$b = \sqrt{47.403\,509\,37}$$
$$b \doteq 6.9$$

The length of side b is 6.9 m.

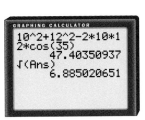

EXAMPLE 2

Packing Confectionery

An ice cream cone manufacturer makes cones with a bottom angle of 30°. The sides of the cones are 14 cm long. The manufacturer would like to pack the cones in cartons with a square cross section 7 cm wide. Will the cones fit in these cartons?

Solution

Let d represent the diameter of the top of a cone. You are given the other two side lengths, both 14 cm, and the contained angle, 30°. Use the cosine law.

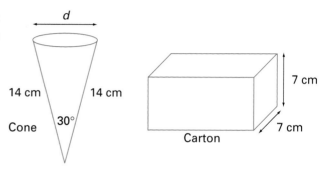

$$d^2 = 14^2 + 14^2 - 2(14)(14) \cos 30°$$
$$d^2 = 52.518\ 041\ 72$$
$$d = \sqrt{52.518\ 041\ 72}$$
$$d \doteq 7.2$$

The diameter of the top of each cone is about 7.2 cm. So, the cones will not fit in the cartons.

The other use of the cosine law is to find an angle, given the three side lengths.

EXAMPLE 3

Angle Measures Using the Cosine Law: CN Tower

On the windiest day, the Skypod in the CN Tower, which is 447 m above ground level, sways through a distance of up to 1 m. Through what angle does the Skypod sway? Give your answer to the nearest hundredth of a degree.

Solution

The diagram shows the information you are given.

Use the cosine law.

$$a^2 = b^2 + c^2 - 2bc \cos A$$
$$1^2 = 447^2 + 447^2 - 2(447)(447) \cos A$$
$$1 = 199\ 809 + 199\ 809 - 399\ 618 \cos A$$
$$399\ 618 \cos A = 399\ 617$$
$$\cos A = \frac{399\ 617}{399\ 618}$$
$$\angle A = \cos^{-1}(0.999\ 997\ 497\ 6)$$
$$\angle A \doteq 0.13°$$

The Skypod sways through an angle of 0.13°, to the nearest hundredth of a degree.

KEY CONCEPTS

• The cosine law relates the three sides and one angle of any triangle.

$$a^2 = b^2 + c^2 - 2bc \cos A$$
$$b^2 = a^2 + c^2 - 2ac \cos B$$
$$c^2 = a^2 + b^2 - 2ab \cos C$$

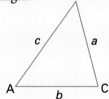

DISCUSS THE CONCEPTS

1. To use the cosine law to find the length of a side, what information do you need to have already?

2. To use the cosine law to find the measure of an angle, what information do you need to have already?

3. What happens to the cosine law when $\angle A = 90°$? Explain why this happens. Also, suggest a way to use this fact as a memory aid when working with the cosine law.

PRACTISE

A 1. Find the indicated quantity, to the nearest tenth.
   a) In $\triangle ABC$, $\angle A = 75°$, $b = 6$ cm, and $c = 7$ cm. Find a.
   b) In $\triangle DEF$, $\angle D = 120°$, $e = 12$ m, and $f = 14$ m. Find d.
   c) In $\triangle GHJ$, $g = 22$ cm, $h = 25$ cm, and $j = 32$ cm. Find $\angle J$.
   d) In $\triangle KLM$, $k = 13.2$ m, $l = 17.5$ m, and $m = 14.3$ m. Find $\angle M$.

2. Find the length of the unknown side in each triangle, to one decimal place.

a)

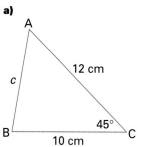

b)

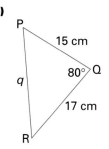

c)

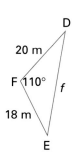

3. Find the measures of all angles in each triangle, to the nearest tenth of a degree.

a)

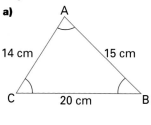

b)

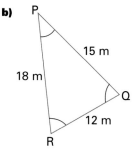

c)

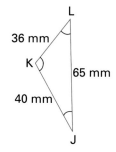

4. Find the measure of each angle and the length of each side in these triangles. Give your answers to one decimal place.

a)

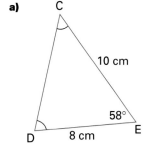

b)

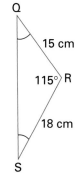

5. **Surveying** A surveyor has measured the distances and angle shown in this map of a lake. To the nearest metre, how long is the lake?

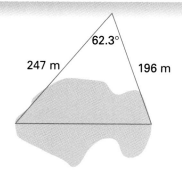

B **6.** **Navigation** A ship is sailing to a destination 250 nautical miles away. Because of an approaching storm, the ship veers off course at an angle of 18° and travels 65 nautical miles in that direction. How far is the ship from its destination after this detour, to the nearest nautical mile?

7. A pendulum is 80 cm long. The width of its swing is 12 cm. What is the total angle through which the pendulum swings? Give your answer to the nearest tenth of a degree.

8. **Peripheral Vision** You can measure the angle of your peripheral vision by extending your arms to the sides until you can barely see the fingers on each hand without moving your eyes.

a) When Jennifer tried this, her opposite fingertips were 80 cm from her eyes, and were 125 cm from each other. What is the angle of Jennifer's peripheral vision, to the nearest degree?

b) Work with a partner to determine your own angle of peripheral vision.

9. In section 1.4, question 8, you assigned measures to ∠ABC and ∠DBC in the design of a pedestrian truss bridge. Using these measures, you calculated the lengths of diagonal beams BC and CD. Determine the lengths of the partial lower chord, AC, and the partial upper chord, BD.

C **10.** Rearrange the cosine law to give an expression for cos A in terms of *a*, *b*, and *c*. In what sort of situation would this form of the cosine law be useful?

Problem Solving With Non-Right Triangles

An interior designer is creating a triangular tile of unusual shape. The designer wants to know the angle measurements so that he can plan designs using tiles of this shape. What are the angles of the designer's new tile?

This chapter has provided you with a toolbox of techniques to use in a range of problems involving triangles. These techniques can be applied most effectively, if you develop the ability to apply the right tool in any situation.

DISCOVER

Choose the Right Strategy

These diagrams represent three different problems. A strategy for each problem is listed, but not in the correct order.

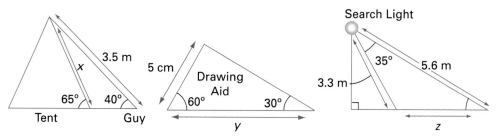

a) Notice that the triangle has a right angle. Use either the sine or cosine ratio.

b) Use the sine rule with two angles and their opposite sides. Solve the resulting equation for the unknown.

c) Use the cosine rule with one angle and two sides to find the remaining side.

1. Match each diagram to a strategy. Explain your reasoning.
2. Apply each strategy to the matched diagram.
3. Design a strategy, similar to those listed, to solve the problem illustrated by this diagram. You are to find both p and q.

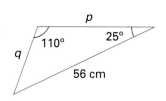

EXAMPLE 1

Firefighting

Two fire towers are located 10 km apart. Exact locations of fires are found by measuring, from each tower, the angle between the fire and the other fire tower. A fire has been located at an angle of 47° measured at tower A and at an angle of 62° measured at tower B. How far is the fire from each tower, to one decimal place?

Solution

Let A, B, and F be the positions of tower A, tower B, and the fire, respectively. Use "opposite sides and angles" labelling: label the distances from the fire to towers A and B as *b* and *a*, respectively.

Start by calculating the third angle:
$$62° + 47° + \angle F = 180°$$
$$109° + \angle F = 180°$$
$$\angle F = 180° - 109°$$
$$\angle F = 71°$$

Because you know one side and all the angles, you can use the sine law to solve for *a* and *b*.

$$\frac{a}{\sin A} = \frac{f}{\sin F}$$

$$\frac{a}{\sin 62°} = \frac{10}{\sin 71°} \qquad \text{Solve for } a.$$

$$a = \frac{10 \sin 62°}{\sin 71°}$$

$$a \doteq 9.3$$

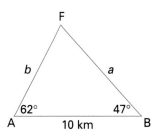

$$\frac{b}{\sin B} = \frac{f}{\sin F}$$

$$\frac{b}{\sin 47°} = \frac{10}{\sin 71°} \qquad \text{Solve for } b.$$

$$b = \frac{10 \sin 47°}{\sin 71°}$$

$$b \doteq 7.7$$

The fire is 9.3 km from tower B and 7.7 km from tower A.

EXAMPLE 2

Pendulum Swing

A 1.5-m pendulum swings through an angle of 6°. How wide is the swing, to the nearest centimetre?

Solution

Let a represent the width of the pendulum's swing. Because you know two sides and the contained angle, you can use the cosine law.

$$a^2 = b^2 + c^2 - 2bc \cos A$$
$$a^2 = 1.5^2 + 1.5^2 - 2(1.5)(1.5) \cos 6°$$
$$a^2 = 0.024\ 651\ 470\ 8$$
$$a \doteq 0.16$$

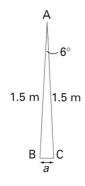

The pendulum's swing is 0.16 m, or 16 cm wide, to the nearest centimetre.

EXAMPLE 3

Design Arts

An interior designer is creating a triangular tile of unusual shape. The lengths of the tile's sides will be as shown. The designer wants to know the angle measurements so that he can plan designs using tiles of this shape. What are the angles of the designer's new tile?

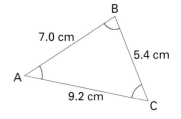

Solution

You know the lengths of all three sides of the triangle. So, you can use the cosine law to find an angle measure. Start with $\angle A$.

$$a^2 = b^2 + c^2 - 2bc \cos A$$
$$5.4^2 = 9.2^2 + 7.0^2 - 2(9.2)(7.0) \cos A$$
$$29.16 = 84.64 + 49 - 128.8 \cos A$$
$$128.8 \cos A = 84.64 + 49 - 29.16$$
$$128.8 \cos A = 104.48$$
$$\cos A = \frac{104.48}{128.8}$$
$$\angle A \doteq 35.8°$$

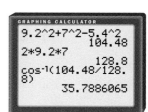

Now that three sides and one angle are known, you can use either the cosine law or the sine law to find the next angle. Because it is easier to work with, use the sine law to find the measure of $\angle B$.

$$\frac{\sin B}{b} = \frac{\sin A}{a}$$

$$\frac{\sin B}{9.2} = \frac{\sin 35.8°}{5.4}$$

$$\sin B = \frac{9.2 \sin 35.8°}{5.4}$$

$$\angle B = \sin^{-1}\left(\frac{9.2 \sin 35.8°}{5.4}\right)$$

$$\angle B \doteq 85.3°$$

GRAPHING CALCULATOR
sin⁻¹(9.2sin(35.8
)/5.4)
                85.27014598

Use the angle sum of a triangle to find the measure of $\angle C$.

$$\angle A + \angle B + \angle C = 180°$$
$$35.8° + 85.3° + \angle C = 180°$$
$$121.1° + \angle C = 180°$$
$$\angle C = 180° - 121.1°$$
$$\angle C \doteq 58.9°$$

The angles of the tile are approximately 36°, 85°, and 59°.

KEY CONCEPTS

- To solve a trigonometric problem, first draw a diagram, and label it. Decide what information is known, and what you are being asked to solve. Then, decide which law to use, based on the following guidelines:
  - The sine law can be used when
    - **i)** two sides and one of the angles opposite them are given—you can find the remaining opposite angle
    - **ii)** any two angles and one side are given—you can find any other side
  - The cosine law can be used when
    - **i)** two sides and the contained angle between them are given—you can find the third side
    - **ii)** all three sides are given—you can find any angle

- If one given angle is 90°, use trigonometric ratios directly, without the sine or cosine laws. In some situations, the Pythagorean theorem is also useful.

DISCUSS THE CONCEPTS

1. Which law, sine or cosine, would you use to calculate the length of side *x*? Explain.

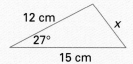

2. Draw and label a triangle in which you would use the sine law to find the measure of an angle. Explain your reasoning.

APPLY THE CONCEPTS

B **Draw or complete a diagram and decide which trigonometric law to use. Round all answers to match the accuracy of the data given.**

1. **Astronomy** The angle formed by the lines of sight from Earth to the stars Sirius and Alpha Centauri is 44°. Sirius is 8.8 light years from Earth. Alpha Centauri is 4.3 light years from Earth. What is the distance between Sirius and Alpha Centauri?

2. **Hockey** A hockey net is 1.83 m wide. Sandra is standing 5.00 m from one side of the net and 6.00 m from the other. Latiyah is standing 6.50 m from one side of the net and 8.00 m from the other. Who has the greater angle of view of the net?

3. **Aviation** Kwok flew his airplane 230 km due east and, after turning 20° northward, flew an additional 75 km before landing at an airstrip. How far is the airstrip from Kwok's starting point?

4. **Golf** The fairway of a "dogleg" golf hole has the dimensions shown. The direct distance to the hole is 359 yards. When drawing a diagram for this fairway, what angle should the artist use for the dogleg?

5. **Athletics** Using video technology, a hurdler measured the distances shown. The distance from the takeoff point to the top of the hurdle is 3.5 m. The distance from the top of the hurdle to the landing point is 2.4 m. The distance from the takeoff point to the landing point is 5.5 m. What is the least angle of inclination the hurdler's trailing leg must make at takeoff?

6. **Forestry** A tree is growing vertically on a hill. The hill slopes at an angle of 15° to the horizontal. When the angle of elevation of the sun is 55°, the tree makes a shadow that extends 2.3 m down the hill. How tall is the tree?

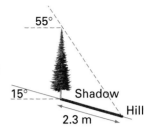

7. **Surveying** To measure the width of a canyon, stakes were driven into the ground at points A, B, and C, with the measurements shown. How wide is the canyon between points A and B?

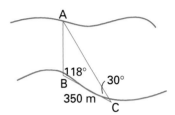

8. **Tracking Stations** A weather balloon has an angle of elevation of 62° from one tracking station and an angle of elevation of 47° from a second tracking station. The tracking stations lie along the same line of latitude, and are 3 km apart. How far is the balloon from each tracking station?

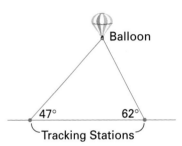

Earth Science Link

Latitude and longitude have been used for centuries to locate positions on Earth's surface. For example, the location of Ottawa is 45.42°N 75.67°W.

9. **Thinking/Problem Solving** Two radar stations that are 120 km apart located an aircraft on their screens. The first station reported the aircraft to the north-east, at an angle of 38.5° with the east-west line formed by the stations. The second station reported a north-westward angle of 56.1° with that same line. How far from each station was the aircraft?

10. **Thinking/Inquiry/Problem Solving** Develop a surveying problem in which you can use the cosine law to determine the distance between two points. Solve your problem.

11. **Astronomy** To an observer on Earth, the diameter of the sun makes an angle of approximately 0.53° across the sky. The distance from Earth to the sun is approximately 150 000 000 km. What is the approximate diameter of the sun?

12. **Construction** A fence 3 m tall is on the side of a hill and tends to lean over. The hill is inclined at an angle of 20° to the horizontal. A brace is to be installed to help stabilize the fence. It will be attached to the fence at a height of 2.5 m, and staked 5 m down the hill from the base of the fence. How long a brace is needed?

13. Archaeology The Mayan pyramid
El Castillo, in Chichen Itza, Mexico,
contains 91 steps on each of four sides.
The base of the pyramid is 55.5 m wide
and each staircase is 36.5 m long.

a) To the nearest degree, what is the
angle of ascent of the stairs?

b) What is the height of the pyramid?

36.5 m

55.5 m

14. Tool And Die A tool and die maker begins with
a rod 50.0 cm long and 0.5 cm wide. At one end,
0.7 cm is to be removed, and at the other end,
0.5 cm is to be removed, as shown. Determine
the angle measures, x and y, of the cuts.

0.7 cm
x
y
0.5 cm

Careers Link

Ergonomics is the
science of tool and
equipment design
for the comfort and
efficiency of the
user.

15. Equipment Design An office equipment designer
is designing a document holder for people who
work at computers. For best viewing, the holder
is to be inclined at an angle between 61° and 65°
to the horizontal. A 16-cm leg is to be attached
14 cm up from the base of the holder. How far
should the foot of the leg be from the base of the
holder? What angle should the leg make with the
holder when it unfolds? Give maximum and
minimum values for your answers.

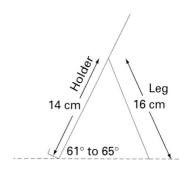

Holder

Leg
16 cm

14 cm

61° to 65°

ACHIEVEMENT CHECK Knowledge/Understanding Thinking/Inquiry/Problem Solving Communication Application

16. Sunfish Apparel has designed a new logo. The company
wishes to make a larger version of it in sheet metal to
place in front of their main store, as shown.
Your task is to find all the exterior
dimensions of the figure, in metres, and the
smallest area of rectangular sheet metal needed
to create the figure as a single piece. Outline the steps
to solve this problem, and then, complete the solution.

x
70 cm
65 cm 110°
135°
y
25°
z

C **17.** **Navigation** At departure, a ship's destination was 500 nautical miles away. The ship travelled 8° off course for 120 nautical miles. By what angle must the ship's course be changed to correct the error? (Hint: First, find the remaining distance the ship must travel, once the captain corrects its heading.)

See page 445 for more information on calculating angles with *The Geometer's Sketchpad®*.

18. **Problem Solving, Application** Snell's law is a formula that calculates the index of refraction when light passes across a flat boundary from one material to a different material. Snell's law can be stated mathematically as

$$n = \frac{\sin A}{\sin B}$$

where $\angle A$ is the angle of incidence and $\angle B$ is the angle of refraction. Use *The Geometer's Sketchpad®* or other software to illustrate some typical angles of incidence and reflection, and calculate the index of refraction.

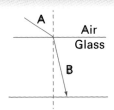

19. **Satellite Tracking** A satellite travelling in a circular orbit 1700 km above Earth is due to pass through point D, directly over a tracking station, at 10:00. The satellite takes 2 h to orbit Earth. Earth's radius is 6380 km.

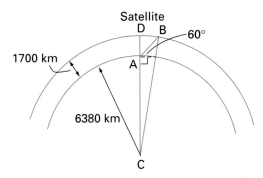

a) The tracking station's antenna is aimed at an angle of elevation of 60°, in the direction of forward travel of the satellite. At what time will the satellite pass through the antenna's beam at point B? (Hint: What fraction of 360° is $\angle C$? How long would it take the satellite to travel along segment DB?)

b) At 10:05, what is the distance between the satellite and the tracking station?

c) At 10:05, what angle of elevation should the tracking station's antenna be given, so as to intercept the satellite's path?

CAREER PROFILE

Careers: Getting Started

Preparing yourself for the work force is not an easy task.
Employers are looking for a range of skills. In addition to
a specific skill, employers look for:
• interpersonal skills
• computer skills
• work experience
• a proven track record in your employment history

Throughout this text, you will discover references to a
number of careers. Each chapter has a Career Profile page
that presents a sampling of careers from both the college
and apprenticeship areas. Some exercises are labelled as
relating to a career. Many chapters also include Career Links
which describe or suggest career options.

The following questions will start you thinking more about the career and
work force you will be entering soon.

1. How do your skills measure up? On a scale of 1 to 5, rank your skill
development in each of the following areas.
  a) Interpersonal skills: Can you get along with co-workers?
  b) Computer skills: Are you confident using e-mail, a word processor,
  and spreadsheets? Do you learn to use new software quickly and
  effectively? Can you type or do you use the "hunt and peck" method?
  c) Work experience: Do you have any work experience related to the
  career of your choice?

2. Discuss how you can improve your skills to better your chance at getting
the job of your choice.

3. The career training you choose may be offered in two ways:
  • courses offered at college
  • apprenticeship (a combination of courses and work term)
  a) Discuss the advantages and disadvantages of both options.

  b) Go to *www.mcgrawhill.ca/links/MPCA12* and follow the links to research
  the colleges and programs you are considering. Make a list, noting the
  advantages and any disadvantages of each of your choices.

CHAPTER PROBLEM WRAP-UP

A pedestrian truss bridge is being designed to cross a creek that is 10 m wide. It will be built to these specifications:

- The bridge should have a straight bottom chord (RS) and two top chords (PB and BQ).
- The bridge should have diagonal support beams between the top and bottom chords.
- The support beams angled to the left should all be parallel, and so should the support beams angled to the right.
- The height at the centre should be 4 m.
- The length of the bridge must be between 10.5 m and 12.0 m.

Your assignment is to choose measures for ∠ABC and ∠DBC, and to determine, based on these, if the bridge will be an appropriate length. If not, you should describe what could be done to change the length of the bridge.

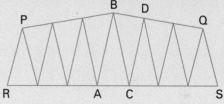

In section 1.4, question 8, you assigned measures to ∠ABC and ∠DBC in the design of a pedestrian truss bridge. Using these measures, you calculated the lengths of diagonal beams BC and CD. In section 1.5, question 9, you found the lengths of partial lower chord AC and partial upper chord BD.

1. Calculate the length of each remaining diagonal brace.

2. Find the lengths of the top half-chord PB and the bottom chord RS.

3. Does your bridge meet the design specifications? Would you change your initial choices for ∠ABC and ∠DBC, and if so, how? Explain your answers.

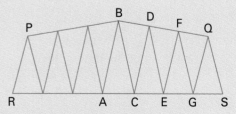

4. Prepare a report on your bridge solution. Include clear diagrams, appropriately labelled. Provide reasoning for your decisions and steps. You could build a model of your bridge to enhance your report.

Review

1.1 Using Trigonometry to Find Lengths, pages 4–9

1. Use the Pythagorean theorem to find the length of the third side in each triangle. Then, write the sin, cos, and tan ratios of the indicated angle.

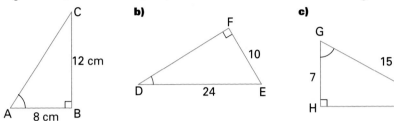

a)

b)

c)

2. **Botany** A botanist needs to determine the height of a tree. She measured the tree's shadow to be 2.3 m long when the angle of elevation of the sun was 78°. What is the height of the tree?

3. **Safety** For safety reasons, a fire ladder must make an angle of 75° with the ground. How high can a 9-m fire ladder safely reach?

1.2 Using Trigonometry to Find Angles, pages 10–14

4. State which trigonometric ratio should be used to find the measure of ∠A. Give reasons. (Do not solve.)

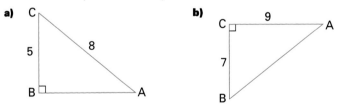

a)

b)

5. Find the measure of the marked angle, to one decimal place.

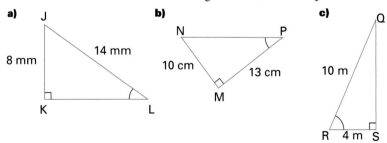

a)

b)

c)

6. **Aviation** An airplane is flying at an altitude of 300 m and is 1000 m by air from the beginning of a runway. At what angle should the pilot begin the descent?

7. **Civil Engineering** A road with an angle of inclination greater than 4.5° is considered steep for trucks. A road rises 65 m in a horizontal distance of 540 m. Is the road considered steep? Explain.

8. Architects and urban planners have to take many factors into account when designing or approving new buildings. For example, the shadow cast by a building may affect the growth of plants and trees in a neighbouring park. During the summer months, the angle of elevation of the sun is at least 50° for several hours in the middle of each day. A new building that is to be 15 m tall should not cast a shadow on a park flower bed directly south of it during these hours. If the flower bed is 13 m away, will the building be approved? Explain your answer.

9. Roberta is standing between two buildings in a townhouse development. The building on Roberta's left is 10 m away and the angle of elevation to its security spotlight, A, is 68°. The building on the right is 7 m away and the angle of elevation to its security spotlight, B, is 74°. Which spotlight is closer to Roberta, and by how much?

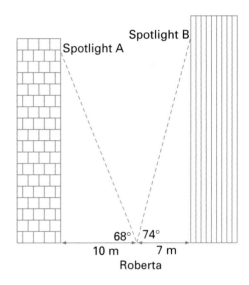

1.3 Trigonometric Ratios of Obtuse Angles, pages 15–20

10. Determine each value to four decimal places.

   a) sin 120°   **b)** cos 75°   **c)** tan 140°

   d) sin 112°   **e)** cos 157°   **f)** tan 165°

11. On the Hammer of Doom midway ride, passengers are strapped into seats at the end of a 30-m arm and hurled toward a solid wall, as shown. The wall lies 25 m behind the mount of the arm. What angle of rotation at the mount will bring the arm to within 1 m of the wall?

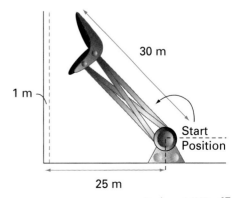

1.4 The Sine Law, pages 21–27

12. Explain why the sine law can be used to find the length of side b in this triangle.

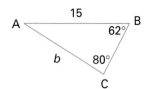

13. Find the measure of each angle and the length of each side.

a)

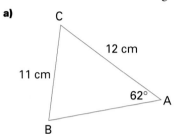

b)

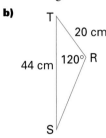

14. Tree Support A 6-m tree is leaning at an angle of 5° to the vertical. To prevent the tree from leaning any farther, a support pole needs to be placed 2 m from the top of the tree, at an angle of 60° with the ground. How long does the pole need to be, to the nearest metre?

1.5 The Cosine Law, pages 28–33

15. Explain how the cosine law can be used to find the measure of ∠P.

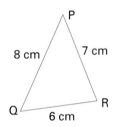

16. Find the length of each unknown side.

a)

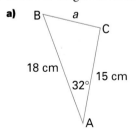

b)

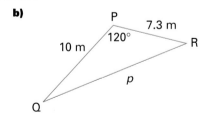

17. Find the measure of the angle indicated.

a)

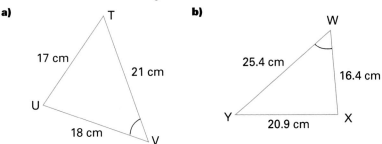

b)

1.6 Problem Solving With Non-Right Triangles, pages 34–41

Language Link

Surveyors often use a process called triangulation to divide plots of land into easily measurable triangular parts.

18. Excavation At an excavation site, points A, B, and C have been marked. The measure of ∠C is 132°. Points A and C are 4.2 m apart. Points B and C are 5.7 m apart. How far apart are points A and B?

19. Forestry A tree is growing vertically on a hillside that is inclined at a 15° angle to the horizontal. The tree casts a shadow up the hill that extends 7 m from the base of its trunk when the sun has an angle of elevation of 72°. How tall is the tree?

20. Mining A mine shaft is dug into the side of a mountain at an angle of depression of 30°. On the other side of the mountain, 6.4 km away, a second shaft is dug at an angle of depression of 45°. How long will each shaft be when they meet?

21. Security A security camera needs to be set so that its angle of view includes the area from a doorway to the edge of a parking lot. The doorway is 18 m from the camera. The edge of the parking lot is 25 m from the camera. The doorway is 30 m from the edge of the parking lot. What angle of view is needed for the camera?

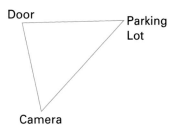

Practice Test

Category	Knowledge/Understanding	Thinking/Inquiry/Problem Solving	Communication	Application
Questions	1–10	6, 7, 9, 10	6, 10	10

1. Find the sine, cosine, and tangent ratios of ∠P.
Then, find the measure of ∠P.

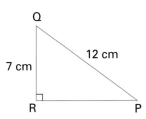

2. **Navigation** A ship is sailing toward a
communications tower that is 50 m tall. The
angle of elevation from the ship to the top of the
tower is 20°. How far is the ship from the tower?

3. A surveyor is standing at the top of a 143 m-high cliff. To isolate a point
that is 86 m from the base of the cliff, what angle of depression must the
surveyor measure?

4. Find the measure of each obtuse angle.
   a) sin A = 0.1657   **b)** cos B = −0.5629   **c)** tan C = −2.5482

5. **a)** Calculate the length of side *b*.   **b)** Calculate the measure of ∠F.

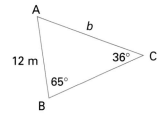   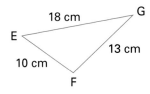

6. **a)** Describe what happens to the value of sin A as ∠A increases from
0° to 90°.
   b) Describe what happens to the value of cos A as ∠A increases from
90° to 180°.

7. For safety reasons, a ladder must be set against a wall at an angle of inclination between 70° and 75°. A 6-m ladder is leaning against a wall so that its foot is 2.8 m from the wall. Is it safe to use? If not, how far does it need to be moved, and in which direction (toward or away from the wall)?

8. An airplane is flying at an altitude of 1500 m. The angle of descent to the airport is 12°. What is the horizontal distance of the airplane from the airport?

9. Sunken treasure lies at an angle of depression of 68° down current from a salvage ship, and at a horizontal distance of 131 m from the ship. A deep-sea diver has been lowered to a point below the salvage ship at an angle of depression of 72° down current.
   a) Make a diagram of the situation.
   b) How deep is the sunken treasure?
   c) How far does the diver need to travel to reach the sunken treasure?

10. An underground water sprinkler system has a sprinkler head positioned at point A. Its angle of rotation must be set to ensure water reaches points B and C. Devise a plan to calculate the rotational angle of the sprinkler. Your plan should use the minimum number of measurements. Explain why you believe your plan is effective.

PROBLEM SOLVING WITH MEASUREMENT

Specific Expectations	Sections
• Solve problems related to the perimeter and area of plane figures, and the surface area and volume of prisms, pyramids, cylinders, spheres, and cones, including problems involving combinations of these objects.	2.1, 2.2, 2.4, 2.5
• Demonstrate accuracy and precision in working with metric measures.	2.1, 2.3, 2.4, 2.5
• Demonstrate an understanding of the use of the imperial system in a variety of applications.	2.1, 2.3, 2.4, 2.5
• Demonstrate a working knowledge of the measurement of length and area in the imperial system, in relation to applications.	2.1, 2.3, 2.4, 2.5
• Perform required conversions between the imperial system and the metric system, as necessary within projects and applications.	2.2, 2.3, 2.4, 2.5
• Use calculators effectively in solving problems involving measurement, and judge the reasonableness of the answers produced.	2.2, 2.3, 2.4, 2.5
• Rearrange a formula to isolate any variable in it.	2.3, 2.4, 2.5

$\mathcal{G}$et Ready

1. **Number skills** Calculate without the use of technology.
   a) $6.5 \div 100$ b) 420×1000 c) 0.05×10
   d) $5678 \div 100\,000$ e) $8 \div 100$ f) 0.003×1.05

2. **Metric measurement** Use the table of metric prefixes to complete these conversions.
   a) 560 mm to metres
   b) 34 000 mL to litres
   c) 78 000 000 mg to kilograms
   d) 5 km to centimetres
   e) 50 L to millilitres
   f) 0.000 05 kg to milligrams

Metric Prefixes			
Prefix	**Symbol**	**Value**	
tera-	T	10^{12}	1 000 000 000 000
giga-	G	10^{9}	1 000 000 000
mega-	M	10^{6}	1 000 000
kilo-	k	10^{3}	1 000
hecto-	h	10^{2}	100
deka-	da	10^{1}	10
deci-	d	10^{-1}	0.1
centi-	c	10^{-2}	0.01
milli-	m	10^{-3}	0.001
micro-	µ	10^{-6}	0.000 001
nano-	n	10^{-9}	0.000 000 001
pico-	p	10^{-12}	0.000 000 000 001

3. **Number skills** What fraction (in lowest terms) is shown in each diagram?
   a)
   b)
   c)
   d)
   e)
   f)

4. **Number skills** Order from least to greatest.
   a) $\dfrac{3}{8}, \dfrac{9}{16}, \dfrac{3}{4}, \dfrac{1}{2}, 1$ b) $1\dfrac{1}{2}, \dfrac{5}{16}, \dfrac{1}{4}, \dfrac{11}{32}$ c) $\dfrac{3}{4}, \dfrac{7}{8}, \dfrac{1}{4}, \dfrac{3}{8}, \dfrac{5}{8}, \dfrac{1}{2}$

5. **Number skills** Simplify.
   a) $\dfrac{1}{16} + \dfrac{3}{8}$ b) $2\dfrac{1}{2} - 1\dfrac{15}{16}$ c) $2\dfrac{1}{4} \div 3$
   d) $3\dfrac{5}{8} \times 4$ e) $2\dfrac{1}{2} + \dfrac{3}{8} + 3\dfrac{1}{2} + 24$ f) $2\left(4\dfrac{7}{16} + 2\dfrac{3}{32}\right)$

6. **Proportion** Solve for x.

   a) $\dfrac{4}{x} = \dfrac{1}{12}$ 　　 b) $\dfrac{1}{16} = \dfrac{x}{80}$ 　　 c) $\dfrac{2.5}{1} = \dfrac{12}{2x}$

7. **Number skills** Calculate. Where necessary, round to the nearest hundredth.

   a) $\pi(3.9)^2(5.4) \div 3$ 　　 b) $2(6 \times 3) + 2\left(4\dfrac{1}{2} \times 3\right)$

   c) $\dfrac{[(1.015 - 0.003) + (1.015 + 0.003)]}{2}$

8. **Pythagorean theorem** Find the length of the side labelled x.

   a)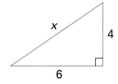

   b)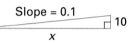

   c)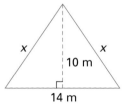

   d)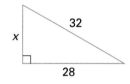

9. **Scientific notation** Convert from standard form to scientific notation.

   a) 0.000 65 　　 b) 12 000 000 　　 c) 0.003

   d) 34 000 　　 e) 0.000 000 1 　　 f) 0.004 07

*C*HAPTER PROBLEM

You have recently purchased a new house and, like many new home-buyers, you are anxious to make your house a home. In particular, you would like to improve the unfinished basement area. There are several home-improvement projects related to the basement that you would like to undertake. Therefore, you need to prioritize, according to need and affordability. To minimize costs, you are going to do-it-yourself.

Throughout this chapter, you will learn how to solve measurement problems in a variety of contexts. You will develop the skills necessary to plan a successful strategy for finishing the basement of your new home.

Measurement Formulas

Geometric Figure	Perimeter	Area
Rectangle	$P = 2l + 2w$ or $P = 2(l + w)$	$A = lw$
Parallelogram	$P = b + b + c + c$ or $P = 2b + 2c$	$A = bh$
Triangle	$P = a + b + c$	$A = \dfrac{bh}{2}$ or $A = \dfrac{1}{2}bh$
Trapezoid	$P = a + b + c + d$	$A = \dfrac{(a+b)h}{2}$ or $A = \dfrac{1}{2}(a + b)h$
Circle	$C = \pi d$ or $C = 2\pi r$	$A = \pi r^2$

Geometric Figure	Surface Area	Volume
Cylinder	$A_{top} = \pi r^2$ $A_{base} = \pi r^2$ $A_{side} = 2\pi rh$ $A_{total} = 2\pi r^2 + 2\pi rh$	$V = \pi r^2 h$
Sphere	$A = 4\pi r^2$	$V = \frac{4}{3}\pi r^3$
Cone	$A_{base} = \pi r^2$ $A_{side} = \pi rs$ $A_{total} = \pi r^2 + \pi rs$	$V = \frac{1}{3}\pi r^2 h$
Square-Based Pyramid	$A_{base} = b^2$ $A_{sides} = 4\left(\frac{1}{2}bs\right)$ $A_{total} = b^2 + 2bs$	$V = \frac{1}{3}b^2 h$
Rectangular Prism (or Cuboid)	$A_{total} = A_{base + top} + A_{front + back} + A_{ends}$ $A_{total} = 2lw + 2lh + 2wh$ $A_{total} = 2(lw + lh + wh)$	$V = lwh$
Triangular Prism	$A_{total} = A_{triangular\ ends} + A_{rectangular\ sides}$ $A_{total} = 2\left(\frac{1}{2}bh\right) + al + bl + cl$ $A_{total} = bh + al + bl + cl$	$V = \frac{1}{2}bhl$

2.1 | Systems of Measure

Many aspects of daily life, from the food you buy to home improvement projects to your chosen career, depend on a good understanding of measurement systems. For most of your studies, you have focussed on the metric system of measure. In everyday life and work, however, the imperial system is also used.

DISCOVER

Associations for the Metric and Imperial Systems

gram   millilitre   centimetre   litre   metre   kilogram
Celsius   kilometre   kelvin   milligram   tonne   millimetre

1. Group these units as measures of mass, volume, length, or temperature.

2. Order each group of measures from least to greatest. Compare your list with a partner.

3. It is useful to associate common objects with measures. Match these items with the appropriate units.
  a) mass of a paper clip
  b) width of a fingernail
  c) mass of one litre of water
  d) amount of liquid in an eyedropper
  e) length of a baseball bat
  f) a ten-minute walk
  g) thickness of a dime
  h) mass of a small car

      1 g   1 mL   1 cm   1 L   1 m
      1 km   21°C   1 mm   1 kg   1 tonne or 1 t

4. Most countries in the world have adopted the metric system of measurement. However, some countries still use the imperial system of measurement for some purposes. Imperial units are widely used in construction and related industries in Canada. For each item in question 3, choose the most appropriate imperial unit for measuring it. Common units in the imperial system are shown in the table. Accepted abbreviations are shown in brackets.

Mass	Volume	Length	Temperature
pound (lb)	quart (qt)	inch (in. or ")	degrees Fahrenheit (°F)
ton (T)	gallon (gal)	foot (ft or ')	
ounce (oz)	pint (pt)	yard (yd)	
	fluid ounce (fl oz)	mile (mi)	

5. Measure each line to the nearest millimetre.

a) ————————————————————————

b) ——————————————————————

c) ——————————————

d) ——————————————————

6. Express the volume in each graduated cylinder in millilitres and in litres.

a)   b)   c)

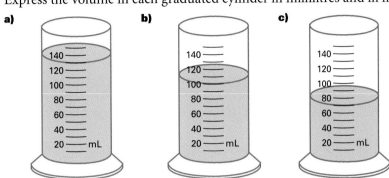

The inch is the smallest measure of length in the imperial system. Whole inches are the tallest lines on a ruler. When measuring lengths that are not a whole inch, fractions of an inch are used.

- Half inches are the next tallest lines shown on the ruler.
- Quarter-inch lines are just shorter than the half-inch lines.
- Eighth-inch lines are just shorter than the quarter-inch lines.
- Sixteenth-inch lines are the shortest lines on this ruler. They are just shorter than the eighth-inch lines.
- Some rulers are accurate to a thirty-second of an inch.

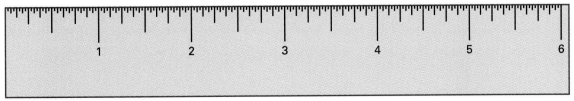

- The average ruler or measuring tape cannot show sixty-fourth inches accurately. However, precision measuring devices and machinery like these Vernier calipers allow materials to be measured to this degree of accuracy.

EXAMPLE 1

Fractions of an Inch

Find the measure of the bolt in inches.

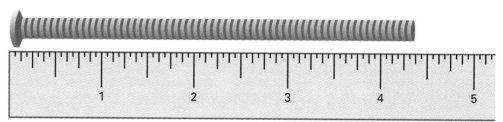

Solution

The end of the bolt is at an eighth-of-an-inch line, 3 eighth-of-an-inch segments after 4″. So, the bolt measures $4\frac{3}{8}″$.

DISCOVER

Unit Relationships Within the Imperial System

This visual tool represents measures of length in the metric system.

| 10 cm | 10 cm | 10 cm | 10 cm | 10 cm | 10 cm | 10 cm | 10 cm | 10 cm | 10 cm |

| 1 dm | 1 dm | 1 dm | 1 dm | 1 dm | 1 dm | 1 dm | 1 dm | 1 dm | 1 dm |

| 1 metre |

Not to Scale

History Link

In ancient times, an inch was the length of three barley corns, a foot was the length of a human foot, and a yard was the distance from the tip of the nose to the tip of the thumb.

1. Using the conversion table, create a similar visual aid to help you remember measures of length, mass, and volume in the imperial system.

Length	Mass	Volume
1 ft = 12 in.	1 lb = 16 oz	1 gal = 4 qt
1 yd = 3 ft	1 T = 2000 lb	1 qt = 2 pt
1 mi = 1760 yd		1 pt = 16 fl oz

2. Use your visual aid to complete the following conversions.
   a) $2\frac{1}{2}$ ft in inches   b) 48 oz in pounds   c) 2 gal in pints

EXAMPLE 2

Convert Length in Imperial Measure

Express 3 ft in inches.

Solution

You know that 12 in. = 1 ft.

So, $\dfrac{12 \text{ in.}}{1 \text{ ft}} = 1$

Units can be converted by multiplying by a ratio that is equivalent to 1. This does not change the actual length, but converts the units from feet to inches.

$3 \text{ ft} = 3 \;\cancel{\text{ft}} \times \dfrac{12 \text{ in.}}{1 \;\cancel{\text{ft}}}$

$3 \text{ ft} = 3 \times 12 \text{ in.}$

$3 \text{ ft} = 36 \text{ in.}$

EXAMPLE 3

Convert Volume in Imperial Measure

Convert 10 qt to gallons.

Solution

$1 \text{ gal} = 4 \text{ qt}$

So, $\dfrac{1 \text{ gal}}{4 \text{ qt}} = 1$

Using this conversion ratio,

$10 \text{ qt} = 10 \;\cancel{\text{qt}} \times \dfrac{1 \text{ gal}}{4 \;\cancel{\text{qt}}}$

$10 \text{ qt} = \dfrac{10}{4} \text{ gal}$

$10 \text{ qt} = 2.5 \text{ gal}$

EXAMPLE 4

Convert Mass in Imperial Measure

Convert 210 oz to pounds and ounces.

Solution

$210 \text{ oz} = 210 \;\cancel{\text{oz}} \times \dfrac{1 \text{ lb}}{16 \;\cancel{\text{oz}}}$ Recall that 1 lb = 16 oz.

$210 \text{ oz} = 13.125 \text{ lb}$

$0.125 \text{ lb} = 0.125 \;\cancel{\text{lb}} \times \dfrac{16 \text{ oz}}{1 \;\cancel{\text{lb}}}$ Convert the fraction of a pound back to ounces.

$0.125 \text{ lb} = 2 \text{ oz}$

$210 \text{ oz} = 13 \text{ lb } 2 \text{ oz}$ Complete the conversion to pounds and ounces.

Length measurements are used to find the distance around an object (**perimeter**), to measure the surface of an object (**area**), and to find the amount of space an object occupies (**volume**).

EXAMPLE 5

Perimeter, Surface Area, and Volume

The skating rink in an arena has the dimensions shown. Assume that the ends are semicircles. The ice has a thickness of $2\frac{1}{2}$ in. Find the following:

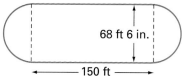

68 ft 6 in.

150 ft

a) the perimeter of the rink, to two decimal places
b) the surface area available for skating, to two decimal places
c) the volume of ice in the rink, to the nearest cubic foot

Solution

In solving this problem, consider the two ends of the rink as forming a whole circle if they were combined.

a) The perimeter (or circumference) of a circle is given by the formula

$$C = \pi d$$
$$C = \pi \times 68.5 \text{ ft} \qquad 68'\,6'' = 68.5'$$
$$C \doteq 215.20 \text{ ft}$$

The total perimeter is

$$P_{total} = 215.20 \text{ ft} + (150 \text{ ft} \times 2)$$
$$P_{total} = 215.20 \text{ ft} + 300 \text{ ft}$$
$$P_{total} = 515.20 \text{ ft}$$

So, the total perimeter of the rink is 515.20 ft.

b) To find the surface area, combine the areas of the circle and the rectangle.

$$A_{circle} = \pi r^2$$
$$A_{circle} = \pi \times (34.25 \text{ ft})^2 \qquad \text{Since } d = 68.5, r = 34.25.$$
$$A_{circle} \doteq 3685.28 \text{ ft}^2$$
$$A_{rectangle} = l \times w$$
$$A_{rectangle} = 150 \text{ ft} \times 68.5 \text{ ft}$$
$$A_{rectangle} = 10\,275 \text{ ft}^2$$
$$A_{total} = 3685.28 + 10\,275$$
$$A_{total} = 13\,960.28 \text{ ft}^2$$

```
GRAPHING CALCULATOR
π*34.25^2
              3685.284532
150*68.5
                    10275
3685.28+10275
               13960.28
```

So, the total surface area of the rink is 13 960.28 ft².

c) The rink is a prism (a three-dimensional figure with identical top and bottom faces). Therefore, to find the volume, multiply the area of the base by the height. First, convert the height from inches to feet:

$$2\frac{1}{2} \text{ in.} = 2.5 \text{ in.} \times \frac{1 \text{ ft}}{12 \text{ in.}}$$

$$2\frac{1}{2} \text{ in.} = 0.208\ 333\ 333\ 3 \text{ ft}$$

V = area of base × height

$V = 13\ 960.28 \text{ ft}^2 \times 0.208\ 333\ 333\ 3 \text{ ft}$

$V \doteq 2908 \text{ ft}^3$

The volume of ice in the rink is 2908 ft^3.

KEY CONCEPTS

- Both the metric and imperial systems of measurement are used in Canada.

- Important relationships within the imperial system are listed in the table on page 58.

- To convert one unit to another, multiply by a ratio equivalent to 1 to eliminate the old units.

DISCUSS THE CONCEPTS

1. For these scenarios, suggest the most appropriate imperial units to use.
  - **a)** laying tile
  - **b)** choosing a size of drill bit
  - **c)** size of a casserole
  - **d)** estimating the capacity of a car's gas tank
  - **e)** painting a house
  - **f)** estimating the space in a home

2. List five common objects that can be described as prisms.

PRACTISE

A **1.** Identify the best metric estimate.

a)	height of a door	500 cm	2 m	110 cm	1.25 m
b)	volume of a car's gas tank	10 L	750 mL	60 L	15 000 mL
c)	mass of a five-year-old	80 kg	20 kg	5000 g	5 kg
d)	volume of a can of soup	2 L	3000 mL	300 mL	0.8 L
e)	diameter of a CD	0.25 m	25 cm	1200 mm	12 cm
f)	mass of a small car	1000 kg	5 t	10 000 kg	10 t
g)	room temperature	10°C	15°C	20°C	25°C

2. Calculate these quantities. Use the prefix table on page 52, when necessary.

  a) the volume of gas, in millilitres, to travel 30 km, if the car's fuel consumption rate is 6.5 L per 100 km

  b) the time, in seconds, between thunder and lightning, if it is measured as 1.95 milliseconds

  c) the size of a computer's hard drive, in terabytes, if it is measured as 20 gigabytes

  d) the speed of a computer, in megahertz, if it is given as 1.5 gigahertz

  e) the voltage of a 9-V battery, in millivolts

  f) the power of a 60-W lightbulb, in kilowatts

  g) the number of building lots, each having 26-m frontage, on a road 1.43 km long

B 3. The top of a table has an area of 1 m^2.

  a) Express the length of each side in centimetres.

  b) Calculate the area in square centimetres.

  c) What is the relationship between square metres and square centimetres?

  d) Use the same process to express

    i) 1 km^2 in square metres

    ii) 1 cm^2 in square millimetres

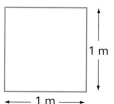

4. Use your calculations from question 3 to complete these conversions.

  a) 0.0067 m^2 to square centimetres

  b) 1485 mm^2 to square metres

  c) 45 268.1 m^2 to square kilometres

  d) 1545 mm^2 to square centimetres

5. A block has a volume of 1 m^3.

  a) Express the length, width, and height in centimetres.

  b) Calculate the volume in cubic centimetres.

  c) What is the relationship between cubic metres and cubic centimetres?

  d) Use the same process to express

    i) 1 cm^3 in cubic millimetres

    ii) 1 m^3 in cubic millimetres

    iii) 1 dm^3 in cubic centimetres

  e) A 1-mL quantity of liquid occupies 1 cm^3 of space. How many litres will a 1-m^3 container hold?

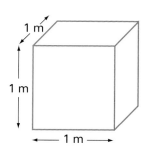

6. What measure is indicated at each letter?

a)

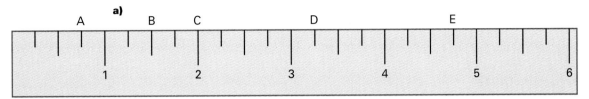

b)

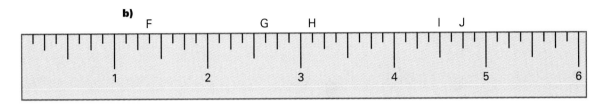

c)

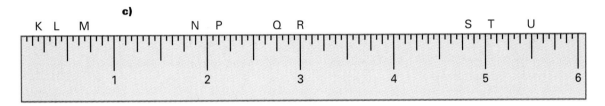

7. Identify the best estimate in imperial measure.

a)	height of a door	10 ft	50 in	7 ft	500 in
b)	volume of a car's gas tank	15 gal	2 qt	10 pt	100 gal
c)	mass of a five-year-old	45 lb	20 lb	100 lb	80 lb
d)	volume of a can of soup	10 gal	10 pt	10 qt	10 fl oz
e)	diameter of a CD	5 in	1 ft	8 in	2 in
f)	mass of a small car	1000 lb	1000 ton	10 000 lb	1 ton
g)	room temperature	60°F	70°F	80°F	90°F

8. Using the scale shown, measure each object, to the nearest sixteenth of an inch.

a) a thumb tack **b)** a house key

c) a bobby pin **d)** a toothbrush

9. Find the perimeter and area of each panel.

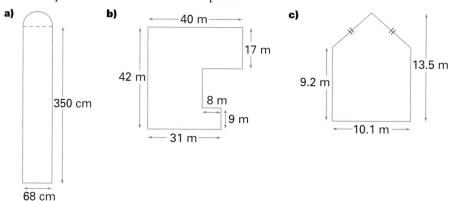

a) 350 cm, 68 cm

b) 40 m, 17 m, 42 m, 8 m, 9 m, 31 m

c) 13.5 m, 9.2 m, 10.1 m

10. Find the perimeter by measuring, to the nearest sixteenth of an inch.

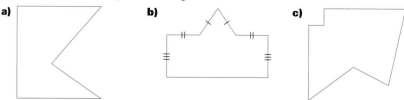

a) **b)** **c)**

11. Find the area of the shape in question 10 b).

12. **Catering sizes** Cakes are sold at a bakery in these sizes:

8″, round   10″, round   full slab (32″ by 8″ rectangle)
half slab (16″ by 8″ rectangle)   quarter slab (8″ by 8″ rectangle)
All cakes are 3″ thick.

a) What is the difference, in cubic inches, in the amount of cake in the two round cake sizes?

b) The 10″ round and the quarter slab are the same price. Which cake is the better buy?

c) If a full slab serves about 80 people, approximately what volume of cake does each person receive?

d) Based on the portion in part c), determine how many people can be served by each size of cake.

13. For each beach item, find the area of plastic and the volume of air.

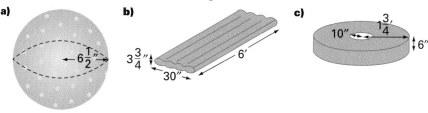

a) $6\frac{1}{2}''$

b) $3\frac{3}{4}''$, 30″, 6′

c) 10″, $1\frac{3}{4}''$, 6″

14. You have just purchased a new semi-detached home in a growing subdivision. You want to make several improvements to your new home, particularly to the unfinished basement. All ceilings in the basement are 7 ft 8 in. high.

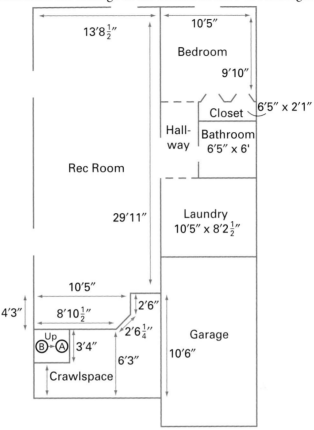

a) Copy and complete the table for this floor plan. The values you enter in the table may be useful for future calculations.

Room	Floor Area (ft^2)	Ceiling Area (ft^2)	Wall Area (ft^2)
Bathroom			
Bedroom			
Hallway			
Laundry			
Rec Room			

b) You want to install flooring in your unfinished basement. The carpet you like is \$1.59/ft^2 and the underpadding required is \$5.50/yd^2. You have also found 12″ by 12″ ceramic tile, which sells for \$19.89 per dozen. Suggest how you would cover the basement floor (keeping the layout in mind) and determine the total cost. Include a sketch to communicate your ideas.

15. Problem Solving Find the surface area of each object.

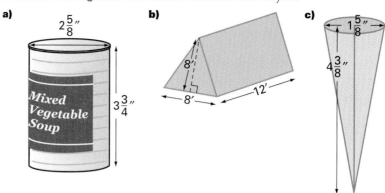

a)

$2\frac{5}{8}''$

Mixed Vegetable Soup

$3\frac{3}{4}''$

b)

$8'$

$8'$

$12'$

c)

$1\frac{5}{8}''$

$4\frac{3}{8}''$

16. Construction materials

a) Find the volume of concrete used to make the brick. Remember to subtract the empty space of the holes.

Diameter

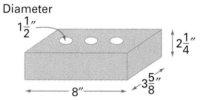

$1\frac{1}{2}''$

$2\frac{1}{4}''$

$8''$

$3\frac{5}{8}''$

b) If the concrete used to make this brick has a density of 140 lb/ft³, what is the mass of a brick, in kilograms?

c) What is the brick's mass, in grams?

d) A rule of thumb in masonry is that three courses (rows) of bricks and three mortar joints produce a height of 8″. Use the brick dimensions in the diagram to determine how much space is reserved for mortar between each course of bricks.

e) Use the information in part d) to estimate the number of bricks needed for a wall 12′ long and 8′ high. Assume that the mortar joints between side-by-side bricks take up the same amount of space as the mortar joints between courses.

Converting Between Metric and Imperial Systems

In some situations, it is important to be able to convert between metric units and imperial units. For example, a North American manufacturer would probably specify dimensions for a product in imperial units. However, key components for the product might be ordered from a European supplier that works in metric units.

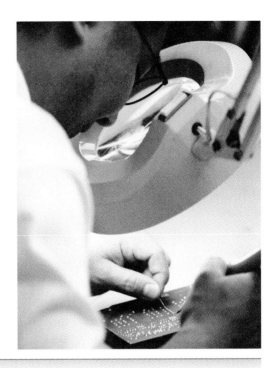

DISCOVER

Building a Conversion Table

Use the information in the questions and your knowledge of ratios to create a conversion table between metric and imperial systems. Note that the conversion table you create, like commonly used conversion tables, will contain approximate values. These values are generated by rounding the precise conversion factors.

Space Link

When converting between different systems, you need to remember which set of measurements is in which system. The U.S. probe *Mars Climate Orbiter* burned up in the atmosphere of Mars in 1999, just because onboard computer software mixed up feet and metres in its orbital calculations.

1. Draw a line that is one inch long. Now measure this line in centimetres. Write your finding in the form "1 in. = … cm."

2. Given that your 5′ tall younger sister measures 152.5 cm, complete the statement: 1 ft = … m.

3. The speedometer on your car shows a speed of 50 m.p.h. as being roughly equivalent to 80 km/h. Calculate a conversion factor for 1 mi.

4. You have seen 454 g listed on the wrapping of a pound of butter. Complete the conversion factor: 1 lb = … g = … kg.

5. A soup label states the amount of soup as 10 fl oz or 296 mL. Give a conversion factor for fluid ounces to millilitres.

6. A quart of oil is labelled 947 mL. Give a conversion factor for quarts to litres.

7. You refill your empty 22-gal gas tank with 85.25 L of gas. Give a conversion factor for gallons (U.S.) to litres.

8. A recipe lists the quantity of cheese as 20 oz or 570 g. Give a conversion factor for ounces to grams.

9. Recall the visual aids created in the Discover on page 58. Create a visual aid to relate metres and feet, quarts and litres, gallons and litres, and miles and kilometres.

EXAMPLE 1

Converting Length Measures

Farad is 5 ft 10 in. tall. Express his height in centimetres. Round your answer to the nearest centimetre.

Solution

First, express 5 ft 10 in. as a single unit (inches).

$$5 \text{ ft } 10 \text{ in.} = 5 \, \cancel{\text{ft}} \times \frac{12 \text{ in.}}{1 \, \cancel{\text{ft}}} + 10 \text{ in.}$$

$$5 \text{ ft } 10 \text{ in.} = 60 \text{ in.} + 10 \text{ in.}$$

$$5 \text{ ft } 10 \text{ in.} = 70 \text{ in.}$$

Now, convert from inches to centimetres, using the conversion factor 1 in. = 2.54 cm.

$$70 \text{ in.} = 70 \, \cancel{\text{in.}} \times \frac{2.54 \text{ cm}}{1 \, \cancel{\text{in.}}}$$

$$70 \text{ in.} = 177.8 \text{ cm}$$

$$70 \text{ in.} \doteq 178 \text{ cm} \qquad 6 \text{ ft} \times 30 \text{ cm/ft} = 180 \text{ cm, so the answer is reasonable.}$$

Notice that the factor 1 in. = 2.54 cm was used. If you were to convert with the less accurate factor 1 in. = 2.5 cm, you would get an answer of 175 cm, a 3-cm difference from the correct answer.

EXAMPLE 2

Fuel Economy

Betty sets the trip odometer on her car to zero at each gas fill-up. When the odometer reads 498 km, it takes 54 L to refill the tank. What is the gas consumption of Betty's car, in miles per gallon?

Solution

Convert both the distance and the volume of fuel to imperial measures.

$$54 \text{ L} = 54 \text{ L} \times \frac{1 \text{ gal}}{3.785 \text{ L}} \qquad\qquad 498 \text{ km} = 498 \text{ km} \times \frac{1 \text{ mi}}{1.61 \text{ km}}$$

$$54 \text{ L} = \frac{54}{3.785} \text{ gal} \qquad\qquad 498 \text{ km} = \frac{498}{1.61} \text{ mi}$$

$$54 \text{ L} \doteq 14.27 \text{ gal} \qquad\qquad 498 \text{ km} \doteq 309.3 \text{ mi}$$

Use your converted quantities to determine the gas consumption in number of miles per gallon.

$$\text{Gas consumption} = \frac{309.3 \text{ mi}}{14.27 \text{ gal}}$$

$$\text{Gas consumption} \doteq 21.7 \, \frac{\text{mi}}{\text{gal}}$$

Note: This text uses U.S. liquid conversion factors; U.S. gallons are smaller than British gallons. 1 gal (GB) = 4.545 L

Betty's car's gas consumption is 21.7 mpg (miles per gallon), to the nearest tenth. This is a reasonable value for fuel economy.

EXAMPLE 3

Floor Dimensions

A rectangular room measures 10.5 ft by 11 ft. What is the floor area

a) in square feet?   **b)** in square metres?

Round your answers to one decimal place.

Solution

a) The dimensions are given in feet, so calculate the area in square feet directly.

$A = l \times w$

$A = 10.5 \text{ ft} \times 11 \text{ ft}$

$A = 115.5 \text{ ft}^2$

The floor area is 115.5 ft^2.

b) To find the area in square metres, the best method is to convert feet to metres *before* doing any area calculations:

$$10.5 \text{ ft} \times \frac{0.3048 \text{ m}}{1 \text{ ft}} = 3.2004 \text{ m} \qquad 11 \text{ ft} \times \frac{0.3048 \text{ m}}{1 \text{ ft}} = 3.3528 \text{ m}$$

Now repeat the area calculation, but in metric units.

$A = l \times w$

$A = 3.2004 \text{ m} \times 3.3528 \text{ m}$

$A \doteq 10.7 \text{ m}^2$

The floor area is 10.7 m^2.

KEY CONCEPTS

- To convert between metric and imperial systems of measurement, multiply by an appropriate factor equal to 1. Set up the factor so that the correct units cancel.

- You can develop your own conversion table using measures of familiar objects.

- Common conversions between the metric and imperial systems are given in the table. All of the equivalent measures shown are approximate.

Imperial to Metric Conversion Factors		
Length	**Area**	**Volume**
1 in. = 25.4 mm or 2.54 cm	$1 \text{ in.}^2 = 6.45 \text{ cm}^2$	$1 \text{ in.}^3 = 16.39 \text{ cm}^3$
1 ft = 0.3048 m	$1 \text{ ft}^2 = 0.0929 \text{ m}^2$	$1 \text{ ft}^3 = 28.32 \text{ dm}^3$ or 28.32 L
1 yd = 0.9144 m	$1 \text{ yd}^2 = 0.84 \text{ m}^2$	$1 \text{ yd}^3 = 0.76 \text{ m}^3$
1 mi = 1.609 km	$1 \text{ acre} = 4047 \text{ m}^2$	
Capacity	**Mass**	**Speed**
1 fl oz = 29.6 mL	1 oz = 28.35 g	1 m.p.h. = 1.609 km/h
1 qt = 0.947 L	1 lb = 0.454 kg	
(U.S.) 1 gal = 3.785 L	1 T = 0.91 t	

DISCUSS THE CONCEPTS

1. Which unit in the imperial system is most similar to each of these metric units?
   a) centimetre   b) metre   c) kilometre   d) litre   e) kilogram
   Justify your choices.

2. Is it true that 1 in.^{2} = 2.54 cm^2? Explain why or why not.

PRACTISE

A 1. Convert each imperial measure, as indicated.
   a) 10″ to centimetres
   b) 120 lb to kilograms
   c) $2\frac{1}{2}$ gal to litres
   d) 18 oz to grams
   e) 6 ft to metres
   f) 8 fl oz to millilitres
   g) 1500 mi to kilometres
   h) 2.7 qt to litres
   i) 1500 lb to kilograms

2. Convert each metric measure, as indicated.
   a) 58 cm to inches
   b) 500 mL to fluid ounces
   c) 40 L to gallons
   d) 500 g to ounces
   e) 40 L to quarts
   f) 25.2 kg to pounds
   g) 62.75 km to miles
   h) 82 mm to inches
   i) 0.3 kg to ounces

APPLY THE CONCEPTS

3. Sam's Place sells sub sandwiches in 6″ and 12″ lengths. What are the lengths of the sandwiches in metric units?

B 4. **Recipe quantities** A family chili recipe measures the ingredients in imperial units. Convert the recipe to metric measures.

4 lb ground beef	48 fl oz tomato juice
8 oz tomato paste	$\frac{1}{4}$ lb chopped onion
28 fl oz baked beans	14 fl oz kidney beans
chili powder to taste	

5. **Food prices** A deli is selling turkey for $2.19 per 500 g and chicken loaf for $1.99 per pound. Which luncheon meat is cheaper?

6. The height of each floor in an apartment building is about 9.5 ft (including the depth of the floor). Estimate the height of the building, in metres, if there are
   a) 10 floors
   b) 16 floors
   c) 21 floors plus a 12-ft recreational floor

7. Plastic cups hold 8 fl oz. How many of these cups can you fill from a 2-L bottle of cola?

8. A standard piece of photocopy paper measures $8\frac{1}{2}$ in. by 11 in.
   a) What is the area of the sheet of paper
   i) in square inches?
   ii) in square centimetres?
   iii) in square millimetres?
   b) A package of 500 sheets is 2 in. thick. What is the thickness of one sheet of paper, in millimetres?

9. Hardware materials A bolt has a diameter of $\frac{3}{16}$ ″. Will it fit into a hole with a radius of 1 mm?

10. Home environment When your house was constructed, a heat recovery unit was installed that has a rating of 185 CFM. This means that this system exchanges 185 ft^3 of the air in your home every minute with fresh air from the outside. If all of your ceilings are 8 ft high, and the floor area of your home is approximately 1600 ft^2, how long will it take for all of the air in your house to be exchanged?

11. A cube has a volume of 1 in.^{3}.
   a) What is the length of each edge, in centimetres?
   b) Calculate the volume of the box, in cubic centimetres.
   c) What is the relationship between cubic centimetres and cubic inches?
   d) Use the same process to complete these statements.
   i) $1\ m^3 = \ldots\ yd^3$ **ii)** $1\ m^3 = \ldots\ ft^3$
   iii) $1\ mile^2 = \ldots\ km^2$ (Note: This is an area, not a volume.)

12. Estimate the amount of cake batter (in cups) that will fit into these baking pans. (Hint: 1 cup is about 250 mL.)
   a) rectangular baking pan: $13'' \times 9'' \times 2''$
   b) round layer-cake pan: $8''$ diameter $\times\ 1\frac{1}{2}''$ deep

13. You are enlarging a $5'' \times 7''$ photograph to an $8'' \times 10''$ print. How many square centimetres larger is the area of your enlarged picture?

14. Problem Solving A German car manufacturer has contacted your tool and die company to produce the part shown.

a) Convert the measurements to inches. Round your answers to the nearest $\frac{1}{16}$".

b) Find the area of the top face of the part.

c) What volume of plastic is required to produce 1000 parts? Explain your reasoning.

All measurements in millimetres
Thickness is 23 mm

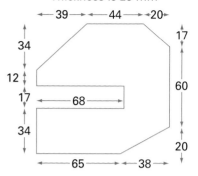

15. Construction Approximate weights per cubic foot of various materials used by construction engineers are shown in the table. Convert each measurement to metric units.

Soil, dry and loose	90 lb
Soil, moist	75 lb to 100 lb
Limestone, uncrushed	160 lb
Concrete	140 lb
Fill dirt, dry and loose	1.2 T
Class 5 road stone	1.25 T

16. The side view of a pool is shown. The pool is 5 m wide.

a) Calculate the volume of water required to fill the pool.

b) How long will it take this pool to drain at a rate of 0.3m^3/min?

12 m
1 m
4 m
3 m
3 m

See page 445 for more information on calculating areas with *The Geometer's Sketchpad*®.

ACHIEVEMENT CHECK Knowledge/Understanding Thinking/Inquiry/Problem Solving Communication Application

17. Retail property costs You are looking for a location to set up a small bookstore. You have narrowed your choices to two options.

Option A: $48.75/m^2 per month

Option B: $5.04/ft^2 per month

a) Calculate the floor area for each option.

b) Explain how you would get both areas in metric units.

c) Explain how you would get both areas in imperial units.

d) Choose one system to represent both floor areas. Which system did you choose? Explain why.

A
15.5 m
4.5 m
15′
B
55′
12.3′

18. Realty survey The survey for a lot you wish to purchase is shown. The realtor has advertised the lot as being more than 2 acres.

a) Given that an acre is equivalent to 43 450 ft^2, do you agree with the realtor's claim? Explain.

b) Property taxes for this area are $0.375 /ft^2 annually. How much will you pay each year in taxes for this lot?

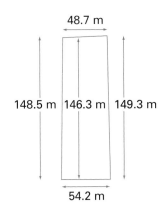

48.7 m

148.5 m   146.3 m   149.3 m

54.2 m

19. The windows in the basement of your new home need covering. You have
- two windows 74 cm deep and 60 cm wide
- one window 74 cm deep and 85 cm wide
- two windows 136 cm deep and 120 cm wide

Use the table to select the blinds you need for these windows. Determine the total cost.

Width	Length	Price
22.5″	58.5″	$32.95
	71″	
34.5″	34.5″	$39.95
	46.5″	
	58.5″	
46.5″	34.5″	$47.95
	46.5″	
	58.5″	
58.5″	34.5″	$55.95
	46.5″	
	58.5″	
70.5″	34.5″	$63.95
	46.5″	
	58.5″	

C 20. Problem Solving, Application Vu wants to sell part of his property, the shaded portion of the diagram. You may wish to use a spreadsheet to complete these questions.

a) Determine the total area, and the area to be sold,
   i) in square feet   ii) in acres

b) Determine the total area, and the area to be sold, in square metres.

c) To determine a fair price, Vu investigates properties being sold in his area. Copy and complete the table, and determine the average cost per square foot.

172′

50′

279′   294′

148′

Lot Dimensions	Area	Price	Cost/ft^2
50′ × 168′		$42 500	
60′ × 145′		$40 000	
55′ × 150′ × 62′ × 182′ irregular		$47 000	

d) What is a reasonable selling price for Vu's lot? Explain your reasoning.

2.3 Measurement for Your Health and Fitness

Various measurements are taken throughout your life as a way to monitor your state of health. People in health occupations use measurement as part of their assessment of patients from the moment a child is born. A healthy lifestyle also involves monitoring your own body, and your own environment, through measurement.

EXAMPLE 1

Body Temperature

One measurement frequently taken to monitor an individual's health is body temperature. Temperature may be measured in degrees Celsius (metric system) or degrees Fahrenheit (imperial system). The oral body temperature of a healthy person is about 37.0°C.

a) Convert 37.0°C to degrees Fahrenheit.

b) Convert 101.2°F to degrees Celsius.

Solution

a) Use the formula

$$T_F = 1.8 \times T_C + 32$$

where T_F is temperature in Fahrenheit and T_C is temperature in Celsius. Substitute.

$$T_F = 1.8 \times 37.0 + 32$$
$$T_F = 98.6$$

So, a healthy body temperature of 37.0°C is equal to 98.6°F.

b) Use the reverse formula to convert degrees Fahrenheit to degrees Celsius.

$$T_C = \frac{(T_F - 32)}{1.8}$$

$$T_C = \frac{(101.2 - 32)}{1.8}$$

$$T_C = 38.4$$

So, a temperature of 101.2°F is equal to 38.4°C. These are reasonable values for a high, but not life-threatening, temperature.

EXAMPLE 2

Calculating Dosage

Medicine is typically administered based on a patient's mass. A 50-lb child is to be given an antibiotic at a dosage of 40 mg per kilogram body mass per day. How much medicine should the child receive in a day? Give your answer to the nearest 10 mg.

Solution

Convert the child's mass to kilograms.

$$50 \text{ lb} = 50 \text{ lb} \times \frac{1 \text{ kg}}{2.205 \text{ lb}}$$

$$50 \text{ lb} = 22.675 \ 736 \ 96 \text{ kg}$$

Determine the amount of medicine required, to the nearest ten milligrams.

$$\text{Dosage} = 22.675 \ 736 \ 96 \text{ kg} \times \frac{40 \text{ mg}}{1 \text{ kg}}$$

$$\text{Dosage} = 907.029 \ 478 \ 5 \text{ mg}$$

$$\text{Dosage} \doteq 910 \text{ mg}$$

So, the child should receive 910 mg of medicine per day.

EXAMPLE 3

IV Drip

A paramedic needs to set up an intravenous line (IV) to deliver 2 drips/s for a period of 15 min. If 10 drips = 1 mL, what volume of solution will she need?

Solution

Start by converting the time to seconds. Then, find the volume as a number of drips. Finally, convert the number of drips to millilitres.

$$\text{Time} = 15 \text{ min} \times \frac{60 \text{ s}}{1 \text{ min}}$$

$$\text{Time} = 900 \text{ s}$$

$$\text{Volume} = 900 \text{ s} \times \frac{2 \text{ drips}}{1 \text{ s}}$$

$$\text{Volume} = 1800 \text{ drips}$$

$$\text{Volume} = 1800 \text{ drips} \times \frac{1 \text{ mL}}{10 \text{ drips}}$$

$$\text{Volume} = 180 \text{ mL}$$

So, the paramedic will require 180 mL of solution.

KEY CONCEPTS

- Monitoring your own body, your environment, and your food intake are important for a healthy lifestyle.

- Health occupations use measurement on a daily basis to check patients' conditions and to determine dosages of medication.

- Temperature is measured in degrees Celsius (metric units) or degrees Fahrenheit (imperial units).

- To convert from degrees Celsius to degrees Fahrenheit, use the formula $T_F = 1.8 \times T_C + 32$.

- To convert from degrees Fahrenheit to degrees Celsius, use the formula $T_C = \dfrac{(T_F - 32)}{1.8}$.

DISCUSS THE CONCEPTS

1. A quick rule of thumb for converting from degrees Celsius to degrees Fahrenheit is to "double and add thirty." Use this method to convert 37°C to degrees Fahrenheit, and compare this to the result when the formula is used. Is this rule of thumb appropriate for
   a) reporting the weather?
   b) measuring someone's temperature?

2. A digital thermometer shows a patient's body temperature to be 99.1°. In which units is the thermometer measuring? How do you know?

APPLY THE CONCEPTS

A 1. At birth, babies are weighed in grams. Parents usually want to know their baby's mass in pounds and ounces. Convert the following to pounds and ounces.
   a) 3325 g   b) 2874 g   c) 4016 g

2. Convert the following baby masses to kilograms and baby lengths to centimetres.
   a) 7 lb 6 oz, 20″   b) 6 lb 8 oz, 19″
   c) 8 lb 3 oz, 21″   d) 9 lb 1 oz, $21\frac{3}{4}$″

3. A respiratory therapist reads these measures of patients' height, mass, and body temperature on the patients' charts.

    a) Convert to metric units.

        i) Ari: 5′10″, 180 lb, 99.1°F   **ii)** Irvana: 5′2″, 143 lb, 100.8°F

        iii) Shar: 5′8.5″, 162 lb, 101.2°F   **iv)** Reid: 6′, 210 lb, 98.7°F

    b) Convert to imperial units.

        i) Jared: 180 cm, 70 kg, 37°C   **ii)** Alexis: 164 cm, 55 kg, 38.2°C

        iii) Lanny: 157.5 cm, 68 kg, 39°C   **iv)** Jasmine: 170 cm, 60 kg, 40.1°C

B **4.** **a)** A person typically has about 75 mL of blood per kilogram of body mass. Convert this volume to fluid ounces per pound.

    b) The total amount of water in the body of an average adult is 39 L. Convert to quarts.

    c) The capacity of the bladder is approximately 1 to 1.5 pt. Convert to litres.

    d) An average adult breathes 2000 to 3000 gal of air per day. Convert to kilolitres.

    e) The average amount of sweat produced per day is 1 L. Convert to pints.

    f) A body temperature of 104°F is considered to be a dangerously high fever. Convert to degrees Celsius.

5. **Medicine dosage** A drug must be administered at a rate of 20 mg per pound of body mass per day. A 50-kg patient is given this medicine in three doses per day. How much should each dose contain?

ACHIEVEMENT CHECK   Knowledge/Understanding   Thinking/Inquiry/Problem Solving   Communication   Application

6. **IV rates** Calculate the volume of 20 drips/mL of IV solution required for the following flow rates.

    a) 2 drips/s for a 12-h period

    b) 2 drips every 5 s for a 6-h period

    c) 15 drips every 20 s for a 48-h period

If each solution were administered in 50-mL bags, which one would run out first? Explain how you know.

---

7. **Problem Solving, Communication** The IV solution used in question 6 is available in 100-mL, 250-mL, 500-mL, and 1000-mL bags. For parts a), b), and c), what quantity of each type of bag would be used to minimize wasted IV solution and also minimize the number of bag changes required? Explain your answers.

8. a) Bacteria in food will grow quickly at temperatures ranging from 60°F to 120°F. What is this range in degrees Celsius?

b) Water freezes at 0°C and boils at 100°C. What are these temperatures in degrees Fahrenheit?

9. Nutrition The food label for a cereal is shown.

a) To be considered a low-fat product, each serving should contain between 0.5 g and 3 g of fat. Does this food qualify?

b) Based on a recommended daily intake of 1500 mg of calcium, how much calcium is in one 30-g serving?

Serving size of 30 g or 125 mL	
Energy	225 calories
Protein	2.8 g
Fat	0.8 g
Carbohydrates	24 g
Sodium	134 mg
Potassium	96 mg
Minerals	**Percent of Recommended Daily Intake**
calcium	7%
phosphorus	7%
iron	29%

c) How many 30-g servings would you require to meet the recommended daily calcium intake of 1500 mg?

d) How many 30-g servings are in a 750-g box of this cereal?

e) What percent (by mass) of a 30-g serving is sodium?

10. A cyclist recorded her distance travelled in kilometres.

a) Devise a simple rule of thumb to convert kilometres to miles.

b) Use your rule to approximate these distances in miles.

i) 10 km   **ii)** 22 km   **iii)** 4 km

11. Dental moulds Dental assistants make moulds of your teeth when you require special dental work. One recipe for mould material requires 16 mL of water for every 6.0 g of powder.

a) How many millilitres of water are required for 42 g of powder?

b) How many millilitres of water are required for 2 oz of powder?

Language Link

People in health occupations use the unit cc to measure volume. The abbreviation cc stands for cubic centimetre. 1 cc is equivalent to 1 mL.

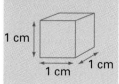

12. Water intake A healthy diet should include at least 2 qt of water daily. A cylindrical water cooler has a base with a 5-in. radius and is 1.5 ft tall.

a) What is the volume of the water cooler, in cubic centimetres?

b) Given that 1 mL = 1 cm³, how many 250-mL cups of water can you fill from the full water cooler?

c) How long will one cooler of water last two adults, if they each consume 2 qt per day?

d) How many cone-shaped paper cups with a 1.5-in. radius and a height of 3 in. can you fill from the full water cooler?

13. Pollutant levels Ontario Standards state the following maximums for pollutants in our drinking water.

Chemical	Maximum Acceptable Concentration (mg/L)
Atrazine (herbicide)	0.025
Dichloromethane	0.05
Carbofuran (insecticide)	0.09
Chloramines	3.0
Triallate (herbicide)	0.23

a) Convert each maximum to ounces per gallon.

b) Based on these standards, what is the maximum safe amount of atrazine you could expect to find in a one-million-gallon reservoir?

c) What is the maximum safe amount of dichloromethane in enough water to fill a bathtub? (Recall: $1 \text{ cm}^3 = 1 \text{ mL}$)

EXTEND THE CONCEPTS

14. Some medicine is prescribed based on the surface area of a patient's body. The following formula can approximate the surface area, in square metres, of an individual based on mass and height.

Surface area $= (\text{mass in kilograms})^{0.537} \times (\text{height in centimetres})^{0.305} \times 0.0242$

a) Find the surface area of a woman who weighs 61 kg and is 158 cm tall.

b) Find the surface area of a 6′ man who weighs 200 lb.

c) Dosage instructions for a medicine are 55 mg/m^2. How much medicine should be administered to a patient with a body mass of 190 lb and a height of 5′8″?

15. Home humidifier To keep the air in your home moist in cold, dry months, you purchase a humidifier. Humidifiers are selected based on the square footage of the home and the rate at which air is exchanged between your home and the outside environment.

Type of Home	Description	Air Exchange
Tight house	Well insulated, snug-fitting doors	One air exchange every 2 h
Average house	Insulated, loose openings	One air exchange per hour
Loose house	No insulation, no storm windows	Two air exchanges per hour

Maximum Square Footage Recommended:

Humidifier Model	Tight	Average	Loose	Price
A	2380 ft^2	1515 ft^2	1085 ft^2	$359.74
B	2850 ft^2	1800 ft^2	1300 ft^2	$447.50
C	3095 ft^2	1970 ft^2	1415 ft^2	$624.72

Select and justify the appropriate model to purchase for the following homes.

a) a 1200-ft^2 home that was built in 1910 and has not had many renovations

b) a 2000-ft^2 home built in 2000

c) a 1900-ft^2 home built in 1970

Measurement for Home Improvement and Construction

The imperial system of measurement is used regularly in the construction industry. Engineers, architects, landscapers, and people in other related professions use the imperial system daily for measurement calculations. Home-improvement projects often require materials that are measured in feet, inches, or gallons.

EXAMPLE 1

Home Improvement

One gallon of paint covers approximately 350 ft². Estimate the volume of paint, in gallons, needed for two coats on the exterior of the garage shown (excluding the doors).

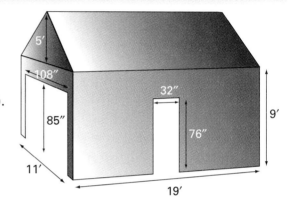

Solution

Start by rounding the measurements to lengths that are easier to work with.
The length of the garage is approximately 20 ft.
The height of the building is approximately 10 ft.
So, the surface area of the two sides of the garage is approximately
20 ft × 10 ft × 2, or 400 ft².

The front and back walls are approximately 10 ft wide by 10 ft high.
So, the surface area of the front and back walls is approximately
10 ft × 10 ft × 2, or 200 ft².

The two triangular roof pieces, if put together, would form a rectangle about
10 ft × 5 ft, or approximately 50 ft².

The total, so far, is about 650 ft². However, you need to subtract the area of the garage and side doors. The measures of these are approximately 10 ft by 7 ft and 3 ft by 7 ft, respectively, giving an approximate surface area of 70 ft² + 20 ft², or 90 ft².

As a generous approximation, a surface area of 600 ft² needs to be covered. For two coats, the surface area to be covered is about 1200 ft².

$$\text{Number of paint cans} = \frac{\text{Area to be painted}}{\text{Area per can}}$$

$$\text{Number of paint cans} = \frac{1200 \text{ ft}^2}{350 \text{ ft}^2}$$

$$\text{Number of paint cans} \doteq 3.5$$

Approximately 3.5 gal of paint is required (or 4 gal, with some left over). This is a reasonable quantity to purchase for painting a fairly large area.

EXAMPLE 2

Concrete Quantity

How many cubic metres of concrete are needed to fill the area shown to a depth of 10 cm?

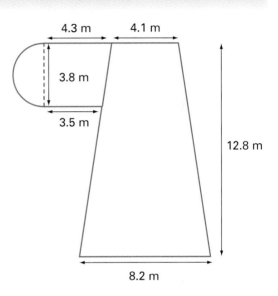

Solution

The diagram can be divided into three shapes: a semicircle and two trapezoids.

$$A_{\text{semicircle}} = \frac{\pi r^2}{2}$$

$$A_{\text{semicircle}} = \frac{\pi(3.8 \div 2)^2}{2}$$

$$A_{\text{semicircle}} \doteq 5.67$$

$$A_{\text{trapezoid A}} = \frac{(s_1 + s_2)h}{2}$$

$$A_{\text{trapezoid A}} = \frac{(4.3 + 3.5) \times 3.8}{2}$$

$$A_{\text{trapezoid A}} \doteq 14.82$$

$$A_{\text{trapezoid B}} = \frac{(s_1 + s_2)h}{2}$$

$$A_{\text{trapezoid B}} = \frac{(4.1 + 8.2) \times 12.8}{2}$$

$$A_{\text{trapezoid B}} \doteq 78.72$$

$$A_{\text{total}} = 5.67 + 14.82 + 78.72$$
$$A_{\text{total}} = 99.21$$

The total area to be concreted is 99.21 m^2.

$V_{cement} = A_{total} \times depth$
$V_{cement} = 99.21 \times 0.10$ Convert 10 cm to 0.10 m.
$V_{cement} \doteq 9.9$

To fill the area to a depth of 10 cm, approximately 9.9 m³ of concrete is needed.

DISCUSS THE CONCEPTS

1. Why are detailed calculations unnecessary for the type of situation in Example 1?

2. Example 1 used approximation techniques. However, in Example 2, precise calculations, based on exact measurements, were made. Describe how you would decide which method, approximate or exact, to use for a particular construction or home-improvement task.

3. Identify and discuss two home-improvement situations where you would be required to calculate
   a) volume   b) perimeter   c) surface area

APPLY THE CONCEPTS

A 1. If a nail is to be driven through a board $1\frac{3}{8}$ ″ thick, and it must extend at least $1\frac{1}{4}$ ″ through the other side, what minimum length of nail should be used?

2. Suggested measurements for a door are shown.
   a) If each hinge is $4\frac{1}{8}$ ″ long, what is the length between the two hinges?
   b) Find the approximate length, in feet, of weather stripping required for the two sides and the top of this door.

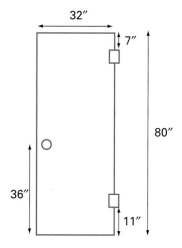

B 3. A 1.5-kg bag of grass seed covers approximately 80 m^2 of new lawn. Estimate the number of bags of grass seed required for the following situations.

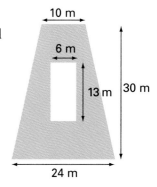

a) a lot 81 m by 39 m

b) a lot 300 ft by 240 ft

c) a 300-m^2 lot with a 50-m^2 house foundation and a 12-m^2 driveway

d) a property with the dimensions shown

4. Thinking/Problem Solving, Application

a) Calculate the cost to put weather stripping around three rectangular windows that measure 7 ft by 3 ft, at a cost of $1.14/ft.

b) Create an equation to determine the total length of weather stripping required for three windows of equal height, but possibly different widths.

c) Use your equation to create a table of quantities and costs for weather-stripping three windows with a range of dimensions. Allow the common height and the individual widths to vary in six-inch steps. Use a spreadsheet, graphing calculator, or other techniques to produce your table.

5. **Problem Solving** Lumber widths are described in whole numbers. In practice, however, they measure to be somewhat less. An 8′-long two-by-four board, when purchased, will still be 8′ long, but will have a thickness and a width approximately $1\frac{1}{2}'' \times 3\frac{1}{2}''$.

Nominal Dimension	Actual Dimension
1″	$\frac{1}{4}$ ″ less
2″ to 6″	$\frac{1}{2}$ ″ less
8″ to 12″	$\frac{3}{4}$ ″ less

Tech Link

When pieces of lumber are first cut, they approach the nominal dimensions. However, weathering and surfacing diminish their size substantially.

a) What are the actual dimensions of

i) a two-by-eight?   **ii)** a one-by-five?

b) You are building an 8′ by 4′ rear deck using two-by-four-by-eights (8′ long two-by-fours) laid flat. How many boards would you be short, if you bought lumber according to the nominal dimensions?

6. Tina wants to cement an area 12 ft by 8 ft to a depth of 4 in.

a) Determine the volume of cement needed, in cubic yards.

b) Cement costs $148.50/yd^3. Estimate the cost of this project.

7. Six equally spaced holes are to be drilled in a piece of oak that is $35\frac{1}{2}$ ″ long. If there is to be $2\frac{3}{4}$ ″ from each end to the centre of the first hole at each end, how far apart should the centres of the holes be spaced?

8. Lumber requirements How many two-by-fours are needed to complete a deck 24′ 9″ wide,

    a) if there are no spaces between the boards?

    b) if there is $\frac{1}{8}$ ″ of space between each pair of boards?

9. Landscape design A landscape technician is designing and installing a circular pond with a diameter of 5 ft 8 in.

    a) How many gallons will the pond hold, if it is 2.5 in. deep?
(Hint: 1 ft^3 = 7.48 gal)

    b) An 8 in.-high stone wall is to surround the pond. How many 4 in.-high stones are required, if each one takes up about 8.5 in. of the pond's perimeter?

10. The top view of a 3 ft 6 in.-deep septic tank is shown.

    a) What is the tank's capacity, in gallons, if 1 ft^3 holds 7.48 gal?

    b) This is the largest area the tank can occupy. By how much should the depth of the tank be increased to handle a capacity of 1750 gal?

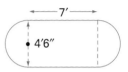

11. Problem Solving, Communication When carpeting a room, you try to minimize both the waste and the number of seams. The carpet you have chosen is sold in 12′ widths.

    a) Draw two plans for carpeting a 21′ by 9′ room.

    b) Determine the amount of carpet, in square yards, needed for each plan.

    c) Explain the advantages and disadvantages of each plan.

12. Decorative red mulch is available in 56-L bags for $3.98 a bag. How many bags will you require to fill the spaces outlined in the diagram with mulch 5 cm deep? What is the total cost?

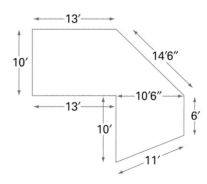

13. Interior decorating Your $9\frac{1}{2}'$ by 12' bedroom has one 2' by 4' window and one 3' by 7' doorway. The height of the ceiling is 9'.

a) What surface area would need to be covered, if you painted the walls and the ceiling?

b) The label on the paint you are buying claims that it has a coverage of 8 to 10 m²/L. How many litres will you need? Give upper and lower estimates.

c) One can of paint holds 3.78 L. How many cans of paint should you buy? Justify your answer.

d) Each can of paint costs $32.98. Rollers, brushes, and other materials will cost about $40. Estimate the cost of painting your room. Include tax at 15%.

14. Problem Solving The label on a new deck stain claims that it covers 400 ft² per gallon on the first coat, and 500 ft² per gallon on the second coat. A 0.90-gal can costs $21.97. Is one can adequate for two coats of stain on the deck shown?

15. Roofing project Your roofing company has been awarded the contract to cover the roof shown. You will need to lay a layer of waterproofing membrane (195 ft² per roll), a layer of asphalt felt (430.5 ft² per roll), and shingles (32.3 ft² per box). Determine the number of rolls or boxes of each material required.

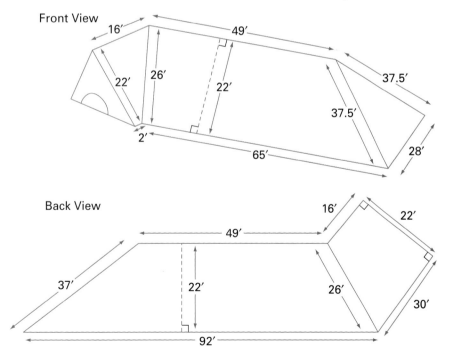

16. The space just under the roof of this single-level home is being used as an attic.

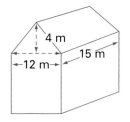

 a) Determine the square footage of attic space.
 b) Find the approximate amount of ventilation needed, if 1 ft² of ventilation is required for every 300 ft² of attic area.

17. **Problem Solving** A carpenter's rule of thumb for building stairs is: the length of two risers and one tread sum to 28 in.

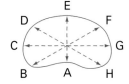

 a) Create an algebraic equation for this model.
 b) Create a table of the desired tread lengths for risers that range from 4 in. to 8 in. high.

18. **Problem Solving** A landscape technician often needs to estimate areas, particularly when working with unusual shapes. A kidney-shaped area is approximated as shown. The letters A to H represent the measurements from the approximated centre to the edges. The average of these lengths is calculated, and substituted into the formula Area = π × (average length)².

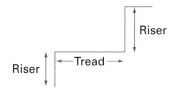

 a) What plane figure is this kidney shape approximating?
 b) How could the average length be calculated in this case?
 c) Determine an estimate for the area of this shape, if A = 1′, B = 4′4″, C = 4′9″, D = 2′6″, E = 2′2″, F = 2′7″, G = 4′8″, and H = 4′3″.
 d) Estimate the amount of edging this region would require.

19. Estimate the materials and costs to drywall the basement of your home, based on these sizes and costs.

½-in. Drywall	Cost ($)	Drywall Compound*	Cost ($)	Drywall Tape	Cost ($)
4 ft × 12 ft	12.48	7 kg ready mix	5.77	500 ft	3.98
4 ft × 10 ft	10.40	20 kg ready mix	10.48	250 ft	2.57
4 ft × 8 ft	8.33	*Use 67 kg per 100 m², or 138 lb per 1000 ft².			

<table>
<tr><td>**2.5**</td><td></td></tr>
</table>

2.5 Measurement for Technology and Design

Advances in technology have improved the speed and accuracy of measurement. Architectural technicians, interior designers, graphic designers, and engineers use computer programs such as *AutoCAD LT®* to complete scale diagrams in a fraction of the time it would take to draw by hand. Designers in many fields create drawings and products to measurement specifications outlined by the client.

EXAMPLE 1

Product Label

A graphic designer is creating the label for a can with a diameter of 9.2 cm and a height of 10.2 cm.

a) Will a circular graphic with an area of 110 cm² fit on the label?

b) If not, by what scale factor does the designer need to reduce the graphic?

Solution

a) The label of a cylindrical can is a rectangle the same height as the can, whose length is the circumference of the can. The limiting dimension is the height of the can, because it is smaller than the circumference. The radius of the circular graphic must be less than half the height of the can. Find the radius of the circular graphic, using the area of a circle formula.

$$A = \pi r^2$$
$$110 = \pi r^2$$
$$\frac{110}{\pi} = r^2$$
$$r^2 = \frac{110}{\pi}$$
$$r = \sqrt{\frac{110}{\pi}}$$
$$r \doteq 5.9$$

The diameter of the circular graphic is approximately 2 × 5.9 cm, or 11.8 cm. So, the graphic is too large to fit on the can.

b) The diameter of the graphic needs to be reduced to less than the height of the can, which is 10.2 cm.

$$\text{Reduction factor} = \frac{10.2}{11.8}$$

Reduction factor $\doteq 0.86$

If the designer reduces the graphic to 85% of its original size, it will just fit. Check whether the answer is reasonable.

Reduced diameter $= 11.8 \times 85\%$

Reduced diameter $= 10.03$

Since 10.03 cm $<$ 10.2 cm, the graphic will fit.

EXAMPLE 2

Theatre Seating Design

A design for movie theatre seating allows two people per square metre of floor area. A side view of the theatre layout is shown. A 1.5-m aisle runs across the theatre, as the side view shows, and another 1.5-m aisle runs from the front of the theatre to the back.

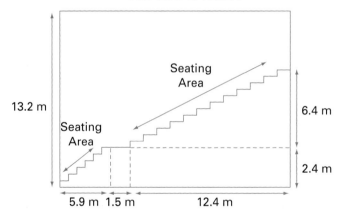

Side View of Theatre

a) About how wide should the theatre be to accommodate 750 people?

b) Find the approximate volume of air space that will need to be cooled by an air conditioning unit installed at this location.

Solution

a) Seating space required $= 750$ people $\times \dfrac{1 \text{ m}^2}{2 \text{ people}}$

Seating space required $= 375$ m^2

The available seating is divided into two areas, both on a slope. The floor space where the seats will be mounted, however, will be horizontal, in a series of steps. So, to calculate the area needed, use the horizontal distances 5.9 m and 12.4 m as the lengths of two rectangles. Combining these rectangles gives a total length of 18.3 m.

To find the width that gives an area of 375 m², substitute into the area formula for a rectangle.

$$A = l \times w$$
$$375 = 18.3 \times w$$
$$\frac{375}{18.3} = w$$
$$w \doteq 20.5 \qquad l \times w \doteq 20 \times 20 \text{ or } 400$$
$$\text{So, the answer is reasonable.}$$

When a 1.5-m aisle is added, the width of the theatre must be 20.5 m + 1.5 m, or 22.0 m.

b) To calculate the air space, divide the volume of the theatre into three prism-shaped sections: A, B, and C. Only an approximate value is required, so round your measures to numbers that are easier to work with. If you calculate the volumes to the nearest 100 m³, this will be an acceptable approximation.

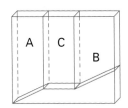

Sections A and B are trapezoidal prisms.

$$A_{\text{base of A}} = \frac{(s_1 + s_2)}{2} \times h_{\text{base of A}} \qquad A_{\text{base of B}} = \frac{(s_1 + s_2)}{2} \times h_{\text{base of B}}$$

$$A_{\text{base of A}} = \frac{(13 + 11)}{2} \times 6 \qquad A_{\text{base of B}} = \frac{(11 + 4)}{2} \times 12$$

$$A_{\text{base of A}} = 72 \qquad A_{\text{base of B}} = 90$$

Notice that the prisms lie sideways, so their "heights" are actually the width of the theatre.

$$V_A = A_{\text{base of A}} \times h_A \qquad V_B = A_{\text{base of B}} \times h_B$$
$$V_A = 72 \times 22 \qquad V_B = 90 \times 22$$
$$V_A \doteq 1600 \qquad V_B \doteq 2000$$

Section C is a rectangular prism, or cuboid.

$$V_C = l \times w \times h$$
$$V_C = 11 \times 22 \times 2$$
$$V_C \doteq 400$$

So, the approximate volume of air that must be cooled is 1600 m³ + 2000 m³ + 400 m³, or 4000 m³.

While approximation allows calculations to be made quickly and easily, there are situations when precision in measurement is critical. In many production-related jobs, you would use special instrumentation to measure materials with a high degree of accuracy. Nonetheless, all measurements contain some degree of error, because of the limitations of the instrument being used.

DISCOVER

Precision in Manufacturing

The table identifies, in both imperial and metric systems, the **degree of precision**, or smallest measurable unit, for various measurement devices.

Instrument	Degree of Precision	
	Imperial	Metric
steel ruler	$\frac{1}{64}$ "	0.5 mm
Vernier calipers	$\frac{1}{320}$ "	0.02 mm
micrometer	$\frac{1}{10\ 000}$ "	0.002 mm
precision gauge blocks	$\frac{2}{1\ 000\ 000}$ "	0.000 06 mm
electronic measuring devices	$\frac{1}{1\ 000\ 000}$ "	0.000 02 mm

1. Match the instruments with the pictures.

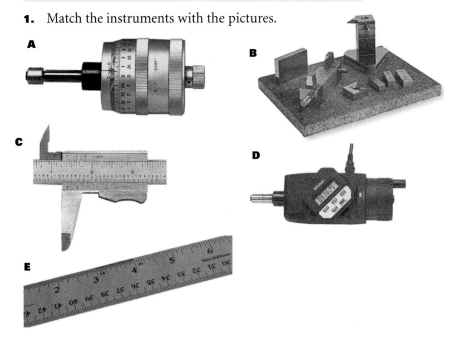

A

B

C

D

E

2. The greatest possible error for a measurement tool is one half of the precision (smallest unit of measure) of the tool. Find the greatest possible error for each measurement tool in the table above.

3. Copy and complete this table using your answers to question 2.

Measure	Instrument	Greatest Possible Error	Range of Actual Length
5.3″	Steel rule	$\pm\dfrac{1}{128}$″	5.292 187 5″ to 5.307 812 5″
$3\dfrac{3}{8}$″	Steel rule	$\pm\dfrac{1}{128}$″	$3\dfrac{47}{128}$″ to …
0.724″	Vernier caliper		
102 mm	Steel rule		
52.79 mm	Micrometer		
11.763″	Electronic measuring tool		
24.015″	Gauge blocks		

4. The degree of precision required in manufacturing depends on the type of item being produced. A rollerblade wheel does not have to be produced with the same degree of accuracy as a component used for the Canadarm on the International Space Station. The amount of variation permitted for a component is known as its **tolerance**. Based on the tolerance given, and the greatest possible errors you have calculated, identify the least accurate measuring instrument that can be used to check each item.
  a) the mould for a CD case, with a tolerance of 0.005″
  b) the metal teeth for a zipper, with a tolerance of 0.0004″
  c) a car engine part, with a tolerance of 0.000 05 mm
  d) a computer part, with a tolerance of 0.000 000 1″
  e) a label for a juice bottle, with a tolerance of 0.05 mm

EXAMPLE 3

Tolerance
A machine part can have a maximum dimension of 44.32 mm and a minimum dimension of 44.30 mm. Determine the mean dimension and the part's tolerance.

Solution
The **mean dimension** of the part is the mean of the maximum and minimum dimensions.

Mean dimension $= \dfrac{44.32 + 44.30}{2}$

Mean dimension $= 44.31$

The mean dimension is 44.31 m.

The tolerance is half the range of acceptable variation.

Tolerance $= \dfrac{44.32 - 44.30}{2}$

Tolerance $= 0.01$

The tolerance of the part is 0.01 mm. Therefore, the part can measure
44.31 ± 0.01 mm.
As a check, the range of values is given by
Mean dimension ± tolerance = 44.31 ± 0.01
So, the minimum dimension should be 44.31 mm − 0.01 mm, or 44.30 mm,
which it is. The maximum dimension can also be used as a check.

EXAMPLE 4

Reading Precision Instruments

a) Read the measure shown on the Vernier calipers on page 90.

b) Read the measure shown on the micrometer on page 90.

Solution

a) Vernier calipers can be read accurately in these five steps:

Step 1: Note the location of the "0" on the bottom Vernier scale. It is just past the 1″ mark, but does not reach the 2″ mark. So, the number of whole inches is 1.

Step 2: The next largest scale measures tenths of an inch. This time, the "0" is after the 3″ mark and just before the 4″ mark. So, add 0.3″ to the previous measure.

Step 3: The 0.1″ scale is divided into four sections, which represent 0.025″ each. In this case, the "0" is just after 3 of these divisions, for an extra measure of $3 \times 0.025″ = 0.075″$.

Step 4: Note the Vernier scale graduation that most closely coincides with a main scale graduation. The 3 mark on the Vernier scale lines up closest with a line on the main scale.
Add $3 \times 0.003\ 125″ = 0.009\ 375″$.

Step 5: Find the sum of all of these measures.
$1 + 0.3 + 0.075 + 0.009\ 375 = 1.384\ 375$
The complete reading on the calipers is 1.384 375″.

b) A micrometer can be read accurately in four steps:

Step 1: Observe the barrel scale. The barrel identifies the number of 0.100 in. In this case, the greatest 0.100 in. division is 0, so the first part of the measure is $0 \times 0.1″ = 0″$.

Step 2: Observe the number of 0.025″ divisions on the barrel scale *after* the last number showing on the barrel. In this case, there are 2. The extra measure is $2 \times 0.025″ = 0.05″$.

Step 3: Add the thimble reading that lines up with the horizontal line on the barrel. In this case, there are 0.6 of a possible 25 thousandths. The final part of the measure is $0.6 \times 0.001″ = 0.0006″$.

Step 4: Find the sum of all the readings.
$$0 + 0.05 + 0.0006 = 0.0506$$
The complete reading on the micrometer is 0.0506″.

KEY CONCEPTS

- Different instruments allow different degrees of precision in measurement. Vernier calipers and micrometers are two instruments that measure very small lengths. They are available in both metric and imperial units.

- Designers and manufacturing engineers are required to follow measurement specifications accurately. Tolerance is the degree to which the actual measurements of a product or component can vary from these specifications.

DISCUSS THE CONCEPTS

1. Why is it important for anyone working in design or manufacturing to be familiar with both the imperial and metric systems of measurement?

2. Discuss the relationship between precision and cost.

PRACTISE

(A) 1. Determine the maximum and minimum measures allowable, based on the given tolerance.

  a) $1.2699″ \pm 0.0009″$ **b)** $2.000″ \pm 0.002″$

  c) $3.0625″ \pm 0.005″$ **d)** $10.203\text{mm} \pm 0.024 \text{ mm}$

  e) $289.005 \text{ mm} \pm 0.007 \text{ mm}$ **f)** $64.86 \pm 0.03 \text{ mm}$

2. Express each of the tolerances in question 1 as a fraction.

Language Link

A "thou" is slang for one thousandth of an inch.

3. Copy and complete the table.

	Dimension	Tolerances	Maximum	Minimum
a)	2.005″	+0.000″, −0.004″		
b)	3.6435″	+0.000″, −0.0012″		
c)	1.0775″	+0.009″, −0.000″		
d)	24.16 mm	+0.00 mm, −0.06 mm		
e)	117.65 mm	+0.07 mm, −0.00 mm		
f)	397.216 mm	+0.009 mm, −0.000 mm		

4. Determine the maximum and minimum allowable wall thicknesses of the steel sleeve shown.

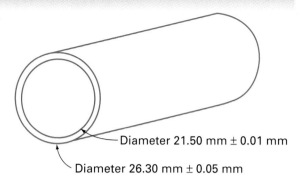

Diameter 21.50 mm ± 0.01 mm

Diameter 26.30 mm ± 0.05 mm

B **5.** **Graphic design** A triangular graphic with an area of $4\frac{1}{2}$ in.^{2} and a base of $2\frac{1}{8}$ in. is to be placed on a square CD cover with a side length of $4\frac{3}{4}$ in.

a) Show how the graphic will fit onto the CD cover.

b) Estimate the number of times that this graphic (in any orientation) can be repeated on the outside back cover.

6. The locations, with tolerances, of two holes that are to be drilled in a piece of iron are shown in diagram B. The machinist drills the holes and then checks for proper locations by taking the measurements shown in diagram A. Are the holes within tolerance? Explain.

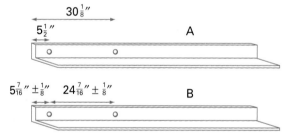

$30\frac{1}{8}''$

$5\frac{1}{2}''$ A

$5\frac{7}{16}'' \pm \frac{1}{8}''$ $24\frac{7}{16}'' \pm \frac{1}{8}''$ B

7. For the part shown,

a) find the area

b) express this area algebraically using π, r, and R

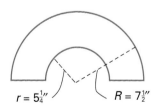

$r = 5\frac{1}{4}''$ $R = 7\frac{1}{2}''$

8. A sine plate is used to create angled cuts in various materials. A 10″ sine plate is tilted on a 45° angle as shown. Use trigonometry to determine the height, x, of the gauge blocks.

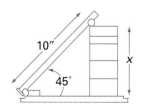

10″

45°

x

9. **Wheelchair access** You are designing a wheelchair ramp to be installed at a local restaurant.

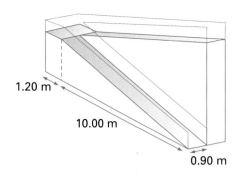

1.20 m

10.00 m

0.90 m

a) If the slope of each section of the wheelchair ramp is 1:12, what is the height of the ramp?

b) Circular cross-section handrails with a diameter of 4 cm are recommended. A person has a hand span of 18 cm. Will their hand go all the way around the handrail?

c) What length of handrail, in feet, is required to cover both sides of the two sloping sections and three sides of the flat section halfway up?

10. The end portion of this piece of aluminum stock is a regular hexagon. Find the following:

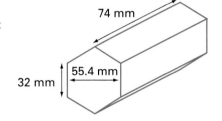

74 mm

55.4 mm

32 mm

a) the cost to coat this stock entirely in a sealant at a rate of $0.02/mm^2

b) the dimensions of the aluminum block from which this stock can be cut

c) the total waste material removed from the aluminum block

d) the mass of the aluminum stock, given that the density of aluminum is 2.6 g/cm^3

11. How many square inches of waste are there in the sheet metal stamping shown? Round your answer to the nearest hundredth.

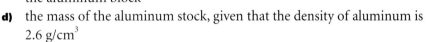

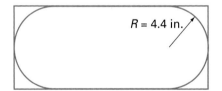

$R = 4.4$ in.

12. The part design shown is sent from Sweden to your tool and die company, and gives all measurements in centimetres. Your machinery, however, is set to imperial measure.

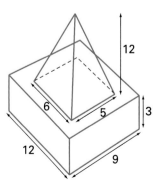

12

6

5

3

12

9

a) What is the total volume of the object, to the nearest hundredth of a cubic inch?

b) If the pyramid is cut from a block 6 cm × 5 cm × 12 cm, how much waste is there, in cubic centimetres?

c) What percent of the original block is wasted?

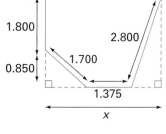

13. Wire-cutting costs In this figure, all measurements are in inches.

a) Determine the length *x*.

b) Calculate the area of the figure.

c) Calculate the perimeter of the figure.

d) Wire-cutting, a process that uses an electric current to cut metal, is being used to cut this piece. The fee charged for wire-cutting is based on the product of the perimeter and the thickness. If this figure is to be $\frac{1}{2}$ in. thick, find the total cost of wire-cutting

    i) at a rate of $11.00/in.^{2} (for a tolerance of 0.001 in.)

    ii) at a rate of $25.00/in.^{2} (for a tolerance of 0.0001 in.)

14. Problem Solving, Application You are arranging furniture in your new home. Your living room/dining room space is rectangular and measures 14′ by 17′. You have been advised to follow these furniture placement guidelines to ensure movement is unrestricted:

- Traffic paths need to be a minimum of 24″.
- Allow 14″ to 18″ between a coffee table and the front of the sofa.
- The distance between the television and the seating should be three times the screen size; for example, for a 30″ screen, you should sit 90″ away from the television.
- In the dining area, allow two feet between the back of the chair and any other piece of furniture or wall.

Draw a scale diagram, showing how you will place the following pieces of furniture in the dining room/living room area:

- couch (rectangular, 80″ by 39″)
- chair (square base, 36″ by 39″)
- entertainment centre (rectangular, 6′ by 23″) plus 28″ screen television
- coffee table (52″ by 20″)
- dining room table (circular, with a radius of 21″)
- four dining room chairs (rectangular, 17.5″ by 18″)

15. Interior zoning The process of situating the elements of a room to meet peoples' needs is sometimes called interior zoning. Interior designers learn about the average working zone for someone in a kitchen, for example.

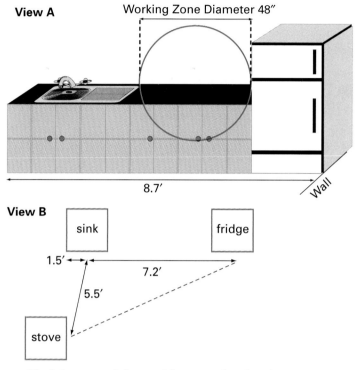

a) Find the area of the working zone in view A.

b) Designers also identify the work triangle in a kitchen as the figure created by the fridge, the stove, and the sink. It is suggested that the perimeter of the work triangle be minimized (ideally, less than about 23′) to reduce travel between key work areas. If the stove for this kitchen is located as shown in view B, does this kitchen meet the work triangle requirements?

16. Problem Solving, Application
  a) Sketch the top views of three homes with a foundation of 2400 ft² each.
  b) Calculate the perimeter of each home in part a).
  c) Compare the ratio of perimeter to area for each of these homes.
  d) If a home has a high perimeter-to-area ratio, then it is proportionally more exposed to the outside environment, reducing energy efficiency. Rank the homes you have sketched from least to most energy efficient.

17. Estimate the number of 3.78-L cans of paint required to paint the bedroom, hallway, and rec room in the basement of your new home (walls and ceiling), if 1 L covers about 130 ft². What is the cost, if each can of paint costs $31.88? Refer to the diagram on page 65.

CAREER PROFILE

Practical Nurse

Practical nurses are on the front line of health care, helping patients in hospitals, care centres, and homes. The profession demands someone who is caring, has good observational and communication skills, and is knowledgeable in health care. A nursing student is expected to develop effective communication and interpersonal skills, problem-solving and decision-making abilities, and a legal and ethical awareness. Nurses must also function as part of a team, to discuss and provide appropriate nursing care for the patient.

Not all community colleges provide nursing programs. All the nursing programs that are available prepare students for examinations held by the Canadian Nurses Association (CNA) for registration in Ontario. Most colleges use a laddering system in which the practical nurse can continue with education and obtain a degree in nursing.

 Go to *www.mcgrawhill.ca/links/MPCA12* and follow the links to find out more about the diploma nursing programs at certain colleges.

Administering medications is an important role for the practical nurse. Answer the following questions on medication dosage.

1. An analgesic (pain killer) is to be given by injection from a 10 cc vial. The medication has a dosage of 3 mg per cc. The patient is to receive 15 mg of the analgesic.
   a) How many cc of the medication should be given to the patient?
   b) How much of the medication remains in the vial?

2. An antibiotic powder containing 2000 units of drug is to be mixed with 17.5 cc of sterile water. The solution of powder and water expands to 20 cc of solution. How much of the drug is in 1 cc of solution?

3. A 250-mL bottle of cough syrup contains 30 mg of active ingredient in each millilitre. How much would you give if 0.9 g were ordered? How much active ingredient is in the bottle?

CHAPTER PROBLEM WRAP-UP

You have recently purchased a new house and, like many new home-buyers, you are anxious to make your house a home. In particular, you would like to improve the unfinished basement area. There are several home-improvement projects related to the basement that you would like to undertake. Therefore, you need to prioritize according to need and affordability. To minimize costs, you are going to "do-it-yourself."

In section 2.1, question 14, you were asked to suggest a plan for covering the basement floor. In section 2.2, question 19, you priced different options for window blinds. In section 2.4, question 19, you estimated the materials and costs to drywall the basement. Finally, in section 2.5, question 17, you calculated the cost involved in painting the walls and ceiling of the basement.

1. Develop an action plan for completing all four home-improvement projects over the next three years, as follows.

   a) Provide a detailed cost description for the projects, and justify the priority given to each.

   b) You currently have $1000 in savings allocated to home improvement, and you expect to commit about $200 monthly for the same purpose. Assume that you are investing your savings at a rate of 5% interest per year, compounded annually. Use a spreadsheet to calculate your available savings at three-month intervals, over the period of the next three years.

   c) Develop a schedule for the three years, showing how you plan to complete each project as your funds allow.

   d) Modify your spreadsheet to show the costs of each project as a debit to your savings account.

2. What factors, other than cost, will have an impact on your action plan?

3. Identify three additional projects that you may choose to undertake in your new home. Suggest how these projects might fit into your existing action plan and savings budget.

Review

2.1 Systems of Measure, pages 56–66

1. Convert the following.

   a) 50″ to feet   **b)** 5 qt to fluid ounces

   c) 245 T to pounds   **d)** 418 ft³ to cubic yards

   e) 500 yd to miles   **f)** 3 lb 4 oz to pounds

2. Find the perimeter and area of each shape.

   a)

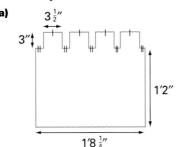

   b)

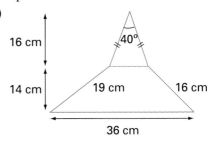

3. Find the surface area and volume of each figure.

   a)

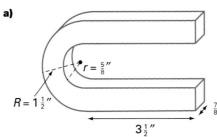

   b)

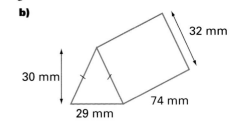

4. **Plant nursery** The region shown is to be filled with topsoil to a 6″ depth.

   a) Determine the volume of topsoil needed, in cubic feet.

   b) The nursery sells topsoil by the cubic yard. Given that 1 yd³ = 27 ft³, how many cubic yards of topsoil are required?

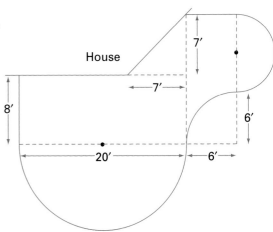

2.2 Converting Between Metric and Imperial Systems, pages 67–73

5. Convert these quantities.

a) 3.5 gal to litres

b) 250 lb to kilograms

c) 18 ft to metres

d) 12.5 oz to grams

e) 6.8 L/100 km to miles per gallon

f) 60 km to miles

g) 17.014 mm to inches

h) 225 g to pounds

6. **Interior decorating** A rectangular room measures 4.2 m by 3.0 m. How much will it cost to carpet the room at $27.50/yd^2?

7. You plan to cover your 34.6 m × 5.2 m driveway with stone to a depth of 5 cm.

a) How many cubic yards of gravel need to be ordered?

b) What is the mass of this gravel, if it weighs 185 lb/ft^3?

8. The following Ontario and American cities are border neighbours. The cost of fuel in each city is listed. If you assume the exchange rate is $1.00 U.S. = $1.38 Canadian, determine the less expensive fuel for each pair of cities.

a) Windsor, Ontario ($0.649/L) and Detroit, Michigan (U.S. $1.29/gal)

b) Sarnia, Ontario ($0.635/L) and Port Huron, Michigan (U.S. $1.35/gal)

c) Fort Erie, Ontario ($0.715/L) and Buffalo, New York (U.S. $1.31/gal)

d) Sault Ste. Marie, Ontario (68.5 cents/L) and Sault Ste. Marie, Michigan (U.S. $1.36/gal)

2.3 Measurement for Your Health and Fitness, pages 74–79

9. Convert the temperatures as indicated.

a) 20.5°C to degrees Fahrenheit

b) 46°F to degrees Celsius

10. **Problem Solving** Your fitness program includes 15 mi of running per week. You run on the indoor track shown.

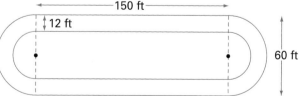

a) What is the difference in length between a lap you run on the inside of the track and a lap you run on the outside?

b) How many laps should you run per week to reach your fitness program goal, if you run on the inside of the track?

c) It takes you 40 s to run one lap on the inside of the track. What is your speed, in metres per second? What is your speed in miles per hour?

11. Drug dosage A drug is to be given at a rate of 45 mg per kilogram of body mass per day. A 145-lb person is to be given two equal doses per day.

   a) How much is each dose?

   b) If the medicine has a concentration of 500 mg/mL in liquid form, what volume must be taken for each dose?

2.4 Measurement for Home Improvement and Construction, pages 80–86

Use the diagram to answer questions 12 and 13.

12. Garden planning

   a) If the radius of the circular flower bed is 5 ft, calculate the amount of soil required for the flower bed at a depth of 8 in. Express your answer in cubic feet and in cubic yards.

   b) If the flowers you wish to plant need to grow in a space of 0.8 ft^2 each, how many should you plant in this bed?

   c) The bricks surrounding the flower bed are 5 in. wide. What is the length of the border if the radius is calculated to the centre of the brick?

   d) Determine the approximate number of bricks required, assuming a brick length of approximately 9 in. and a brick thickness of 4 in. How many bricks are needed so that they are flush with the top of the soil?

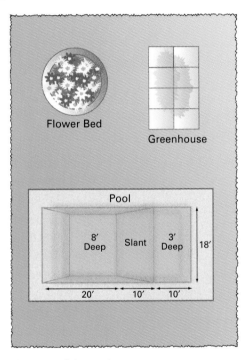

13. Swimming pool equipment

   a) What area should the solar blanket have for it to be flush with all four sides of the pool shown above?

   b) How many litres of water are in the pool, if it is filled to 6 in. below the top edge?

   c) If the hose you use to fill the pool flows at a rate of 30 L/min, how long will it take to fill up the pool in the spring?

   d) If the concrete patio surrounding the pool is to have a uniform width of 5 ft and a thickness of 6 in., how much concrete must be ordered, in cubic yards?

   e) Draw a side view of the pool, labelling all dimensions.

14. Greenhouse construction

  a) How many square feet of glass are required to build the greenhouse shown?

  b) To ventilate the greenhouse, a fan is needed. Fans are sold according to the number of cubic feet of air moved in a minute. How many cubic feet of air are in the greenhouse?

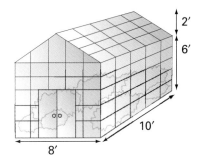

2.5 Measurement for Technology and Design, pages 87–97

15. a) What is the measure shown by each of these instruments?

i)　　　　　　　　　　　　　　**ii)**

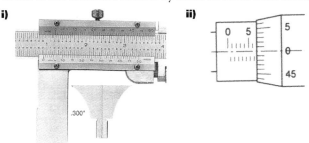

  b) What is the measure of error for each of the above instruments?

16. Copy and complete the table.

	Dimension	Tolerances	Maximum	Minimum
a)	1.245 "	±0.002"		
b)	$3\frac{1}{64}$ "	$\pm\frac{1}{128}$ "		
c)	34.725 mm	+0.000 mm, –0.005 mm		
d)	42.593 cm	+0.004 mm, –0.006 mm		

17. Product design

  a) A television is described as having a 19″ screen, which means that it has a diagonal measure (from bottom corner to opposite top corner) of 19″. If the screen's width is $15\frac{1}{4}$″, what is the angle of inclination of the diagonal?

  b) A larger television has a 30″ screen. If the width of the television screen is 24″, is the angle of the diagonal significantly different than for the smaller television? Justify your answer.

  c) Measure the dimensions of various sizes of television screens. What is the angle of the diagonal of each screen? Are the angles the same? What would happen to the view of the screen if you increased or decreased the angle?

  d) How do the dimensions of a television screen differ from a computer monitor?

Practice Test

Category	Knowledge/Understanding	Thinking/Inquiry/Problem Solving	Communication	Application
ACHIEVEMENT CHART CATEGORY CONNECTIONS				
Questions	1–6	1, 3, 6	6	6

1. Calculate the surface area and volume of each of the following.

a) This square-based pyramid, at the Louvre museum in Paris, France, has a base with an area of 900 m². You estimate the height to be about 20 m.

b) The Cinesphere at Ontario Place in Toronto has a radius of 62 ft 4 in.

2. Complete each of the following conversions.

a) A car has a fuel consumption of 6.3 L/100 km. What is this in miles per gallon?

b) A project requires the use of a socket to loosen a $\frac{3}{16}$″ bolt. Your metric socket set includes the following sizes: 4 mm to 15 mm (in 1 mm increments), 17 mm, 19 mm, 20 mm, and 22 mm. Which socket could you use?

c) You must take a medication if your temperature exceeds 101.5°F. Your digital thermometer reads 38.8°C. Is medication required?

d) You are purchasing blinds for a window that measures 180.3 cm by 139.7 cm. At the store, the blinds are measured in inches. What size blinds should you purchase?

e) If a 1.36-kg bag of nuts sells for $11.99, what is the price per pound?

3. **a)** What is the total surface area of the shaded portions?
  b) Express this area in square inches.
  c) Determine the total volume of material used, in cubic millimetres and cubic inches.
  d) What is the difference in material costs for this component, comparing aluminum ($0.50/in.^{3}) and stainless steel ($2.50/in.^{3})?

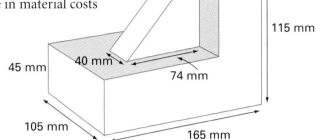

4. What is the name of each instrument?

a)

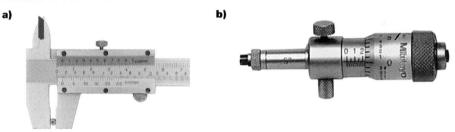

b)

5. Find the maximum and minimum allowable measures.
  a) 12.048″, ± 0.010″   **b)** 13.586 mm, +0.0003 mm, −0.0007 mm

---

ACHIEVEMENT CHECK   Knowledge/Understanding   Thinking/Inquiry/Problem Solving   Communication   Application

6. The layout of a property is shown. The dotted line represents the location of a fence that must be just within the property line. Fence posts are to be 6′ to 8′ apart, from centre to centre. A gate 5′ wide will be installed onto side E.

  a) For aesthetic reasons, you would like the fence posts for each side to divide that side into equal sections. What is the length between the fence posts on each side?
  b) How many fence posts will you need?
  c) What is the total length of fence?
  d) If the cost of sod is $0.55/ft^2, find the cost to sod the fenced area. Explain your calculation.

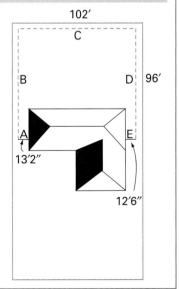

GEOMETRY IN DESIGN

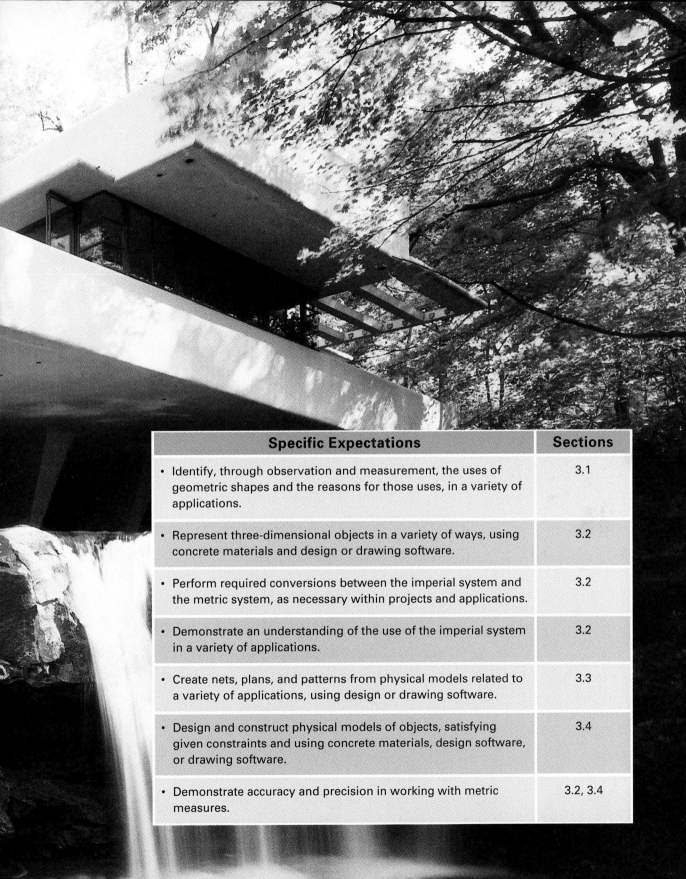

Specific Expectations	Sections
• Identify, through observation and measurement, the uses of geometric shapes and the reasons for those uses, in a variety of applications.	3.1
• Represent three-dimensional objects in a variety of ways, using concrete materials and design or drawing software.	3.2
• Perform required conversions between the imperial system and the metric system, as necessary within projects and applications.	3.2
• Demonstrate an understanding of the use of the imperial system in a variety of applications.	3.2
• Create nets, plans, and patterns from physical models related to a variety of applications, using design or drawing software.	3.3
• Design and construct physical models of objects, satisfying given constraints and using concrete materials, design software, or drawing software.	3.4
• Demonstrate accuracy and precision in working with metric measures.	3.2, 3.4

Get Ready

1. **Geometric shapes and solids**
   a) Identify each of these polygons.

   b) The sum of the interior angles of a polygon with *n* sides is given by the formula $(n - 2) \times 180°$. Determine the sum of the interior angles of each regular polygon in part a).

See Appendix B, page 442, for more information on *The Geometer's Sketchpad®*.

2. **Geometric shapes and solids** Accurately draw each regular polygon, up to and including an octagon. These figures can be drawn with *The Geometer's Sketchpad®* or using a pencil, protractor, and compasses.

3. **Geometric shapes and solids** A Platonic solid, also called a regular polyhedron, is a particular, highly symmetric type of three-dimensional object with these properties:
   • The faces are all identical regular polygons.
   • Each vertex joins the same number of polygons.
   Each Platonic solid has the same appearance, no matter which face is at the front. Only five Platonic solids exist.

   | Tetrahedron | Cube | Octahedron | Dodecahedron | Icosahedron |

   a) Identify which polygon is the face type of each polyhedron.
   b) For each of the five Platonic solids, determine the number of faces, edges, and vertices. Copy and complete the table.

Platonic Solid	Number of Faces	Number of Edges	Number of Vertices
Tetrahedron			
Cube			
Octahedron			
Dodecahedron			
Icosahedron			

4. Geometric shapes and solids A typical house has the structure of a cube with a triangular prism on top. Design a futuristic house using different polyhedron shapes.

5. Geometric shapes and solids
   a) Draw a square. From the square, create a cube.
   b) Draw a circle. From the circle, create a sphere.
   c) Draw a sector of a circle. From the sector, create a cone.
   d) Describe what you did in each step to create a three-dimensional appearance from the first shape.

6. Perimeter and area
   a) Determine the dimensions of the foyer, and find its area.
   b) Find the perimeter of the foyer.
   c) Find the area of the entire living area.

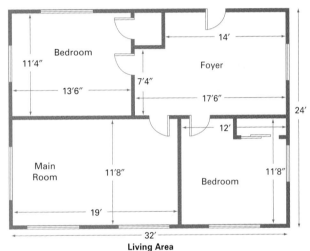

Living Area
Allow 2 in. for thickness of interior walls.

7. Geometric shapes and solids Look around the room. List the two- and three-dimensional geometric shapes in the objects that you see.

CHAPTER PROBLEM

The more electronic gadgets we acquire, the less space we seem to have! You need to solve Chris's problem. She is moving to a college dorm and needs to organize her television, stereo system, and DVD player to fit the space she has in her room.

3.1 Geometric Shapes in Design

Most objects in our day-to-day lives are created in a process that moves from an idea to a sketch, then to a technical drawing, and finally to production. Geometry is the basis for the design of many objects. Geometry can improve these aspects of a design:

- the stability of the *structure* through the distribution of forces
- the *strength* of the finished product
- the *function* of the object as fulfilled by its design
- the *aesthetics* of the design

Geometry is also essential to creating realistic scenery and characters in computer games. Computer game objects and levels are created using geometric modelling software. You can sometimes be aware of this software when you move through a game level and you see an "edge."

DISCOVER

Finding Geometry in Design

These are examples of the different uses of geometry in design. For each design,

a) describe the basic shape or combination of shapes used in the design

b) discuss, perhaps with a partner, the reason you think the shapes were used in the design

c) describe how geometry improves any or all of these design aspects: structure, strength, function, and aesthetics

1. Geometry in engineering: The Confederation Bridge spans 13 km to link Prince Edward Island and New Brunswick.

2. Geometry in music: The electric guitar combines ease of playing, support for the strings, and a dynamic appearance.

Language Link

A geodesic line on a sphere is the shortest distance over the surface between any two points. In the geodesic dome invented by Buckminster Fuller, triangle vertices in different parts of the dome are linked by strong geodesic lines made from triangle edges. Buckminster Fuller developed a geometry he called synergetics, based on the patterns he saw in nature. The geodesic dome is one example.

3. Geometry in architecture: In the geodesic dome of the Cinesphere at Ontario Place, the triangular form provides strength when braced against other triangles. This arrangement gives the maximum strength for a spherical shell structure, using a minimal amount of material.

4. Geometry in fine arts: Geometric forms can be used to analyse and (as a beginning art student) create beautiful drawings and paintings. In this drawing by Leonardo da Vinci, the shapes in each figure (for example, the torso of the animal) have been modelled as ovals or other curved forms.

DISCOVER

Natural Design: The Golden Ratio

In nature, as in human-made design, geometry's role is aesthetic, strengthening, structural, and functional. The Golden Ratio is a ratio of dimensions that is often found in nature. It can be observed in the growth of seeds in a flower, the spiral of a seashell, and in the proportions of a human body. The Golden Ratio is approximately 1:1.618. An object with Golden Ratio proportions is 1.618 units long for every unit it is wide. Thus, a golden rectangle 36 cm wide is 58 cm long, to the nearest centimetre.

1. Determine the ratio of each pair of measurements by dividing the larger measurement by the smaller.
   a) A dog standing 40 cm tall at the shoulder, and measuring 65 cm from shoulder to hindquarters
   b) The Parthenon, an ancient temple in Athens, Greece, measuring 18 m high by 31 m wide
   c) The length and width of this nautilus seashell

2. For each ratio in question 1, find the percent difference between this value and the Golden Ratio. For example, for the ratio 1.513,

$$\frac{1.513}{1.618} \times 100\% = 93.5\% \text{ and } 100\% - 93.5\% = 6.5\%$$

So, 1.513 is 6.5% less than the Golden Ratio.

3. With a partner, measure pieces of furniture in your classroom, or objects in the school. Do any of these items conform to the Golden Ratio? Use percent difference calculations to support your answers.

4. Explore facial proportions. With a clear plastic ruler, take your partner's facial measurements. Measure the vertical or horizontal distance between each of the two locations listed.
   • Use the schematic below as your guide.
   • For each measurement, measure "flat" to the plane of an imaginary window between you and your partner's face.
   • Measure to the nearest tenth of a centimetre.

   a = top of head to chin
   b = top of head to pupil (centre of eye)
   c = pupil to tip of nose
   d = pupil to lip
   e = width of nose
   f = outside distance between eyes
   g = width of head
   h = hairline to pupil
   i = tip of nose to chin
   j = lips to chin
   k = length of lips
   l = tip of nose to lips

Language Link

Anthropometry, from the ancient Greek words *anthropos* (human) and *metron* (measure), is the measurement of the human body in terms of dimensions, proportions, and ratios.
Anthropometry is used in the manufacture of clothing to plan size ranges.

5. Calculate the following ratios, using your measurements from question 4.

$$\frac{a}{g}, \frac{b}{d}, \frac{i}{j}, \frac{i}{c}, \frac{e}{l}, \frac{f}{h}, \frac{k}{e}$$

How close are these ratios to the Golden Ratio?

6. Switch roles with your partner, and repeat questions 4 and 5.

7. **a)** Suggest other natural proportions, or proportions in architecture, that might be in the Golden Ratio.

   b) Do an Internet search on the Golden Ratio to test your ideas.

EXAMPLE 1

Architectural Form

Frank Lloyd Wright's architecture is widely known for his use of clean geometric lines. During the 1920s and 1930s, at a time when others were creating ornately decorated buildings, Frank Lloyd Wright stripped his buildings of ornamentation and expressed them in their geometric form.

His buildings, such as Fallingwater at Bear Run, Pennsylvania, (pictured here and on page 106) often appear simple, like a child's building block sculpture.

a) Describe and comment on the main shapes used in this house.

b) Explain how Frank Lloyd Wright used geometry to enhance his design.

Solution

a) The basic shapes are shallow rectangular prisms. The layout and look of the house are determined by the way these prisms fit together.

b) Geometry enhances the design in the following ways:

- Structure: The components of the building are relatively easy to construct. The low profile of the building enhances its stability.
- Function: Frank Lloyd Wright believed that architecture should be tailored to its use by humans. In this building, the rectangular prisms are designed to create pleasant and practical open-plan living spaces.
- Aesthetics: Repeated use of the same shape throughout the design provides symmetry and balance for greater aesthetic appeal. Frank Lloyd Wright used these geometric forms in a naturalistic arrangement to echo the staggered horizontals in the natural landscape.

Find more information on this beautiful house and its architect on the Internet. Go to *www.mcgrawhill.ca/links/MPCA12* and follow the links.

EXAMPLE 2

Computer Game Graphics

The best computer games feature the most realistic graphics. What makes this figure look convincing?

Solution

The appearance of three-dimensional depth is very convincing. It is created in this scene by the use of perspective (you will study this further in section 3.2). During game play, your point of view shifts as the character turns, runs, and jumps, adding to the feeling of "virtual reality." From the Designer screen, new figures can be created. During this process, the geometry is obvious, as shown here.

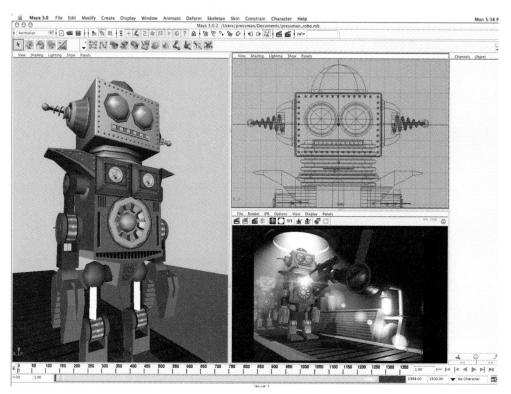

EXAMPLE 3

Computer Ergonomics

Ergonomics is a science that studies people at work. It applies information about human characteristics to the design of tools, machines, systems, tasks, jobs, and environments for productive, safe, comfortable, and effective human use. Ergonomics draws on a variety of diverse disciplines: physiology, engineering, psychology, biomechanics, environmental studies, and anthropometry.

Keyboards and mice are the main sources of computer input, and should be comfortable to use. Discuss how these pieces of hardware have changed during the history of computers.

Solution

The keyboard on the right is straight. Your hands and body must conform to the keyboard and screen arrangement. The natural keyboard below conforms to the natural angles the hands form when brought together.

Likewise, with mouse design, the more modern design fits the hand's natural shape and requires less movement.

KEY CONCEPTS

- Geometry is used in design to improve structure, strength, function, and aesthetics.

- The Golden Ratio is often found in natural dimensions and is aesthetically appealing.

- Geometric design can create convincing computer graphics, improve the ergonomics of computers and other tools, and improve many other aspects of everyday life.

DISCUSS THE CONCEPTS

1. a) Describe the different geometric shapes you often observe in the course of a day. These could include landscape forms, people, animals, plants, product packaging, buildings, and vehicles.

b) Discuss how geometry influences the design of any manufactured objects you describe.

2. Think of three areas of design where geometry is used as the main tool. Give an example of each, describing how the geometry affects the design.

APPLY THE CONCEPTS

Ⓐ 1. Visit a grocery store, or use grocery store advertisements, and look at the containers of packaged food. Compare two containers that show the same volume or mass, but look different in size. Explain what was done to the design to make one look larger.

2. What are the similarities and differences between polygons and polyhedra?

3. **Fine arts** Artists sometimes use wooden mannequins to model the human form when drawing. Name or describe the main geometric shapes that are used to create this model.

Ⓑ 4. **Computer games** Choose a favourite computer game.
a) Consider the objects, or players. Describe the shapes used to create these characters.
b) Now, look at a particular level, possibly in a navigational map. What do you notice? Draw the main elements of a level, using regular polygons and polyhedra.

5. **Inquiry, Communication** Consider the chair you are sitting in now. Are you comfortable? Are your lower legs at a 90° angle with your thighs? Are your thighs at a 90° angle with your torso? These are indicators of good ergonomic arrangement and design in your desk and chair. On the other hand, some innovative desk chair designs produce a kneeling posture that is also good ergonomically, and can help reduce back strain.

   a) Choose a desk chair design (perhaps from a furniture catalogue) that you believe would be ergonomically effective. Analyse its design in terms of structure, strength, function, and aesthetics.

   b) In groups of three or four, measure anthropometric data for each group member. Copy the table and record the data for each person.

   c) Calculate the averages for your group.

   d) What additional measurements would you need to design a desk chair using the average data, and why? Make these additional measurements.

Measurements	Person 1	Person 2	Person 3	Person 4	Average
Shoulder to elbow					
Hip to knee					
Knee to foot					
Back of knee to lower back					

6. **Wallpaper design** Frieze designs for wallpaper feature a pattern, or combination of patterns, with symmetry about one or more lines. Translational symmetry is the repetition of a pattern along the length of the band. If the pattern is flipped about a horizontal line, it also has reflection symmetry. A pattern can also have reflection symmetry about a vertical line.

   a) Describe the translation and reflection symmetries in the design shown below.

Translation or slide

Horizontal reflection followed by slide or translation

Vertical reflection

Horizontal reflection followed by vertical reflection

Horizontal reflection followed by vertical reflection

Horizontal reflection followed by horizontal translation followed by horizontal reflection

Horizontal reflection followed by vertical reflection followed by horizontal reflection

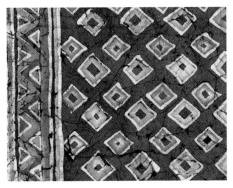

b) Many patterns also have rotational symmetry—a pattern is rotated about a central point. Describe any rotational symmetries you can find in the design.

c) Create a wallpaper design that incorporates one or more symmetries. Then, analyse your design for any extra symmetries, besides the symmetries you planned to include.

7. **Inquiry, Application** Quilt blocks are made of a variety of simple shapes joined to form other interesting patterns.

a) Find some quilt blocks in a book or catalogue, and describe how the shapes are used to form patterns and symmetries.

b) Use regular polygons in a range of colours to create a quilt block of your own. If you have access to geometry or drawing software, you may use it in creating your design.

8. This spiral construction models the nautilus shell shown on page 112.

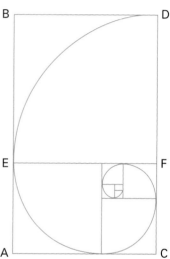

a) Analyse the construction in terms of similar rectangles and the Golden Ratio.

b) Suppose the smallest rectangle in the centre has a width of one unit. What is the width of the smallest square, which is beside it? What is the width of the next larger square? What is the ratio of these two dimensions?

c) Explain how the width of each larger square is related to the next two smaller widths. Does this remind you of a particular sequence of numbers?

d) Draw this construction and reproduce the spiral. Use *The Geometer's Sketchpad*® or a set square and a pair of compasses. (Hint: Start from the outer rectangle.)

e) Suggest a reason why this spiral models the nautilus shell so effectively.

9. Refer to your list of shapes from question 7 on page 109.

a) Classify the shapes as
i) regular polygons **ii)** other two-dimensional shapes
iii) Platonic solids **iv)** other three-dimensional shapes

b) Which shapes appear to be most widely used? Why?

c) Do any of these shapes involve the Golden Ratio?

d) Based on your geometric analysis, suggest ways to improve the design of the room
i) ergonomically **ii)** aesthetically

10. **Graphic design** Imagine that you are starting a new company. You will need a logo to make your products easily identifiable. Design a logo for your company using drawing software, *The Geometer's Sketchpad®*, or drawing tools and paper.

11. A Chinese tangram is a famous puzzle that uses polygons for its pieces.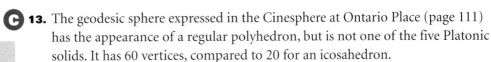

    a) Create a tangram by cutting a square piece of card into the pieces shown in the diagram.

    b) Identify the shapes that are used in this puzzle.

    c) Form as many of these shapes as possible, by laying some or all of the tangram pieces next to each other. You are not allowed to overlap pieces.

       i) triangle   **ii)** trapezoid

      iii) pentagon   **iv)** hexagon

       v) human figure

    d) Which, if any, of the shapes could not be formed? Suggest reasons why.

*C*HAPTER
PROBLEM

12. Chris's television is a 21″ model with a built-in VCR player. The television is 22″ wide, 19″ deep (front to back), and 19.5″ high. It cannot support anything stacked on top of it. The stereo components are all 17″ wide and 15″ deep. The tuner is 6″ high, and the CD player and DVD player are each 3″ high. Chris's speakers have their own stands, and can be easily placed anywhere in the room, so they do not need to be on the entertainment unit. There are two places in the dorm where Chris can place the entertainment unit. One is approximately 30″ wide and the other is approximately 45″ wide. Using a variety of shapes, sketch a few designs for each of the two spaces available.

C 13. The geodesic sphere expressed in the Cinesphere at Ontario Place (page 111) has the appearance of a regular polyhedron, but is not one of the five Platonic solids. It has 60 vertices, compared to 20 for an icosahedron.

Science Link

Buckminster Fuller's geodesic sphere is often called a Buckyball. Buckyballs are even found in nature in the form of C_{60}—a rare carbon molecular structure with a carbon atom at each of the 60 vertices.

    a) Fit five equilateral triangles together around one vertex. Is the result curved or flat? Now try fitting six equilateral triangles together in the same way. Is the new arrangement curved or flat? Explain your findings.

    b) The triangles in an icosahedron are equilateral, and five of them fit around each vertex. Are the triangles in a geodesic sphere actually equilateral? Why or why not? Compare your answer with the definition of a regular polyhedron or Platonic solid on page 108. From this, explain why the geodesic sphere is not a regular polyhedron or Platonic solid.

Representing Three-Dimensional Objects

To construct an object, whether it is a simple box, a camera, a dress, or a house, you need to have a plan. In design and engineering, plans are expressed in technical drawings, which provide the necessary information to make the object.

There are different types of technical drawings and representations:
- orthographic drawings, which show an exactly scaled front, top, or side view
- views that are angled to show more than one side of a structure simultaneously
- perspective views and renderings, based on angled views, which are technically exact but are also more realistic-looking
- scale models, which represent the structure in three dimensions

This section will focus on orthographic drawings, perspective views, and scale models.

DISCOVER

Orthographic Drawings

Orthographic drawings show the front, side, and top views of a structure. They are sometimes called multi-view or three-view drawings. Each view represents the size and shape of the structure's features as accurately as possible. All the key measurements are clearly marked. The scale used is clearly stated. These orthographic drawings of a component show the front and left-side views. The scale is 1:5; so, each dimension is drawn at one fifth of its actual measure. The lower the second number in the scale is, the closer the drawing is to the actual size. A small component might be magnified in a scale drawing. In this case, the first number in the scale is larger than the second; for example, 2:1.

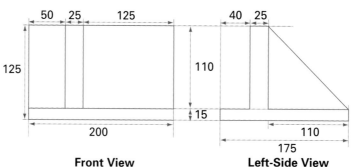

Scale = 1:5
All measurements are in millimetres.

Front View   **Left-Side View**

1. Copy the views accurately, using fine pencil. Make each measure exactly to the scale of 1:5. For example, the overall height of 125 mm should measure 25 mm on your drawing. Use a set square or protractor to create accurate right angles. (If you have access to drawing or geometry software, you may recreate the views on a computer.)
2. Lightly draw a vertical line downward from each vertical feature in the front view. These vertical lines will guide your construction of a top view.
3. Draw a horizontal line to represent the back of the component. The side view shows that the next feature of the component, the tall ridge, is 40 mm forward from the back of the component. Since the scale is 1:5, draw another horizontal line 8 mm below the first one, for the back of the ridge.
4. Use the other measurements from the side view to place the remaining horizontal lines in your top view.
5. Draw vertical lines for the left and right edges of the component and the sloping flange. Use your vertical guide lines and the horizontal lines already drawn as guides.
6. The design for this component includes four screw holes, one at each corner of the base of the component. These are visible in the top view only. Each screw hole is 5 mm in diameter, and is positioned 10 mm horizontally and 10 mm vertically from each edge of the base. Draw the screw holes, using short guide lines to position the centres. Remember to draw the dimensions to scale.
7. To complete the top view, neatly and accurately draw and label arrows to show each measurement. Show the actual dimensions, not the scaled-down lengths you used to create the drawing. Include the measurements for the screw hole placements and diameters.

DISCOVER

One-Point Perspective and Scale Model

Orthographic drawings provide a complete and accurate guide for building a structure. They are also the starting point for creating realistic representations. **Perspectives** are drawings that use vanishing points to give the feeling of distance and depth. The simplest perspective drawings, such as this one, use a single vanishing point.

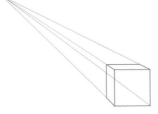

One-Point Perspective

A scale model provides the manufacturer with a physical "mock up" of the object made to scale (a smaller copy of the object), which the manufacturer can use as a reference. A scale model allows the client to see a building or product before it is actually built to size.

Draw a perspective of the component you drew in the Discover on pages 120 and 121. Then, plan and build a scale model.

1. Turn a piece of paper so that its short side is vertical. Accurately copy the side view of the component, near the bottom left corner of the paper. Omit the sloping flange from this copy, which represents only the left end of the component.

2. Mark a single vanishing point near the top right corner of the paper. Lightly draw diagonal lines connecting this point to each point in the side view where two or more lines meet. The lines represent the direction in which the object seems to recede from your point of view.

3. The placement of the right side of the component should give a realistic impression of the component. Place the bottom front right corner on the lowest diagonal, 20 mm from the bottom front left corner. Compared to the orthographic drawing scale, this scales down the measurements along the diagonals, as you would see them if you looked at the actual component from this angle.

4. Starting from this point, draw another copy of the side view, but this time, include only the edges that would be visible from this angle.

5. To draw the flange, you need to position its corners. The bottom front edge of the component is half as long in this drawing as it is on the original front view. The flange corners are 10 mm and 14 mm from the left of the component, using the scale of the front view. So, divide these distances by two, and mark points at 5 mm and 7 mm along the lowest diagonal from the bottom front left corner. Lightly draw vertical lines from here to the next diagonal up to give the starting points for the flange. Locate the other corners of the flange by lightly drawing horizontally and vertically. Draw the edges of the flange that you would see at this angle.

6. Use the diagonals to draw the remaining edges (or parts of edges) of the component that would be visible at this angle.

7. Using a material of your choice, construct a scale model of the component. Use a scale of 1:2 so that the model is half the size of the actual component. Carefully plan the construction of your model so that each dimension is produced according to scale.

For complex objects, such as large buildings, a highly realistic image, called a rendering, is usually created. This helps the purchaser or client to visualise and approve or modify the design before the expensive construction process begins. A rendering is developed from a perspective. This view of a street corner restaurant has been rendered from a two-point perspective, with vanishing points placed along the two streets that meet at the building.

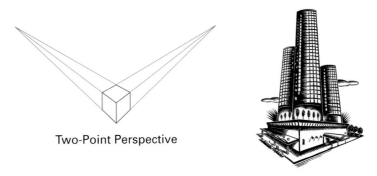

Two-Point Perspective

Appendix B has more information on both *CorelDRAW®* (page 446) and *AutoCAD LT®* (page 450).

Designers today can use a range of drawing software tools to create plans and images. *CorelDRAW®* is one of the most common programs, in schools and on some home computers. *AutoCAD®*, a more powerful software package, is widely used by professional designers and is available on some school networks. Even if you do not have access to programs like these, the most important aspects of electronic design are the same ideas you discovered in hand-drawn design work.

EXAMPLE 1

Wood Moulding: Orthographic Drawings

Create the three orthographic drawings that would be used in the production of this piece of wood moulding.

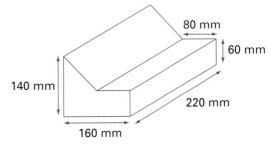

Solution

Method 1 Pencil and Paper

Use a set square or protractor. Based on the dimensions, a scale of 1:8 is easy to work with and produces drawings of an appropriate size. Start with the front view, showing the two levels of the wood moulding as two stacked rectangles. Use horizontal guide lines to develop a side view, showing the slope from the back down to the lower front level. Use vertical guide lines to draw the top view, which is similar to the front view, but with different vertical measurements. Finally, add dimension labels and a scale legend.

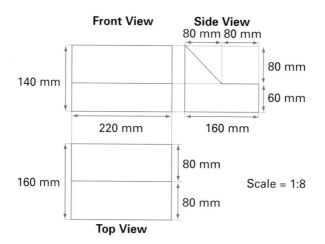

Front View

140 mm

220 mm

Side View

80 mm 80 mm

80 mm

60 mm

160 mm

Scale = 1:8

160 mm

80 mm

80 mm

Top View

For help with drawing and modifying an object in *CorelDRAW®*, see page 448 and 449.

Method 2 *CorelDRAW®*

The side view is the most complex.

Step 1: Using the **Rectangle Tool**, draw a rectangle.

Step 2: If the rectangle is not already selected, select it using the **Pick Tool**.

Step 3: Select the **Shape Tool** and position it on the midpoint of the top edge of the rectangle. Right-click and select **Convert to Curves**. Right-click again and select **Add**. This will add a new node, or vertex, to your rectangle.

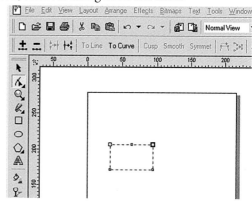

Step 4: Still using the **Shape Tool**, click on the new node and drag it to the middle of the rectangle.

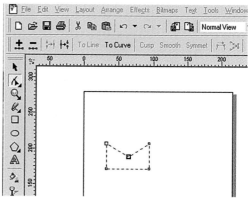

Step 5: With the **Shape Tool** still selected, click on the top right vertex and drag it halfway down toward the bottom right vertex. This completes the side view of the moulding.

The front view of the moulding looks like two long rectangles. Create this view using the **Rectangle Tool**, and draw a line midway down using the **Line-Connect Tool**. Create the similar-looking top view in the same way.

EXAMPLE 2

Wood Moulding: Perspective View

Create a one-point perspective of the wood moulding in Example 1.

Solution

Method 1 Pencil and Paper

Accurately copy your side view from Example 1. Place a vanishing point a good distance from the side view, positioned so that the front of the moulding will be visible in the finished perspective. Lightly draw lines from each vertex to the vanishing point.

Choose a realistic-looking position for the other end of the moulding. Draw the edges of the moulding that will be visible from your viewpoint. Along the diagonal guide lines, draw the lines that represent the horizontal lines of the object.

Method 2 *CorelDRAW*®

For more information on the Extrude Tool in *CorelDRAW*®, see page 449.

Step 1: Use the **Pick Tool** to select the side view you created in Example 1.

Step 2: Select the **Extrude Tool**, click on one of the vertices, and drag it to create a perspective view.

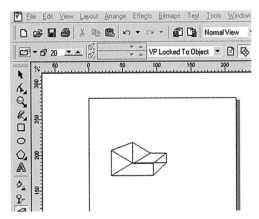

For more information on drawing with *AutoCAD LT*®, see page 450.

Create a Three-Dimensional Cube using *AutoCAD LT*®

1. Select the **Rectangle Tool** (or use the menu **Draw, Polygon, Rectangle**), and draw a square. You can draw the square by clicking to place the first vertex of the square, then move diagonally and click to place the other vertex. You should see a square on the screen.

2. Using the **Copy Tool** (or use the menu **Modify, Duplicate**), click on one of the lines to select it. The lines will turn to dotted lines. Key **Enter** to tell the program you have finished making the selection.

3. Next, move the cursor to anywhere on or off the lines of the square, click to pick up the square, and click to drop the square. You will see a copy of the square created.

4. Using the **Line Tool** (or use the menu **Draw, Line**), draw lines between the two squares to give the impression of three dimensions in your drawing. With the line tool, click on the first vertex, click on the second vertex, then key **Enter** to complete the line. Then, select the **Line Tool** again for the next connecting line.

5. When you have completed this diagram, go to **File, Save as**, and save the file to the directory your teacher has assigned.

You can find more help with *CorelDRAW*® and *AutoCAD LT*® on the Internet. Go to *www.mcgrawhill.ca/links/MPCA12* and follow the links.

Orthographic drawings, perspectives, and scale models are important to the success of the design and construction process. For more complicated designs, you might want to draw a cross-section view. This is a special type of orthographic drawing where you imagine the object being sliced parallel to your plane of view. In Chapter 2, you studied several floor plans, which are examples of cross-section top views. An outside top view of a cottage would show only the roof. You would have no information about the layout of the rooms inside. The floor plan opposite shows the cottage as though the top has been sliced off at window height. It shows all interior walls, several fixtures, and the positions of the doors and windows.

Floor plans can also be classified as schematics. Schematic drawings give a full description of a design using the top view only. Other examples of schematics include maps, line markings in sports arenas, and electrical drawings.

EXAMPLE 3

Floor Plan: Scale and Measurements

To complete the floor plan shown,

a) calculate the scale of the drawing

b) find the dimensions of each room

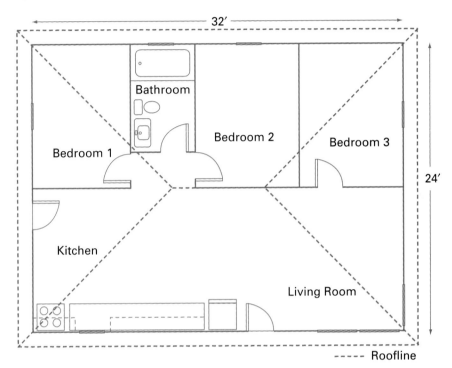

Solution

a) The outside dimensions of the cottage are given, 24 ft by 32 ft. On the drawing, these dimensions measure exactly 3 in. and 4 in., respectively. The scale is, therefore, given by the following ratio:

Scale: 3 in. represents 24 ft

$\therefore$ 3 in. represents $\left(24 \text{ ft} \times \dfrac{12 \text{ in.}}{1 \text{ ft}} \right)$

3 in. represents 288 in.

1 in. represents $\dfrac{288}{3}$ in. = 96 in.

As a check, the other dimension should give the same scale:

4 in. represents 32 ft

$\therefore$ 4 in. represents $\left(32 \text{ ft} \times \dfrac{12 \text{ in.}}{1 \text{ ft}} \right)$

4 in. represents 384 in.

1 in. represents 96 in.

So, the scale is 1:96.

b) The table shows the conversion of measurements on the drawing to the actual dimensions.

Dimension	Drawing Measure	Actual Measure		
		Inches	Feet	Feet and Inches
Living room width	$1\frac{1}{2}''$	$96 \times 1\frac{1}{2}'' = 144''$	$144'' \times \frac{1'}{12''} = 12'$	12'0''
Length of bedrooms	$1\frac{1}{2}''$	$96 \times 1\frac{1}{2}'' = 144''$	$144'' \times \frac{1'}{12''} = 12'$	12'0''
Width of bedroom 1	$1\frac{1}{16}''$	$96 \times 1\frac{1}{16}'' = 102''$	$102'' \times \frac{1'}{12''} = 8.5'$	8'6''
Width of bedroom 2	$1\frac{1}{8}''$	$96 \times 1\frac{1}{8}'' = 108''$	$108'' \times \frac{1'}{12''} = 9'$	9'0''
Width of bedroom 3	$1\frac{1}{8}''$	$96 \times 1\frac{1}{8}'' = 108''$	$108'' \times \frac{1'}{12''} = 9'$	9'0''
Length of bathroom	$1''$	$96 \times 1'' = 96''$	$96'' \times \frac{1'}{12''} = 8'$	8'0''
Width of bathroom	$\frac{11}{16}''$	$96 \times \frac{11}{16}'' = 66''$	$66'' \times \frac{1'}{12''} = 5.5'$	5'6''

KEY CONCEPTS

- In order to construct an object or to assess its design, accurate drawings are needed.

- Orthographic drawings are used to give top, side, and front views, with dimensions.

- Perspective drawings are used to give a realistic impression of the object.

- Scale models are used to see what the finished object will look like.

DISCUSS THE CONCEPTS

1. Look at the Homes section of your weekly newspaper. It has many advertisements for new houses.
   a) How do the advertisements show you what the house looks like? Cut out an advertisement, and write your response in your notebook.
   b) Discuss your findings with a classmate.

2. Self-build furniture is supplied with instructions to help you assemble the product from the separate pieces. Describe how the types of views shown in the instructions help the purchaser to assemble the furniture correctly.

A **1.** **a)** In groups of two or three, measure the dimensions of your classroom.

b) Draw a floor plan of your classroom. Use a set square or protractor, graph paper or orthographic dot paper, or drawing software.

2. Sketch the front view of three different structures you see daily.

3. A side view is to be drawn of a 32-m passenger airplane. If the scale is 1:50, how wide will the drawing be, to the nearest centimetre?

4. **Map scales**

a) Which scales are commonly used on road maps? Explain why each scale is used.

b) On any map that shows its scale, choose a starting point and a destination. Calculate the straight-line distance, in kilometres, between the two points.

c) On a map of Ontario, London is 9.5 cm from Toronto. Find the actual distance between the two cities, and determine the scale used for the map.

B **5.** A house is 30′ wide, 23′ deep (front-to-back), and 25′ high. The orthographic views are to be drawn so that each one fits on an $8\frac{1}{2}''$ by 11″ sheet of paper, allowing 1″ of space around the edge of the paper. All three views are to be drawn to the same scale. What is the largest scale that can be used?

For help with drawing software, see page 442 (*The Geometer's Sketchpad*®), page 446 (*CorelDRAW*®), or page 450 (*AutoCAD LT*®).

6. **Application** Use drawing software to create an accurate scale floor plan of a basketball court. Find the regulation lengths of all the court lines, and show these dimensions on your plan. If you do not have access to drawing software, make a scale drawing by hand.

7. **Design** You have been commissioned to design a garage to fit a medium-sized car. Determine the information you will need and then, obtain it. Make sketches based on this information. From the sketches, prepare orthographic drawings.

*C*HAPTER PROBLEM

8. Before Chris can build the entertainment unit, she must make scale drawings. A perspective drawing will give her an idea of how the unit will look, and orthographic drawings will be used when the unit is built. Of the designs sketched in section 3.1, question 12, choose one for each space, and draw perspective and orthographic views.

9. **Application** An isometric view of an object looks similar to a perspective, but it does not use vanishing points. Instead, three axes are used to represent each of the three spatial dimensions, as shown. The same scale is used on each axis, and the angle between the axes is 120°.
Construct an isometric view of the wood moulding from Example 1 on page 123. Use a set square or protractor, or drawing software.

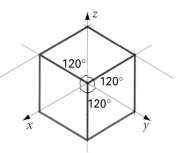

10. a) Using toothpicks and modelling clay, make a model of each of the five regular polyhedra.

b) Record, in a table, the number of edges (E), the number of faces (F), and the number of vertices (V) for each polyhedron. What do you notice about the number of faces of a dodecahedron and the number of vertices of an icosahedron? Explain how to draw an icosahedron inside a dodecahedron.

c) Record the sum $F + V$ for each polyhedron. Express any pattern you notice as an equation in terms of E, F, and V.

ACHIEVEMENT CHECK　Knowledge/Understanding　Thinking/Inquiry/Problem Solving　Communication　Application

11. The drawing shows a stationery holder that must be able to hold 50 sheets of $8\frac{1}{2}$ in. × 11 in. paper and 25 envelopes, 10.4 cm × 24.1 cm.

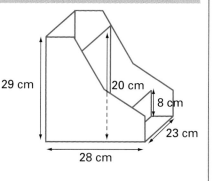

a) Using drawing software, or drawing tools, design a different stationery holder that will also hold 50 sheets of paper and 25 envelopes.

b) Prepare orthographic drawings of your design.

c) Describe the special features of your design.

C **12.** **Thinking/Inquiry/Problem Solving, Application** Design a package for a company that is planning to sell regular-sized pop cans in packs of ten. Packaging material should be kept to a minimum. Explain how the shapes used in your design save on packaging material.

Creating Nets, Plans, and Patterns

Nets, plans, and patterns provide a flat, two-dimensional representation of a three-dimensional object.

DISCOVER

Objects Manufactured Using Nets

The **net** of a three-dimensional object can be seen by unfolding the shape, possibly cutting or separating some of the edges. You can make a net by opening an empty cereal box at both ends and cutting along one edge. All the surfaces of the box are now laid flat, side by side, at the same time.

1. Gift boxes are often sold in flattened form. One of the layouts shown is a net that can be made into a gift box, but the other cannot. Predict which net can be folded into the gift box, and explain why.

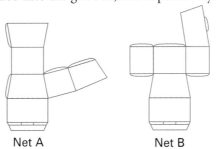

Net A   Net B

2. Verify your answer by copying each net, cutting it out, and folding it along the edges.

3. Create another net that is not a reflection or rotation of the net for the box above, but which also makes a gift box of the same shape and design. Assess your own net and the true net from question 1 as designs for the gift box. Which is the more effective design, and why?

4. Origami, the Japanese art form in which a sheet of paper is folded into a three-dimensional shape, has many similarities to net-folding. Compare the process you completed in question 2 to the process involved in making an origami piece like the bird shown. How are these processes similar, and how are they different?

DISCOVER

Nets and Patterns in Design

In a pattern, the surfaces are not attached by common edges, as they are in a net. A **pattern** is a net that has been divided into individual shapes. The pieces of a pattern are usually separate from one another, and are joined together by sewing, gluing, carpentry, or other joining techniques.

Arts Link

You may remember using a pattern in a Family Studies course to make a stuffed animal, a piece of clothing, or a pillow. A pattern can also act as a set of directions, as with knitting and macramé.

1. What is the basic shape of the cosmetics bag shown?
2. Can you create a net for this bag? Explain.
3. Create a pattern that can be used to make this bag.
4. Name three other items of this shape that you see on a daily basis.
5. Home furnishings change according to fashion. Many people update the look of their homes by painting the walls a new colour, making slipcovers for their sofas, or renovating. Changing the lampshades is a simple home decorating change. Purchasing new lampshades can be expensive, but with some creativity, fabric, scissors, and glue, you can make new lampshades with any fabric you choose. Describe how you could create a pattern to make a lampshade.

EXAMPLE 1

Creating Product Patterns

Many portable CD players do not come with a case. Create a pattern for a cloth carrying case shaped to hold a small portable CD player.

Solution

Step 1: Measure the width, depth, and height of a portable CD player. If you do not have access to one, estimate the dimensions.

Step 2: Conceptualize a design for the CD player carrying case. One possibility is a rectangular prism with a zippered top.

Step 3: Draw two rectangles—one for the front of the case and one for the back. Use the width and depth measurements from step 1, but allow a little extra space along each edge for the seam where the cloth will be sewn.

Step 4: Use the width, depth, and height measurements, with seam allowance, to draw three rectangles for the sides and bottom of the carrying case.

Step 5: Will your carrying case have a zipper, a flap with velcro, or some other fastening? Describe your choice and adjust your pattern, if necessary.

Step 6: If you would like your carrying case to have a strap, include it in your pattern.

Step 7: Draw the completed pattern. Include the seams. The zipper is to be sewn in last, with the foldover shown.

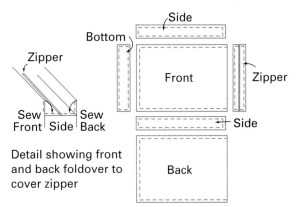

Detail showing front and back foldover to cover zipper

In order to build a house, the architect must prepare the drawings, or plans, which give all the details of the house from the foundation up to the roof. **Plans** are expressed as a series of orthographic drawings. The cottage floor plan in section 3.2 (page 127) is an example. Once these plans are prepared, they serve as the blueprint for the construction of the house. Similarly, furniture, vehicles of all kinds, and many other products require accurate plans for manufacturing.

Language Link

The word "blueprint" comes from an early, chemical form of photocopying. The original plans, drawn on translucent paper, were placed onto blueprint paper sensitized with certain chemicals, and exposed to light. The areas of blueprint paper not covered by the ink of the original drawing reacted, and turned blue when washed in water, revealing the plan copy as unexposed white lines.

EXAMPLE 2

Replicating Antique Furniture

A cabinetmaker would like to recreate this antique table design.

a) What dimensions are needed to create the plans?

b) Neatly, but not to scale, sketch a set of plans for recreating the design.

Arts Link

The table in the photograph was designed by George Walton, a Scottish architect and designer. Walton's more famous colleague Charles Rennie Mackintosh, whose furnishings and buildings are recognized as landmarks in modern design, created the chairs in the photograph.

Solution

a) The cabinetmaker would need the following measurements:
  - height and thickness of legs
  - height at which the lower supports connect the legs
  - diameter and thickness of the circular table top
  - curve and thickness of the upper supports
  - curve and thickness of the lower supports
  - angle and height of bend at the base of the legs

  The curves could be specified by a pattern.

b) To determine which views to draw, consider the best angles from which to show all the parts and dimensions. These three views are an effective choice.

Top View

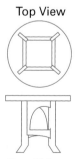

Front View
(showing supports)　　Front View
(showing legs)

KEY CONCEPTS

- A net is the shape of an object that has been taken apart and flattened.

- Patterns are two-dimensional pieces used in creating three-dimensional objects. Patterns can be created from three-dimensional objects by taking them apart and separating all joining pieces. A net is a single-piece pattern.

- Plans are orthographic views used to describe an object that must be constructed to specific dimensions.

DISCUSS THE CONCEPTS

1. Give two examples, different from the examples mentioned in this section, of design or manufacturing uses of
   a) nets   **b)** patterns   **c)** plans

2. Describe the process involved in creating your athletic shoe, in terms of nets, patterns, and plans.

APPLY THE CONCEPTS

B **1.** **Kites** Research the history and practice of kite design. Create the net for a simple kite. Mark all the dimensions.

For help with drawing software, see page 442 (*The Geometer's Sketchpad*®), page 446 (*CorelDRAW*®), or page 450 (*AutoCAD LT*®).

2. **Fashion** Consider the pieces used to make a T-shirt. Draw, or sketch, the pattern pieces by flattening a T-shirt and tracing around the edges. Draw the pattern pieces, using drawing software, if available.

3. **Computer case** Measure or estimate the dimensions of a portable computer, and draw plans for making a carrying case.

4. **Nets for polyhedra** If you were to trace, cut, and fold each of these nets, identify the solid you would construct.

a)  **b)** 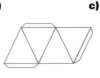 **c)** 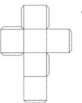 **d)**  **e)**

5. **Thinking/Inquiry, Communication** 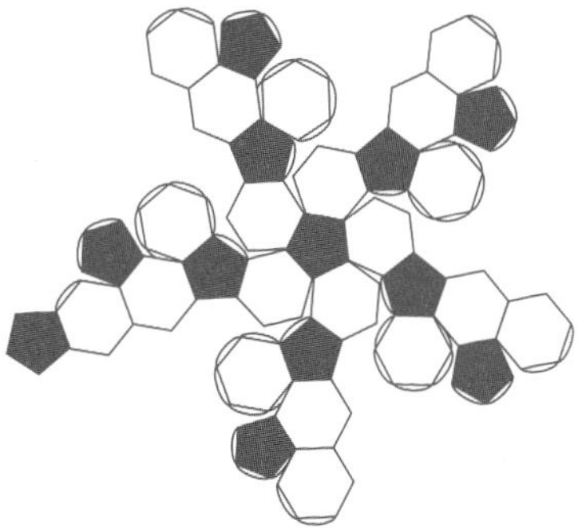
   a) Is the object formed by this net a Platonic solid? Why or why not?
   b) Archimedean solids, like Platonic solids, are comprised of regular polygons, but are made up of more than one type of regular polygon. A soccer ball can be compared to one Archimedean solid, the truncated icosahedron. Describe two more of these solids, in terms of the type and number of faces. You could use your local library or the Internet as a resource.
   c) Draw a net for one of the solids you described in part b). Construct your solid to show that the net works.

6. **Construction** Examine how each framing feature is designed for houses. Model each feature, using drawing software, if available.
   a) roof support frames
   b) wall framing

7. Two designs are being considered for a cardboard package. The nets for both are shown.

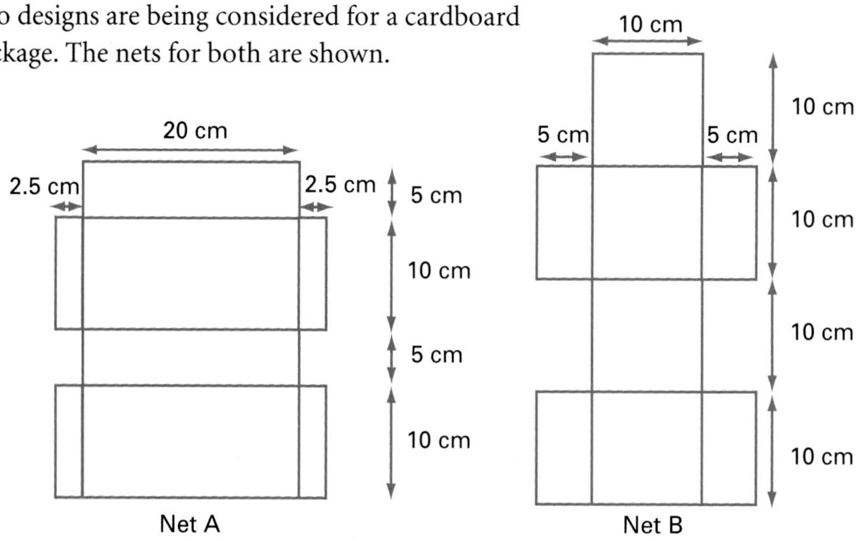

Net A

Net B

   a) Verify that both designs have the same volume.
   b) Calculate the surface area of each design when the net is folded into a box. Which design has the smaller surface area?
   c) Suppose the net is to be cut out of a rectangular piece of card. For each design, state the smallest possible dimensions of the rectangle, and find its area. Which design uses the smaller area of card?
   d) Recommend, with reasons, one of these designs for
      i) a cereal box
      ii) a gift box

8. Walkway design Obtain and examine catalogues from companies that design and make bricks and blocks for driveways and walkways. Design a pattern for a walkway that is 2 m wide and 10 m long, using drawing software, technical drawing equipment, or graph paper or orthographic dot paper.

9. Chris has decided to use medium density fibreboard (MDF) to construct her entertainment unit. Select one of the designs developed in section 3.2, question 8, and draw the individual pieces that must be cut to build the unit.

C **10.** An interesting geometric form, the Klein bottle, was created by extending the idea of a Möbius strip.

a) Create a Möbius strip by putting a half-twist in a strip of paper and taping the ends together. Draw a line along the middle of the strip until you return to your starting point. How many sides does a Möbius strip have?

b) Make a sequence of sketches to illustrate the geometric creation of a Klein bottle, following these steps:
• A square of material is made into a tube by joining two opposite sides.
• The tube is bent into a U-shape.
• One end of the tube is pushed through the material near the other end.
• The two circular ends are joined together.

c) Add orientation arrows along the sides of the square in your sketches. On the unfolded square, two arrows should point up, and two should point to the right. In your final sketch, which two arrows now point in opposite directions?

d) Based on the photograph and your sketches, design a pattern or net for a Klein bottle. Use your design to construct a Klein bottle. Is the inside of your model separate from the outside?

e) Predict what would happen if you tried to roll up a Möbius strip by joining its edges together. Test your prediction with your Möbius strip from part a).

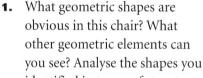

3.4 Designing and Constructing Physical Models

Design, in the most basic terms, is a plan for making an object. The object can be anything: a building, a machine, a shoe, or a piece of cutlery. Ideas for designs usually stem from a need for an object with certain features or attributes. The designer may have an idea, or will look for sources of inspiration to trigger an idea for creating a design to meet the identified need. Once an idea for a design is established, the designer uses the techniques you have studied so far to plan within the given constraints for the object, such as dimensions, materials, volume, area, and budget. Designers often simulate this process by constructing a physical, three-dimensional model.

DISCOVER

Furniture Design

Many geometric elements are integrated into furniture design. Historically, architects used furniture design as a way of experimenting with different design features, most of which are a combination of geometric shapes. As you saw in section 3.1, shapes can play structural, strengthening, functional, and aesthetic roles. A model is an effective test of the shapes chosen.

1. What geometric shapes are obvious in this chair? What other geometric elements can you see? Analyse the shapes you identified in terms of structure, strength, function, and aesthetics.

2. In the previous section, Charles Rennie Mackintosh was noted as an influential architect and designer. Use the Internet or your local library to research these questions.

   a) Besides Charles Rennie Mackintosh and George Wilson, who else was influential in the modernism movement, and what contributions did they make?

   b) Find some current furniture designers. What do you like about their designs?

3. Choose a variety of geometric shapes that you believe would be effective in your own design for a piece of furniture.

4. Create plans for your piece of furniture. Select some simple materials (such as cardboard, wire, and glue) and construct a model of it.

EXAMPLE 1

CD Shelf

Design a shelf to hold 30 CDs. Make a full-scale model out of cardboard.

Solution

Take all the measurements of a CD case. A typical CD case is 14.3 cm wide, 12.6 cm tall, and 1.0 cm deep. To fit 30 CDs, you need to make the shelf's inside width at least 30 cm, and its inside depth enough to support the CDs securely. The corresponding outside dimensions will depend on the thickness of the material used. In the case of cardboard, the difference can be ignored.

- A relatively simple design requires a back, two sides, and a bottom.
- It is sensible to make the back and bottom width 1 cm wider than the width of 30 CDs, to accommodate fingers when a CD is pulled from the shelf.
- Making the depth of the bottom piece 1 cm *less* than the width of a CD will also make CDs easier to remove and replace. However, this is a design choice.
- The sides of the shelf should be tall enough to support the CDs. In this sample design, a circular curved shape has been chosen for the side pieces.

A pattern to create this CD shelf includes tabs to join the sides. Notice that the tabs on the sides fit under the bottom and behind the back, to help support the bottom and the CDs resting on it. Using this pattern, cut and glue the cardboard pieces together to test your full-scale CD shelf.

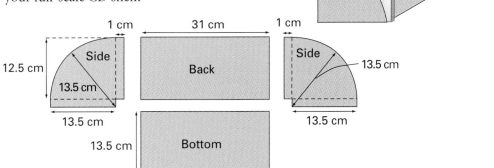

The sides could be further shaped, perhaps with cut-outs, to make the CD shelf more appealing. It could be painted or decorated and placed on a stereo stand.

EXAMPLE 2

Telephone Stand

Design a telephone stand for a small entry hall in a city apartment. The stand must be able to hold a telephone and two telephone books, with an area for paper and pens for taking messages.

Solution

Since a specific size of telephone is not given, allow enough space for a range of sizes. However, you must keep in mind that the entry hall is small. So, you should try to make the table as efficient as possible. The dimensions 27 cm × 23 cm × 8 cm are realistic for a Toronto telephone book, for example.

- The most efficient design for a small space is a surface for the telephone, paper, and pens, with a shelf mounted underneath for the telephone books to lay on. Since telephones are considerably smaller than telephone books, the size of the telephone book should be considered first in deciding the width and depth of the stand.
- A width of 35 cm and a depth of 30 cm is a generous area for telephone books, allows plenty of space on the stand top, and is not too big for a restricted space.
- For two telephone books, allow 20 cm below the top of the stand. This will give some flexibility for increase in the thickness of the books.
- To create a space for the pens, pencils, and message pads on top of the stand, you can use a shallow groove. In this example, a 15-cm width has been allowed, leaving space for a telephone, with spare surface to rest on while writing a message.
- For the height of the table, use 70 cm. This is just below most people's waist height and is quite comfortable for dialling or for writing messages.

So, the dimensions of the telephone stand will be 70 cm high × 35 cm wide × 30 cm deep, with the book shelf 20 cm below the top. For a simple design, use a rectangular top and four legs. To give some aesthetic appeal, the legs could be in the shape of triangular prisms. The plan shows the stand.

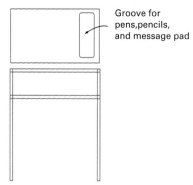

Groove for pens, pencils, and message pad

EXAMPLE 3

Aesthetic Furniture Design

Design a chest of drawers where the width to height ratio is the Golden Ratio. The width of the chest should be 80 cm.

a) To the nearest centimetre, what is the height of the chest?

b) If as many drawers as possible are also in the Golden Ratio (height of drawer to width of drawer), how many drawers with these dimensions can you fit in the chest?

c) To the nearest centimetre, what is the height of the remaining drawer?

d) Make a 1:10 scale drawing of the front view of the chest of drawers.

Solution

Method 1 Paper and Pencil

a) As you saw in section 3.1, the Golden Ratio is 1:1.618. Since the width required is 80 cm, the height is calculated as follows:

$$\text{Height} = 80 \times 1.618$$
$$= 129.44$$
$$\doteq 129$$

The height of the chest is 129 cm.

b) The Golden Ratio height of a drawer 80 cm wide is

$$\text{Drawer height} = \frac{80}{1.618}$$
$$\doteq 49$$

Two drawers 49 cm tall fit into the chest.

c) The last drawer is 31 cm tall.

d) The front view shows the chest of drawers.

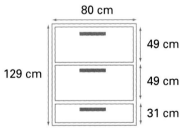

(Note: For this example, the distance in from the sides and between the drawers (the thickness of the wood) has been ignored. If you were designing this chest of drawers for manufacture, these design considerations would have to be taken into account.)

For help with drawing and modifying an object in *CorelDRAW*®, see page 448 and 449.

Method 2 *CorelDRAW*®

Parts a) to c) from **Method 1** give the dimensions to be used.

d) Select the **Rectangle Tool,** and draw a rectangle 8.0 cm wide by 12.9 cm tall. Use the rulers and the position readout in the lower left-hand corner to help you.

Select the **Connector Line Tool** and draw a horizontal line 4.9 cm down from the top of the chest. Then, draw a second horizontal line 3.1 cm up from the bottom of the chest. Use the middle drawer as a check; it should be 4.9 cm high. Your final drawing should look like the screen to the right.

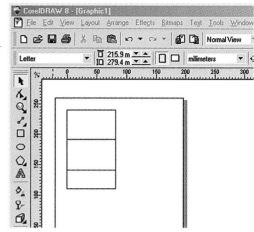

KEY CONCEPTS

- In order to successfully design and create an object, specific information is required, such as purpose or use, dimension constraints, and material(s).

- When creating models, the material is light and is used only to represent the form the final object will take.

- When considering material for the final product, the drawings must allow for the added dimensions of the material.

DISCUSS THE CONCEPTS

1. Discuss the design elements you would incorporate in the design of a book bag, hockey bag, laptop computer case, or a carrying case of your choice.

2. Consider what you like or dislike about the bag or case you use now. For what specific purposes was it designed? Is it big enough, or too big? Is it comfortable to carry in the way you choose to carry it (over one shoulder, in your hand, or as a backpack)? How can you improve the design?

A 1. **Child's playhouse** Design and construct a child's playhouse. It should be able to hold three small children. Create a 1:8 scale model of the playhouse using found material.

2. **Computer equipment** Create a container to hold computer game disks and games magazines.

B 3. **Fashion** Design a pair of sunglasses. Sketch and draw the plans for your design. Make a full-scale model of your design out of cardboard and transparency plastic.

4. **Stereo unit** Find the average size of stereo components and design a unit that will accommodate most stereo systems. Try to incorporate some curved shapes into your design. Draw the front view of your design.

5. **Thinking/Inquiry/Problem Solving, Communication** Design a closet unit for storing sweaters during the summer. The unit should be able to hold at least ten sweaters, and be durable for storage anywhere in the house. Consider a tight-fitting lid. Draw all three orthographic views. Describe your procedure, measurements, and design solution.

*C*HAPTER
PROBLEM

6. Chris is still undecided about which design to choose. She would like you to create a scale model of the design so that she can see what the finished unit will look like. Using the design and drawings you selected in section 3.3, question 9, create a 1:8 scale model out of cardboard.

EXTEND THE
CONCEPTS

7. For the chest of drawers design in Example 3 on page 141, wood pieces 1 cm thick are to be used.
   a) The height and width of the chest, allowing an extra 1 cm for the top, bottom, and each side, are still to be in the Golden Ratio. Calculate the new height of the chest, based on a width of 82 cm.
   b) Subtract 2 cm for the top and bottom to get the internal height of the chest. Is this still in the Golden Ratio with the original 80-cm width?
   c) Wooden support pieces 1 cm thick are to be inserted between the drawers. If two of the drawers have the same Golden Ratio dimensions as before, what is the new height of the third drawer?
   d) Create a new front view that allows for the thickness of the material.

C 8. **Dwelling** Design an outdoor structure made of waterproof cloth and PVC pipe, which can be connected with a variety of joining pieces to create corners. The structure should fit two people sleeping, or four people seated, with room for a small table. The structure should be strong enough to survive wind and rain, yet allow the occupants to create a window for light. Consider one of the regular polyhedra for a basic form.

*C*AREER PROFILE

Graphic Design Artist

Books are produced in four main stages: writing, editing, designing, and printing. The design stage begins with manuscript, sketches of diagrams, and discussions with the publisher, and ends with printer-ready files that deliver the pages to the printer exactly as they will appear in the book. Thus, a graphic design artist is a key player in the process of publishing a book—such as the one you are holding now.

Graphic design has undergone many dramatic changes since the introduction of the computer. Before computerization, graphic design artists were expected to assemble the type, pictures, and graphics into a layout that would be photographed and then, processed for the printing presses. Today's graphic design artist must learn how to handle computer files, often converting them from one format to another.

Graphic design work includes designing packaging for a range of products, creating advertising signage on billboards, and developing attractive and easy-to-use textbooks. Proportional scaling of objects is a vital mathematical concept used in this industry.

A graphic design program is generally three years in length, and qualifies the graduate to be a member of the association of Registered Graphic Designers of Ontario (RGD Ontario). A graphic design program includes courses such as Computer Graphics, Creative Thinking and Problem Solving, Three-Dimensional Graphic Design, Advertising, and Marketing. Go to *www.mcgrawhill.ca/links/MPCA12* and follow the links to find colleges that provide this program.

1. A photograph whose dimensions are 3.5" × 8" needs to be enlarged onto a layout page that is 8.5" × 11".
   a) What percent of the page does the photograph occupy before it is enlarged?
   b) If the enlargement is to fill as much of the page as possible, without any cropping or reorientation of the image, by what percent can it be enlarged?
   c) What percent of the page will the enlargement occupy?

2. A few of the courses that graphic design artists are expected to take are mentioned above. Research a college that provides a graphic design program, and identify other courses that are included in their plan of study.

CHAPTER PROBLEM WRAP-UP

The more electronic gadgets we acquire, the less space we seem to have! You need to solve Chris's problem. She will be moving to a college dorm and needs to organize her television, stereo system, and DVD player to fit the space she has in her room.

- Chris's television is a 21" model with a built-in VCR. The television is 22" wide, 19" deep (front to back), and 19.5" high. It is not built to support anything stacked on top of it.
- The stereo components are all 17" wide and 15" deep. The tuner is 6" high, and the CD player and DVD player are each 3" high.
- There are two places in the dorm where Chris can place the entertainment unit. One is approximately 30" wide and the other is approximately 45" wide.

1. Sketch a few designs for each of the two spaces available.
2. Choose one design for each space, and draw perspective and orthographic views.
3. Select one of the designs, and draw the individual pieces that must be cut to build the entertainment unit.
4. Create a 1:8 scale model out of cardboard.
5. Chris likes your design, based on your scale model. To complete this project, the following tasks remain.

   a) The medium density fibreboard (MDF) will require staining to match Chris's other furniture. Research types of stain and calculate the amount needed for this unit.

   b) Chris will need a storage area for her CDs and DVDs. Modify your plan to allow storage. Chris currently has about 100 CDs and 20 DVDs, but will likely acquire more in the future.

   c) Now that the design has been determined, research the cost of MDF and calculate how much this project will cost. Include the cost of the joining hardware and the stain.

   d) Prepare a design package for Chris, consisting of
   - the perspective drawing
   - the orthographic drawing
   - your design rationale
   - the breakdown of costs

eview

3.1 Geometric Shapes in Design, pages 110–119

1. Using only regular polygons,
   a) create a design that would be suitable for an auto-part manufacturing company's letterhead
   b) create a piece of art in the style of this painting by Piet Mondrian

2. **Packaging design** Choose three products of the same type. Consider the designs of the packaging.
   a) For each packaging design, suggest aspects of the product the designer was trying to convey.
   b) In a table, list the products, main shapes, and design characteristics particular to each package.
   c) Choose one package, and discuss its design in terms of structure, strength, function, and aesthetics.

3. **a)** Describe the polygons used in a pyramid.

   b) The Great Pyramid at Giza, Egypt, has geometry based on the Golden Ratio. Use your local library, or go to the web site *www.mcgrawhill.ca/links/MPCA12* and follow the links to learn more about the Great Pyramid. Write a brief report on the Golden Ratio in its architecture.
   c) Describe the use of the Golden Ratio in another architectural example.

3.2 Representing Three-Dimensional Objects, pages 120–130

4. Represent your initials as a three-dimensional object.

5. Draw perspective views of these objects, using drawing software, if available.
   a) a cube van   **b)** a chocolate mint in the shape of a rhomboid

For help with drawing software, see page 442 (*The Geometer's Sketchpad*®), page 446 (*CorelDRAW*®), or page 450 (*AutoCAD LT*®).

6. **a)** Using building blocks, create a model of a house.
   b) Draw what you have built in the three orthographic views: front, top, and side.
   c) Draw a perspective based on the front view.
   d) Imagine the model sliced in half in any plane of your choice. Draw a cross section of the house.

7. A camera is made of simple shapes. Sketch the basic shapes of this camera. Create a perspective of the camera on an angle. Smooth out the edges to give a more realistic look.

3.3 Creating Nets, Plans, and Patterns, pages 131–137

8. a) Create a net for a 200-mL juice box.
   b) Enlarge the pattern to create a net for a juice box that holds 750 mL of juice. Use the cube root of the volumes to determine the scale factor for the linear dimensions.
   c) What are the dimensions of the new juice box?
   d) Explain why you needed to use the cube root scale factor.

9. Refer to Discuss the Concepts in section 3.4 (page 142). Draw the pattern pieces required to make your book bag or backpack.

10. Create a pattern for a pair of snow boots. Write a brief report, describing the process you used to create this pattern.

11. Create the plans to build a simple rectangular jewellery box 5 cm deep, 10 cm wide, and 15 cm long, with a hinged lid. Use drawing software, if available.

3.4 Designing and Constructing Physical Models, pages 138–143

12. Design a desk organizer to hold pencils, pens, computer disks, paper clips, and message pads. Describe some of the materials that would be suitable for this type of organizer.

13. Find a picture of a tent with it dimensions. Choose a scale and create a model of the tent, using materials of your choice.

14. Survival dwelling On a popular survival television show, contestants are required to make a dwelling out of indigenous material and whatever fabric they have. Imagine you are stranded in the rainforests of South America.
   a) Research the rainforest environment, and list the constraints and needs a rainforest shelter would have to meet.
   b) Design a dwelling that would be suitable for this environment.

Practice Test

ACHIEVEMENT CHART CATEGORY CONNECTIONS

1. Name three different polygons you see in traffic sign shapes. Explain how these shapes are used as visual cues for the type of information in the sign.

2. Examine the headgear shown. What are the basic shapes used? What information, or measurements, do you think a hat designer would need to create these designs?

3. Using this floor plan, calculate
   a) the area of the whole living space
   b) the amount of carpet required to cover the two bedroom floors and the living room floor, with 5% additional carpet for waste allowance

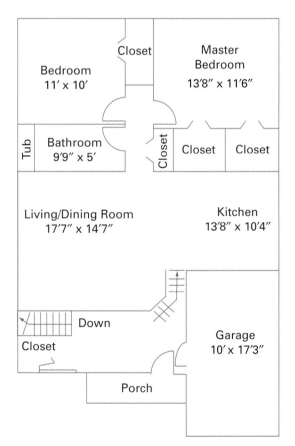

4. Draw a one-point perspective of the armchair shown in these orthogonal views.

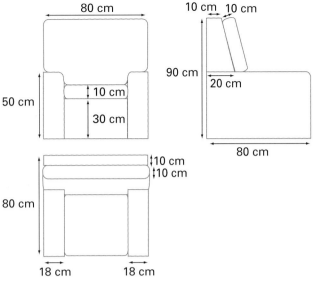

5. When this net is folded together, what object does it form? What are the object's dimensions? Suggest a specific use for the object.

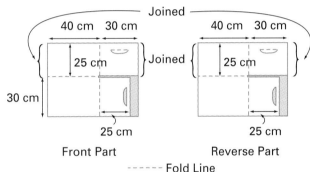

6. Design and sketch an art portfolio, including dimensions. The portfolio should allow the artist to carry paper as large as 55 cm by 75 cm and have a place for storing pencils or charcoal while travelling.

ACHIEVEMENT CHECK Knowledge/Understanding Thinking/Inquiry/Problem Solving Communication Application

7. Kitchen Gadgets wishes to introduce a new series of baking pans, all with the same volume. Designers have decided to produce an open rectangular box design that is 13 in. long by 9 in. wide by 2 in. high. They will also produce a design in the shape of an open circular pan that is also 2 in. high.

a) Prepare orthographic drawings for the circular pan using drawing software or drawing tools.

b) Prepare a new design and make a model from a suitable material for a third baking pan. (Hint: Look at the different kinds of pans available in kitchenware stores and magazines for ideas.)

c) Describe the features of your design with reasons for your choices.

Cumulative Review

Chapters 1–3

1. **Surveying** A surveyor measures the angle of elevation to the top of a building that is under construction. When the surveyor stands 10.0 m from the front of the building, the angle of elevation to the top is 67°.
   a) Find the height of the unfinished building.
   b) A further three floors are to be added to the building, adding 10.5 m in height. The surveyor is to check that the building will conform to planning regulations. Calculate the angle of elevation from the surveyor's position to the top of the completed building.

2. The backyard of a house has a sloping portion, which extends from the back of the house at an angle of depression of 20°. The owners would like to put in a flowerbed near the base of the slope, but want to be sure the flowers will receive enough sunlight. The height of the roofline at the back of the house is 8 m. The angle of elevation from the base of the slope to the roofline must be at most 50°. What is the shortest distance the flowerbed must be down the slope?

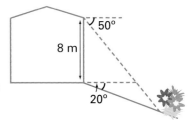

3. **Automotive design** A design engineer for a car manufacturer is working on the "dip" function for a car headlight. The road-illumination requirement is shown in the diagram. Calculate ∠A, the angle the dipped headlight beam needs to cover to produce this area of illumination.

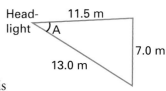

4. A house is built at the intersection of two streets. The house plot is surveyed as shown.
   a) Calculate the lengths of the sides of the plot labelled v and w.
   b) Calculate ∠A and the lengths x, y, and z.
   c) Find the length of fencing required to enclose the backyard, which has the layout shown.

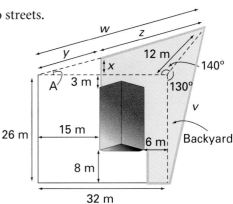

5. Convert the measures as indicated.
   a) 6.2 ft to inches
   b) 8 qt to ounces
   c) 1.4 T to pounds
   d) 1.5 mi to yards
   e) $6\frac{1}{4}$ lb to pounds and ounces
   f) $1\frac{1}{4}$ yd to inches
   g) 1 mi to inches
   h) 56 fl oz to pints and fluid ounces

6. Find the volume of each object. Remember to convert all measures to the same units.

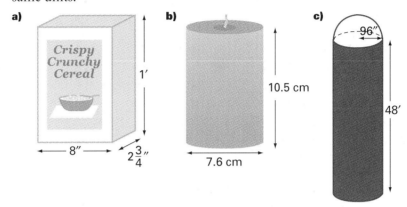

a)
Crispy Crunchy Cereal
1′
8″
$2\frac{3}{4}$″

b)
10.5 cm
7.6 cm

c)
96″
48′

7. **Home improvement** You have measured your rectangular kitchen floor to be 2.5 m by 6 m. However, when you arrive at the flooring store, the tiles you like are in imperial measure, $11\frac{7}{8}$ ″ by $11\frac{7}{8}$ ″. With grout, each tile will cover an area of 1 ft².
   a) How many tiles will you need? Draw a plan to explain your answer. Assume you can cut the tiles cleanly to fit along the floor edges.
   b) In each box, there are a dozen tiles. How many boxes should you buy?
   c) Each box of tiles costs $21.49. You will also need a bag of mortar for $16.79 and a bag of grout mix for $15.99. How much will the materials cost altogether, with 15% tax?
   d) How many wasted tiles do you expect? Explain your thinking.

8. **Hardware purchasing** The home you have just purchased is missing screens on three of its upstairs windows. You decide to purchase the materials at a hardware store and construct the screens yourself. The dimensions of the three windows are 3′7″ × 4′2″, 2′5″ × 3′1″, and 2′6″ × 2′.
   a) Screen is available in 50′ rolls, 3′, 4′, and 5′ wide. Which type of roll should you buy? Explain.
   b) Screen framing is available in 4′, 6′, and 8′ lengths. What frame lengths should you buy to minimize waste? Explain your answer.

9. **Nutrition** It is recommended that adults get 1500 mg of calcium in their daily diet. The calcium content in some foods is shown.

Food	Calcium Content (mg)
150 g canned salmon	150
400 g fruit yogurt	310
250 mL skim milk	298
one slice whole wheat bread	23

a) How much calcium does 300 mL of fruit yogurt contain?

b) What is the calcium intake of a lunch consisting of a salmon sandwich, made with 75 g salmon and two slices of whole wheat bread, and 500 mL of skim milk?

10. All measurements in the diagram of a machine part are in inches.

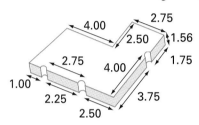

a) Find the perimeter of the top face.

b) Find the cost to wire-cut this material at $11.00/in.

c) What is the volume and mass of this part, if it is made from brass with a density of 0.3 lb/in.^{3}?

d) If the part is to be mounted on a rectangular base with a minimum of 6 in. clearance from each edge, how large must the base be?

e) What will the base cost, if it is 2 in. thick and the material costs $0.60/in.^{3}?

11. **Apartment floor plan** Design a one-bedroom apartment with an area of approximately 60 m^2. The living room should be separate from the bedroom. The kitchen should be no smaller than 2.75 m by 2.75 m, and the bathroom should be no smaller than 1.5 m by 2.5 m.

a) Draw a floor plan to illustrate your design, including dimensions and a scale legend.

b) Your apartment design is to be modified for a person who uses a wheelchair. Describe the design modifications required.

c) Modify your design to these requirements.

12. **Marketing appeal** Cool Cola Company wants to change the look of their bottles. They have asked you to create a new bottle to hold 380 mL of cola. They would like the bottle to look full and tall. They want to capture a young market. Design and sketch a bottle. Once you have decided on a design, draw a front view with the dimensions. Explain the inspiration for your design, and create a logo or catch phrase.

13. **Modernist design** Charles Rennie Mackintosh and George Walton helped create the influential Glasgow Style, a prominent part of the modernist design movement in the early twentieth century. The photograph on page 134 shows some of their work. On the Internet, search on Charles Rennie Mackintosh, the modernist movement in design, and art deco. Identify key design aspects of design in this period that explain why the designs have kept their modern appearance, and do not look dated.

14. **Child's toy** Design a child's riding toy, or a scooter. Create the plans, using drawing software, if available.

15. **Furniture design** A typical project in first- or second-year post-secondary design courses is to design a chair. Commonly, the material of choice for modelling this project is corrugated cardboard.
    a) Consider what shapes are necessary to make the cardboard chair stable and strong enough to hold the weight of an average man.
    b) Draw the three orthographic views for your cardboard chair.
    c) Create the pattern for your cardboard chair.
    d) Using corrugated cardboard, build your design of a chair. You may not have enough material to build this to full scale, so use light cardboard and build to 1:4 scale.

16. **Doghouse project** Design a wooden doghouse that could be sold in pieces for the consumer to put together in a few simple steps. Prepare the following details.
    a) drawings of the pieces, with all measurements labelled, that would be given to the lumber company that is to cut the wood—if hardware is needed to build the doghouse, include details about those parts
    b) an instruction sheet, with plans, for the purchaser, stating how the pieces are to be put together
    c) a detailed description of the cardboard box that the parts are to be packaged in for sale—include the design to be shown on each face of the box

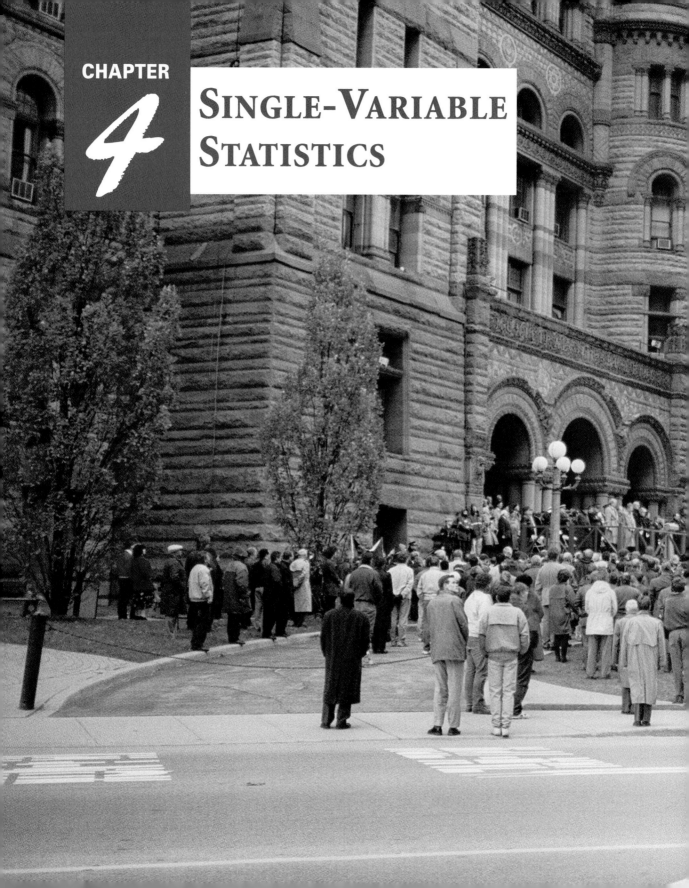

CHAPTER 4

SINGLE-VARIABLE STATISTICS

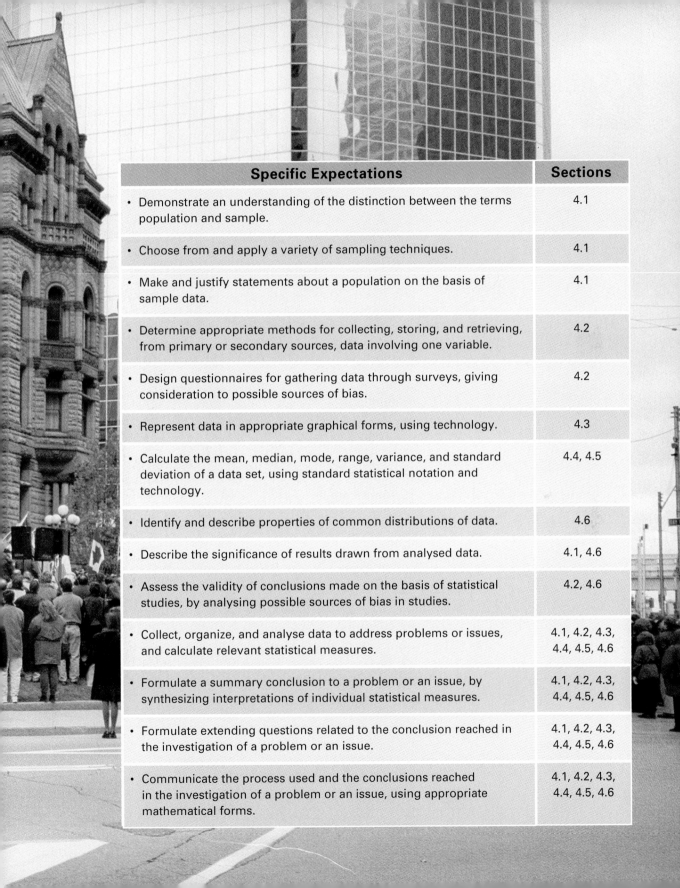

Specific Expectations	Sections
• Demonstrate an understanding of the distinction between the terms population and sample.	4.1
• Choose from and apply a variety of sampling techniques.	4.1
• Make and justify statements about a population on the basis of sample data.	4.1
• Determine appropriate methods for collecting, storing, and retrieving, from primary or secondary sources, data involving one variable.	4.2
• Design questionnaires for gathering data through surveys, giving consideration to possible sources of bias.	4.2
• Represent data in appropriate graphical forms, using technology.	4.3
• Calculate the mean, median, mode, range, variance, and standard deviation of a data set, using standard statistical notation and technology.	4.4, 4.5
• Identify and describe properties of common distributions of data.	4.6
• Describe the significance of results drawn from analysed data.	4.1, 4.6
• Assess the validity of conclusions made on the basis of statistical studies, by analysing possible sources of bias in studies.	4.2, 4.6
• Collect, organize, and analyse data to address problems or issues, and calculate relevant statistical measures.	4.1, 4.2, 4.3, 4.4, 4.5, 4.6
• Formulate a summary conclusion to a problem or an issue, by synthesizing interpretations of individual statistical measures.	4.1, 4.2, 4.3, 4.4, 4.5, 4.6
• Formulate extending questions related to the conclusion reached in the investigation of a problem or an issue.	4.1, 4.2, 4.3, 4.4, 4.5, 4.6
• Communicate the process used and the conclusions reached in the investigation of a problem or an issue, using appropriate mathematical forms.	4.1, 4.2, 4.3, 4.4, 4.5, 4.6

*G*et Ready

1. **Number skills** Evaluate. Round to three decimal places where necessary.

   a) $\sqrt{81}$

   b) $\sqrt{169}$

   c) $\sqrt{45}$

   d) $\sqrt{245}$

   e) $\sqrt{40 + 80 - 5}$

   f) $\sqrt{291 + \dfrac{35}{2}}$

2. **Proportions** Solve each proportion.

   a) $\dfrac{x}{15} = \dfrac{48}{90}$

   b) $\dfrac{n}{8} = \dfrac{6}{100}$

   c) $\dfrac{k}{5} = \dfrac{18}{150}$

3. **Interpreting graphs** Bill and Jan spent their week's vacation touring their home province. The graph represents the daily number of kilometres that they travelled.

   a) On which day did they drive the fewest number of kilometres?

   b) How much farther did they drive on Tuesday than on Thursday?

   c) What was the total number of kilometres driven?

   d) What type of graph is this?

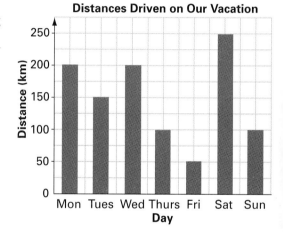

Distances Driven on Our Vacation

4. **Interpreting graphs** The graph represents how much time Lana spent in each of the countries on her 20-day vacation to Europe.

   a) How many countries did Lana visit?

   b) How many days did she stay in France?

   c) How many days did she stay in Portugal?

   d) What percent of the time did Lana stay in countries other than England?

   e) What type of graph is this?

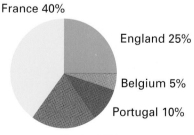

Countries Visited

France 40%   England 25%   Belgium 5%   Portugal 10%   Spain 20%

5. Interpreting graphs The mass of a patient is being carefully monitored. The graph shows the record between June 5 and July 31.

Patient's Mass

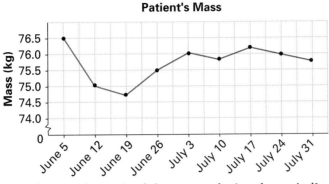

a) What was the patient's least mass during the period?

b) During which week did the patient lose the greatest amount? How much was lost that week?

c) During which week did the patient gain the most? How much was gained that week?

d) The patient's birthday is July 12. What was this person's mass on that day, approximately?

e) What type of graph is this?

*C*HAPTER PROBLEM

Most statistics start with a problem. A need is identified for which statistics will provide insight, give a better understanding of the problem, and hence, help in solving the problem. To do this, the statistician along with the client must clearly identify what exactly is to be studied. The statistician must then collect, organize, and analyse the data that will lead to an answer to the problem.

Consider the training program you hope to enter after high school, or the career to which you are aspiring. List any concerns or questions you have about this future plan. For example, if you are interested in a particular college course, perhaps you would like to know the answer to one of the following questions:

• What percent of students successfully complete the course?
• What percent of students find employment in the field within six months of completion?
• How much will it cost to complete the course, including all living expenses?

Record the problem that you plan to investigate. Work with a partner to ensure that each of you has a clear statement of what you need to research.

4.1 Collecting Data: Sampling Techniques

Statistics involves collecting, organizing, analysing, and interpreting numerical information. The numerical information collected is called the **data**. Familiar data include sports statistics, such as baseball, hockey, or Olympic records. Data can be collected from a group of individuals or from the whole **population**. One example of surveying the whole population is the census, which is collected every five years by Statistics Canada. More often, statistics are gathered from a **sample** of a population. A sample is a smaller group that generally represents the whole population.

To provide an accurate representation of the population, the sample needs to be selected carefully. There are several methods of choosing a representative sample. Among these methods are random selection and stratified selection.

 In **random sampling,** any member of the population has an equal chance of being selected. For example, to choose a random sample of 10 jelly beans from 50 mixed colours, put the jelly beans in a bag, and remove 10 without looking in the bag.

In **stratified sampling,** a population is divided into groups with similarities, or strata, and then, a sample is chosen from each of those groups. For example, a survey on shopping characteristics of families could have samples grouped by household income.

EXAMPLE 1

Choosing Teams

In physical education class, a simple method of grouping players for drills is to assign a number from 1 to 5 to each student. Then, all the students with the same number form a group. Is this a random selection of teams? Is there a fairer method?

Solution

This is a crude interpretation of a random sampling. The person numbering the students could easily bias the selection. For example, numbering all the strong players 3 and all the weak players 4. The team made up of all the 4's would be the weakest in skills.

A fairer random sampling would be to use cards numbered from 1 to 5, with sufficient repetitions to cover the number of students in the class. Shuffle the cards and hand them out, face down, one to each student. This method gives each student an equal chance of obtaining any one of the numbers.

EXAMPLE 2

Choosing a Stratified Sample

A school principal is researching the relationship between mathematics and science marks. She wants to use a sample of 60 students who are taking both math and science. In a school of 800 students, 300 are taking both math and science. The number of students taking each subject by grade are shown. How should the principal select a representative sample?

Grade 9	100
Grade 10	80
Grade 11	75
Grade 12	45

Solution

The principal should use a stratified sample, chosen so that the sample represents all grades fairly. An appropriate number of the sample should be chosen from each grade.

For the 100 grade 9 students taking math and science, use a proportion to find the number that should be selected for the sample.

Let n represent the number of grade 9s in the sample.

$$\frac{n}{60} = \frac{100}{300}$$

$$n = 60 \times \frac{100}{300}$$

$$n = 20$$

Let t represent the number of grade 10s in the sample.

$$\frac{t}{60} = \frac{80}{300}$$

$$t = 60 \times \frac{80}{300}$$

$$t = 16$$

Similarly, by proportional reasoning, 15 grade 11s and 9 grade 12s should be selected. The principal should then form her sample by randomly choosing 20 grade 9s, 16 grade 10s, 15 grade 11s, and 9 grade 12s from the 300 students taking math and science.

You can use a random number generator to select a sample without bias. Most graphing calculators have a random number generation feature. Check your calculator's user manual to learn how this feature works.

For more information on generating random numbers using a graphing calculator, see Appendix B, page 428.

To generate random whole numbers, first set the mode to zero decimal places.
Press (MODE).
Move the cursor down to **Float** and select **0**.
Press (ENTER).

EXAMPLE 3

Using a Graphing Calculator to Generate Random Numbers

The principal, from Example 2, might decide to select students from each grade level randomly by number. Use the random number generator of a graphing calculator to choose 20 of the 100 grade 9 students who take math and science.

Solution

Make a list, numbered from 1 to 100, of the grade 9 students.
Press (MATH).
Select **PRB**, for the Probability menu.
Press **1: rand**.
Press (ENTER).

Now enter the upper limit of random numbers wanted.
Here numbers from 1 to 100 are eligible.
Key in 100.
Press (ENTER).
Each press of (ENTER) will show the next random number.
Press (ENTER) until twenty different random numbers are generated.

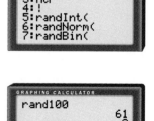

(Note: Your random numbers will be different. The first number generated in this example was 61, the second was 8, the third 2, and so on.)
The principal would then match the random numbers to the list of students to select the twenty students from the grade 9 group.

KEY CONCEPTS

- Population refers to the whole group being studied.

- Sample refers to a smaller portion, or subset, that is representative of the population.

- In random sampling, all participants have an equal chance of being selected. In stratified sampling, participants are grouped according to some common trait or characteristic.

- Random numbers can be created using a graphing calculator or using a computer random number generator.

DISCUSS THE CONCEPTS

1. Describe a situation where random sampling would be an appropriate method of data collection.

2. Describe a situation where it would be appropriate to use stratified sampling to collect data.

3. a) Is it possible to produce the following results in generating five random numbers from 1 to 10 using a random number generator? Explain.

   3 3 3 3 3

   b) If you were using this method to choose five representatives from a group of ten students, what would you need to do to select your sample?

APPLY THE CONCEPTS

A 1. Describe how a sample might be chosen to gather data in each case. Describe what the population is and indicate whether it would be feasible to survey the whole population.
   a) 30 teenagers for a new health study to be conducted by a local hospital
   b) in a city, the percentage of people aged 18–24 attending college
   c) the average number of televisions in the homes of students at your school
   d) the most popular car colour in your neighbourhood

2. In a study of favourite clothing stores, what is the population? How would you select a survey sample? How would you collect the data?

3. What are the problems with the following types of polls?
   a) A radio station is conducting a telephone poll on the favourite musical group.
   b) The number of people needing the services of the local charity food bank is being surveyed on a randomly chosen day at a randomly chosen time.

4. Music companies frequently study their audience and their share of the market. One such company would like to see how their latest "boy-band" is doing. They need a stratified sample of the listening population. Describe how they could choose the sample for this type of survey. Explain.

5. Children are selected from a large elementary school for a study on the effects of a new reading program. The study would like to choose 30 students from each of grades 1, 2, and 3. Describe how a sample could be selected to represent the students. Explain.

6. An educational research company is studying students' attention span during lectures. The researchers need to select two samples of 20 students each from a college with an enrolment of 650. Describe how they could select the two sample groups.

For help with generating random numbers using a graphing calculator see Appendix B, page 428.

7. A coach is choosing a team. To help make the selection, she decides to choose randomly. The coach numbers each player from 1 to 20.
   a) Generate random numbers to help the coach select 10 players.
   b) Describe possible problems with this method of selecting the team.

8. **Communication** For the next week, pay attention to the media's reporting of statistics. Is any information given on how the data were collected, or how the sample was chosen? Discuss this with a classmate. In your notebook or journal, record your findings.

CHAPTER PROBLEM

9. Consider the problem you are examining.
   a) Decide what type of sample would best represent the population you wish to study. Justify your choice.
   b) Choose your sample individuals.

10. **Communication, Application** Go to *www.mcgrawhill.ca/links/MPCA12* and follow the links to Statistics Canada's website, Student Resources. Research and write a report summarizing how samples are selected by Statistics Canada. Give examples to illustrate your description.

4.2 Methods of Collecting Data

Data collection is usually done using some sort of survey, questionnaire, or poll. In designing a survey and collecting your own data, consideration must be given to the questions that will be asked. If the survey is poorly designed, bias can occur, making the data collected invalid. An easy rule to remember is to start with the five Ws. Consider the Who? What? When? Where? and Why? These questions may not be suitable for all problems studied, but they provide a guide to the process of designing a survey.

Bias will also occur when a sample is poorly selected. A sample should represent the whole population as closely as possible. If not, sampling bias occurs and the results are invalid.

A survey that you conduct yourself is a **primary source** of data. Data that have been collected by someone else is called a **secondary source**.

DISCOVER

Work in a small group.
1. Design an informal survey to determine what kind of music your friends like. Is one question sufficient? Consider whether there are other questions that might be helpful in learning what specific kinds of music they like, not only singers or groups. Write down the questions.
2. Survey several friends by asking your questions. Did you get the type of response you were looking for? If not, explain what is wrong with the question(s). Rewrite your questions, if necessary, and try the survey again with different individuals.
3. Continue the revision process until you are satisfied that you have designed good survey questions. Compare your questions with those of other groups.

EXAMPLE 1

Designing a Survey

A snack food company wants to know what snacks teenagers like to eat. Create a survey to collect relevant data.

Solution

You may feel that young teenagers like different snacks than do older teenagers. One way to test this theory would be to divide the survey into two age groups: 13 to 15, and 16 to 18. Choose your sample from each of these groups, thus creating a stratified sample. Ensure ages are relatively evenly distributed in each group, genders are balanced, and the two groups are equal in number. This will allow for an unbiased selection of a sample.

Next, create a survey. Here are possible questions:
- Do you like to snack after school?
  (Answers "When" teenagers like to snack.)

- Do you prefer sweet or salty snacks?
  (Answers "What" kinds of snacks teenagers like.)

- Do you snack because you are hungry or bored?
  (Answers "Why" teenagers snack.)

- Do you
  a) buy your snacks on your way home?
  b) make your own snacks?
  c) have pre-packaged snacks at home?
  (Answers "Where" teenagers get their snacks.)

Other questions might include the following:
- Do you usually have enough money to buy a snack?
- How much do you usually spend on snacks each week?

EXAMPLE 2

Recognizing a Biased Survey

Stepan is a newspaper reporter. He likes playing computer games. He is assigned to write an article on the latest and most popular computer games. To research the topic, he asks all his friends what game they like best. He reports from his collection of data that Star Daze–Unearthed is the favourite computer game of 97% of the population.

a) What is wrong with the data?

b) How might his readers interpret these results?

Solution

a) The data are biased since Stepan only asked his game-playing friends. They may play Star Daze—Unearthed together, so it is not even an independent choice. Thus, the conclusion is not valid.

b) The population is not specified. Readers might believe that 97% of the whole Canadian population like Star Daze–Unearthed or that 97% of people who play computer games like Star Daze–Unearthed. Since we do not have the full article, we do not know whether the author reported that he asked only his friends. This is an example of how some statistics are presented as a matter of fact without proper investigation or supporting material. Reader beware!

KEY CONCEPTS

- Data that are collected by an individual in a survey or experiment are a primary source of data.

- Data that have been gathered by someone else and re-used or re-published are a secondary source of data.

- Biased information can be collected by improper selection of a sample or by creating a poorly worded survey.

- Data must be gathered in an unbiased way to ensure that valid conclusions or inferences can be drawn.

DISCUSS THE CONCEPTS

1. Interpret the following sentence. What does it mean?
   60% of all school-aged children drink ABC Cola.
   Do you have enough information to draw a conclusion from this statement? What else would you like to know?

2. Station QPTV claims they are the most popular television channel. They report, "39% of viewers were tuned into QPTV, while 22% watched ZYTV, and 49% tuned into other networks." What is wrong with this report?

APPLY THE CONCEPTS

A 1. A newspaper article reports that a province has 2 094 001 voters, and the turnout at an election was 60.2%.
   a) How many people voted?
   b) What type of data did the article use in the report, primary or secondary? Explain your reasoning.

2. Describe how you would collect data on each topic. Classify your source of data as primary or secondary.
   a) the top ten musical performers by income
   b) the country of origin of immigrants to Canada in 2000
   c) the most popular type of cookies of 5-year-olds in your neighbourhood

3. Find an article in a newspaper that quotes statistics.
   a) Discuss what information you can draw from the data presented in the article.
   b) Apply the terms data, bias, primary source, secondary source, and validity of conclusions in your discussion.
   c) What questions or assumptions are not explicitly stated?

B 4. **Communication** Working with a partner, discuss what questions should be asked when you are reading or listening to a report that includes statistical information. Create a checklist.

5. **Communication, Application** Find two examples of biased or misleading statistical reports in newspapers, magazines, or on-line reports. Explain why they are misleading.

6. a) Design a survey to collect information from students in your school on one of the following topics:
   • Favourite fast-food restaurant
   • Favourite TV show
   • Favourite school subject
   b) Describe how you will collect responses to your survey. Will you poll the whole student population? Will you use a sampling technique, and if so, which type?
   c) Test your survey on a small number of students. Revise the wording of the questions, if necessary.
   d) Carry out your survey.
   e) Are your data from a primary or a secondary source?

7. **Communication** Write an article that includes some fictitious statistics. Test how believable your article is by asking a classmate to interpret your data.

8. **Communication** Design a poster that misleads the reader with some statistics. Be prepared to explain what is wrong with the presentation.

9. In a Statistics Canada report on the waste management industry, the following was reported:

Generation from residential sources amounted to 330 kg per Canadian, of which 100 kg, or 30%, was diverted from landfills or incinerators.

a) Explain what you think the statement means.

b) Go to *www.mcgrawhill.ca/links/MPCA12* and follow the links to read the whole report. How did what you read differ from what you interpreted from the excerpt? Determine whether the report used primary or secondary sources of data.

c) Use the **Data Bank**, page 459, which gives information on Waste Management in Ontario. Create a truthful but misleading headline.

CHAPTER PROBLEM

10. Design an appropriate survey that will help research your issue. Consider carefully the questions and possible responses. List the most important questions you want to ask. Decide how you will choose a sample to survey.

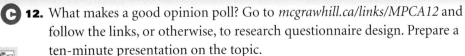

ACHIEVEMENT CHECK Knowledge/Understanding Thinking/Inquiry/Problem Solving Communication Application

11. A large high school wishes to study the destination paths of grade 9 students to predict the class sizes in each of three streams. Out of a grade 9 student population of 455, a questionnaire will be given to 46 students.

a) Design a suitable survey question.

b) The sample of students who will be asked the survey question is to be chosen using one of the following methods. For each method,

i) identify the type of sample obtained

ii) identify any possible bias

iii) explain why you would or would not choose this method

Method A: Using board-issued student numbers, every 46th student will be chosen

Method B: Using board-issued student numbers, a computer will randomly select 46 numbers

Method C: Students will be grouped into academic or applied, and a proportional number of students will be chosen from each stream

Method D: Students will be listed by their birthdays and the first 46 will be chosen

EXTEND THE CONCEPTS

C 12. What makes a good opinion poll? Go to *mcgrawhill.ca/links/MPCA12* and follow the links, or otherwise, to research questionnaire design. Prepare a ten-minute presentation on the topic.

4.3 Representing Data

To analyse data, they must be presented in a form that can be easily read and interpreted by the statistician and the clients. Statistical data can be represented in many graphical forms, including pictographs, circle graphs, broken-line graphs, bar graphs, and histograms.

You have used bar graphs in the past. They are good for displaying data that involve criteria that do not involve any measurement, such as favourite colour. A **histogram** is similar to a bar graph, but it has no separation between the bars, and the criteria become intervals of measurement. To draw a histogram, the data are first organized into a frequency table. Then, the data are grouped using **class intervals**, so the data are well represented with no gaps between the class intervals.

DISCOVER

Drawing a Histogram

One hundred six golfers participated in a charity golf tournament. The number of golfers who obtained a particular score is known as the frequency of the score, as indicated in the table.

Score	Frequency
64	1
66	2
67	1
69	11
70	18
71	14
72	26
73	18
74	6
75	5
76	2
78	2

1. Make a new table for the data by grouping the data using class intervals as started here. Each class interval should be the same width. The notation [60, 65) means 60 is included in this class interval, but 65 is not.

Class Interval	Frequency
[60, 65)	1
[65, 70)	14

2. On grid paper, mark the class intervals on the horizontal axis and frequency on the vertical axis. Draw adjacent bars to represent the data. Give your histogram a title.

EXAMPLE 1

Drawing a Histogram Manually

A report of English marks must be prepared for the principal. One teacher has the following marks recorded. Prepare a frequency table and present the information in a histogram.

52 66 75 80 52 48 95 85 84 68 86 82 63 78 75 64 79 81 66 53 76 75 69 65

Solution

The range of the data is 48 to 95, so suitable class intervals are $[40, 50)$, $[50, 60)$, and so on. A frequency table can be made with the help of tallies.

Class Interval	Tally	Frequency
[40, 50)			1
[50, 60)	|||	3
[60, 70)	⊬⊬ ||	7
[70, 80)	⊬⊬		6
[80, 90)	⊬⊬		6
[90, 100)			1

Draw the graph.

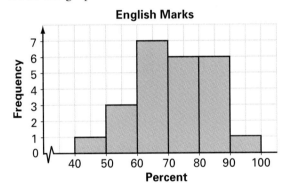

English Marks

EXAMPLE 2

Drawing a Bar Graph Using Spreadsheet Software

Show the data given in Example 1 as a bar graph using spreadsheet software.

Solution

Make a frequency table.

Marks	Frequency
40–49	1
50–59	3
60–69	7
70–79	6
80–89	6
90–100	1

Enter the data into a spreadsheet. Enter the Titles in cells **A1** and **B1**. Enter the data, with the marks in column A, cells 2 through 7, and the frequency in column B, cells 2 through 7.

For help with graphing data in spreadsheets, see page 432 (Microsoft® Excel) or page 436 (Quattro® Pro).

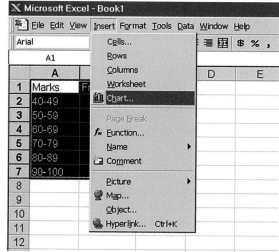

To create a graph, highlight all the data (including titles) that you want in the bar graph, starting at cell A1 going to the bottom right corner of the input data, cell B7.

Click on the Chart icon.

Follow the wizard options, clicking on the instructions to produce the bar graph. Remember to change the names of the Chart Title, *x*-axis, and *y*-axis.

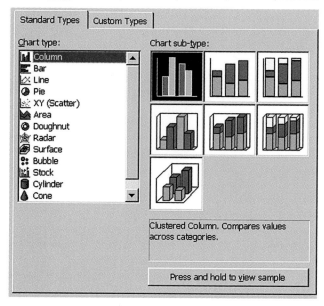

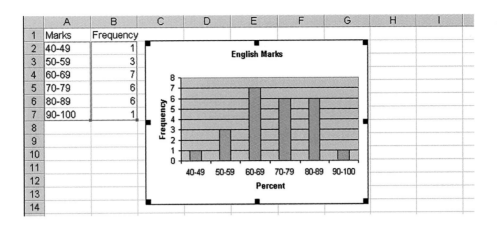

	A	B	C	D	E	F	G	H	I
1	Marks	Frequency							
2	40-49	1							
3	50-59	3							
4	60-69	7							
5	70-79	6							
6	80-89	6							
7	90-100	1							
8									
9									
10									
11									
12									
13									
14									

EXAMPLE 3

Drawing a Histogram Using Fathom™

Consider the many family movies that have been made over the years.

Movie Title	Release Date	Length (min)
Fantasia	1940	120
Swiss Family Robinson	1940	93
Dumbo	1941	88
Pinocchio	1940	88
Bambi	1942	70
The Jungle Book	1942	94
Song of the South	1946	75
Cinderella	1950	74
Alice in Wonderland	1951	75
Peter Pan	1953	76
Lady and the Tramp	1955	75
Old Yeller	1957	83
Sleeping Beauty	1959	75
101 Dalmations	1961	68
In Search of the Castaways	1962	98
The Sword in the Stone	1963	79
The Love Bug	1968	107
The Aristocats	1970	78
Teenage Mutant Ninja Turtles	1990	93
Beauty and the Beast	1991	84
The Lion King	1994	89
Pocahontas	1995	81
Toy Story	1995	81
The Land Before Time V	1997	74

a) Use the data to create a frequency table and display the data in a histogram using Fathom™.

b) What inferences might be made based on the data? What is the typical length for a family movie? How does the length of a family movie seem to have changed with time?

Solution

For an introduction to Fathom™, see page 438 in Appendix B.

a) Open Fathom™.

On the menu bar, click on **File**, **New** to create a new document.

Open an empty case table by clicking on the table icon, or by going to **Insert** and then, selecting **Case Table**.

Click in the cell with <**new**> and type in the attribute name, the column label Length_min. Then, enter the data in that column.

To enter a title, use the **Data** menu. Select **Rename Collection** and key in the title Length of Family Movies.

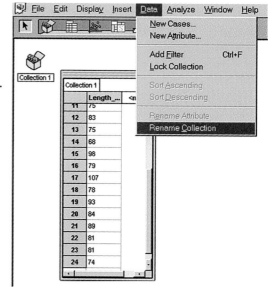

To create a histogram:

Use the **Insert** menu. Choose **Graph**.

An empty graph will appear.

Click on the column of data. A small hand will appear. Drag the column of data onto the x-axis of the graph. By default, a dot plot will appear.

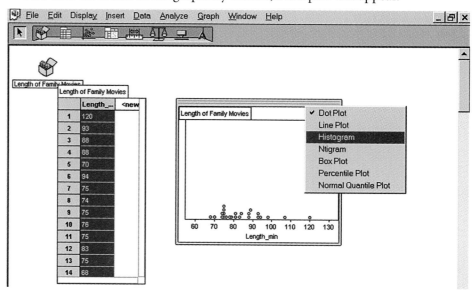

Use the pull-down menu in the upper right-hand corner. Select **Histogram**.

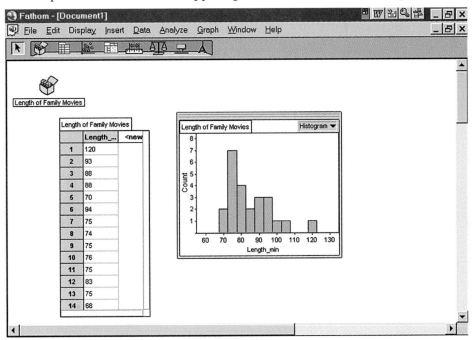

Language
Link

Fathom™ and other recent applications use the term *bin width* instead of class interval.

Fathom™ automatically chooses suitable bin widths.

Use the **Graph** menu, **Show Graph Info** to see how the histogram is drawn.

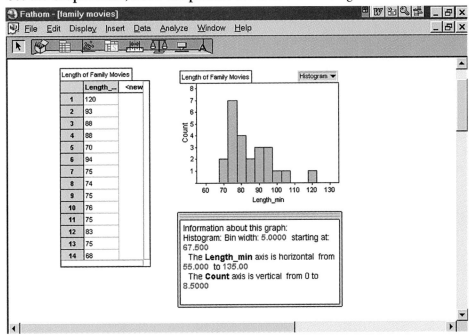

To alter the bin widths, select the blue bin width and key in your own choice. Similarly, you can change the starting value.

Click on the graph and explore other features. For example, if you click inside a bar, the data corresponding to that bar are highlighted in the case table.

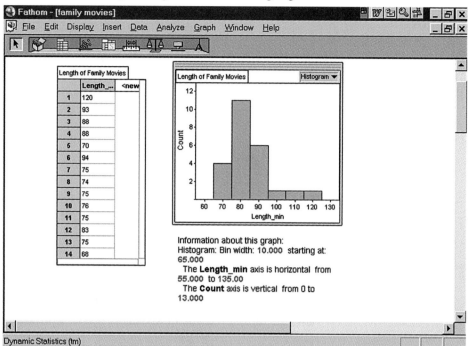

b) Many family movies seem to be between 75 and 85 min in length. The oldest movie listed is the longest, the shortest was a 1961 release, but the most recent movie listed is also quite short. There is no recognizable pattern of change with time. More data would need to be gathered to be certain of these inferences.

KEY CONCEPTS

- A **histogram** presents data, based on frequency of occurrence, grouped in intervals of equal width, usually with no gaps between the intervals.

- A **bar graph** presents data grouped according to categories, with equal spaces between the bars.

- Many tools can be used to help create graphs, including spreadsheet software and Fathom™.

DISCUSS THE CONCEPTS

1. There is very little difference in appearance between a bar graph and a histogram. How would you explain the differences to a younger student?

2. Certain types of graphs are more appropriate for some data than others. Describe a set of data that each type of graph is most appropriate for displaying.
   a) circle graph   **b)** bar graph   **c)** pictograph   **d)** histogram

3. Data are to be collected from a sample of students in your school to provide an answer to each question. Which results would you display in a bar graph and which would you display in a histogram? Explain.
   a) What is your favourite colour?
   b) How much do you on entertainment each month?
   c) How do you usually travel to school?
   d) How many music CDs do you own?

APPLY THE CONCEPTS

A 1. What class intervals would be appropriate to use in organizing the data described to display them in a histogram?
   a) heights of ten-year-olds, where the range of data collected is 123 cm to 176 cm
   b) daily maximum temperatures, where the range of data collected is 0°C to 30°C
   c) length of recordings, where the range of data collected is 3:06 min (3 min 6 s) to 6:42 min

2. The mass, in kilograms, of each player on a college football team was recorded.

78 72 80 85 76 73 79 75 76 88 69 71 80 76 68 75 70 72 67 61

   a) Organize the data into a frequency table using class intervals [60, 65), [65, 70), and so on.
   b) Display the data in a histogram.

B 3. Following a stroke (an interruption of blood flow to the brain), patients must go through a rehabilitation period. An occupational therapist recorded the average number of weeks of therapy for a random sample of 30 patients.

8 21 6 9 4 15 10 9 7 9 6 17 8 9 10
9 2 8 3 10 16 13 5 3 2 1 9 4 13 7

   a) Create a frequency table and prepare a histogram to display the results.
   b) What is the most common number of weeks of therapy?

4. **Application** Retail businesses spend a lot of money advertising their products. They do a great deal of research to promote their products effectively. One thing they research is the appeal of their packaging. Which shape should a business use for a logo on the label of their product? These four shapes are being considered.

a) Predict which shape you think will be the most appealing to the majority of people.
b) Decide on a method to collect data using a sample. Conduct the survey.
c) Display your data in a suitable graph.
d) Do your data confirm your prediction?

5. **Application** As a school project, Adam researched the caloric value of nutrition bars. He collected the following data.

130 219 230 250 368 170 220 230 252 440 170 220
230 280 465 180 220 234 296 490 200 240 225 310

Create a frequency table and prepare a histogram to display the data.

6. Discus throw has been part of the Olympics since ancient times. Since 1928, women have also participated in this event.

Year	Winning Country	Winning Distance (m)
1928	Poland	39.62
1932	United States	40.58
1936	Germany	47.63
1948	France	41.92
1952	Soviet Union	51.42
1956	Czechoslovakia	53.69
1960	Soviet Union	55.10
1964	Soviet Union	57.27
1968	Romania	58.28
1972	Soviet Union	66.62
1976	East Germany	69.00
1980	East Germany	69.96
1984	Holland	65.36
1988	East Germany	72.30
1992	Cuba	70.06
1996	Germany	69.95
2000	Belarus	68.40

a) Which country has won the most gold medals in women's discus throw?
b) Draw a bar graph to show the number of gold medals by country.
c) What other information can you infer from these data?

7. The length, in minutes and seconds, of 30 songs by the group The Tragically Hip are as follows.

3:06 5:30 3:17 3:59 3:21 4:55 3:32 4:10 5:53 4:08
4:57 4:38 3:43 4:06 4:25 3:28 3:20 3:28 4:12 4:10
4:23 4:36 3:41 3:54 4:16 3:37 4:40 3:56 4:30 3:46

a) Display the data in a histogram.

b) What is the typical length of a song by this group?

For help with graphing data in spreadsheets, see Appendix B, page 432 for Microsoft® Excel or page 436 for Quattro® Pro.

8. **Application** Find a local team's season scoring data. Use technology to organize the data in a spreadsheet and then, display it in a histogram.

9. **Data Bank** Canada is abundant in open spaces and parks. Use the Ontario Provincial Parks data, on page 464.

a) Which park has the greatest area? Which park has the least area?

b) Which park has the greatest area per campsite?

c) Which park has the least area per campsite? Express the area in square metres per campsite.

d) Create a histogram to show the number of campsites available.

10. **Data Bank** Use the Passenger Aircraft Operating Statistics, on page 462. Create a histogram for each of the following data sets.

a) wingspan **b)** length

11. **a)** Using the data you have collected for your problem, decide which type of graph would best represent the data. Explain your choice.

b) Draw the graph.

12. Use the data on Craters.

a) Sort the information and draw a graph to show the number of craters by continent. What type of graph is most appropriate for this data?

b) Organize the data by diameter. What type of graph is most appropriate for displaying this data? Draw the graph.

c) Write a paragraph describing what you have learned about craters on Earth's surface from these statistics.

4.4 Measures of Central Tendency

In everyday communication, the term *average* is often used. In statistics, three measures of central tendency of a set of data are identified: the mean, the median, and the mode. The mean is the statistic that is commonly referred to as the average. The median is the middle value, and the mode is the most frequently occurring value.

The **mean** is calculated by taking the sum of the data and dividing by the number of data. The mean is sometimes denoted by $\bar{x}$.

$$\text{Mean} = \frac{\text{sum of all data}}{\text{number of data}}$$

The **median** is found by sorting the data in ascending order and finding the middle value. If there is an odd number of data, there is a single middle value. If there is an even number of data, then the median is the average of the two middle values.

The **mode** is also found using sorted data. It is the value that occurs most frequently.

DISCOVER

Finding Measures of Central Tendency

One Saturday, a store sold forty pairs of running shoes in the following sizes.

7	5	8	6	6	9	10	8	12	6	7	10
9	6	11	8	10	9	8	10	4	9	8	9
8	10	8	6	9	10	7	10	8	13	10	9
8	6	9	10								

1. Arrange the data in ascending order.
2. What is the mode size? Can there be more than one mode in a set of data? Can there be no mode?
3. What is the median size?
4. Describe a short cut for finding the sum of all the data when they are arranged in ascending order. Calculate the mean size.
5. Which measure of central tendency is of most interest to the store manager?

EXAMPLE 1

Finding Measures of Central Tendency

The scores on a college entrance test are shown.

| 62 | 75 | 90 | 71 | 67 | 69 | 62 | 82 | 47 | 66 | 64 | 57 | 74 | 65 | 69 |

Find each measure of central tendency.

a) mean **b)** median **c)** mode

Solution

Method 1: Using a Calculator

a) Mean $= \dfrac{\text{sum of all data}}{\text{number of data}}$

$= \dfrac{62 + 75 + 90 + 71 + 67 + 69 + 62 + 82 + 47 + 66 + 64 + 57 + 74 + 65 + 69}{15}$

$= \dfrac{1020}{15}$

$= 68$

On a scientific calculator, use the summation key $\boxed{\Sigma+}$ and the mean key $\boxed{\bar{x}}$:

$\boxed{\Sigma+}$ 62 $\boxed{\Sigma+}$ 75 $\boxed{\Sigma+}$ 90 … $\boxed{\Sigma+}$ 65 $\boxed{\Sigma+}$ 69 $\boxed{\bar{x}}$ 68

The mean score is 68%.

b) Arrange the scores in ascending order.

47, 57, 62, 62, 64, 65, 66, 67, 69, 69, 71, 74, 75, 82, 90

Since there are 15 scores, the median score is the eighth score.
The median score is 67%.

c) Scores of 62 and 69 both occur twice.
There are two mode scores, 62% and 69%.

Method 2: Using a Graphing Calculator

First, clear any data lists using $\boxed{\text{STAT}}$ **4:ClrList**.
Enter the list of scores in L1.

See page 429 of the Technology
Appendix for an introduction to the
statistical functions of the TI-83 Plus.

a) Go to the home screen.
Press $\boxed{\text{2nd}}$ $\boxed{\text{STAT}}$.
Move the cursor over to **MATH**
and select **3:mean(** .

Press (2nd) 1 to enter L1.
Enter) and press (ENTER).

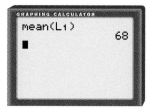

The mean score is 68%.

b) Use the same steps, but from the MATH menu,
select **4:median(**.

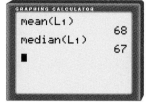

The median score is 67%.

c) The calculator can sort the data to scan for the
mode(s).
Press (2nd)(STAT) and select **OPS** 1:SortA(.

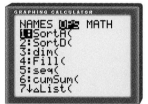

Press (ENTER)
Enter L1).
Press (ENTER).

Press (STAT) and select **1:Edit**.
Press (ENTER).
Move the cursor up to the top of L1. An
ordered list is shown. You can move the cursor
down to identify the modes.

The modes are 62% and 69%.

EXAMPLE 2

Finding Measures of Central Tendency From a Frequency Table

The selling prices of running shoes were recorded in a frequency table.

Number Sold	Price ($)
4	29.95
10	39.99
12	48.95
6	55.69
5	68.90
3	85.00

a) Determine the mean, median, and mode selling price.

b) Which of these measures would the store manager probably choose to report to his head office? Why?

Solution

a) Mean $= \dfrac{\text{sum of all data}}{\text{number of data}}$

$= \dfrac{4 \times 29.95 + 10 \times 39.99 + 12 \times 48.95 + 6 \times 55.69 + 5 \times 68.90 + 3 \times 85.00}{40}$

$= \dfrac{2040.74}{40}$

$= 51.0185$

The mean selling price is $51.02.

There are 40 sales, so the median is between the 20th and 21st selling price.
In ascending order of price, both the 20th and 21st prices are $48.95.
The median selling price is $48.95.
The most frequently occurring price is $48.95.
The mode selling price is $48.95.

b) The store manager would probably report the mean selling price. This measure is the greatest. The other measures might give the impression that sales were lower.

KEY CONCEPTS

The mean, median, and mode are known as measures of central tendency.

- The mean, $\bar{x}$, is Mean $= \dfrac{\text{sum of all data}}{\text{number of data}}$.
- The median is the middle value when the data are arranged in ascending order. If there is an even number of data, the median is the average of the middle two values.
- The mode is the value that occurs most frequently.

DISCUSS THE CONCEPTS

Use examples to support your answers.

1. Is it possible for a set of data to have the same value for the mean, median, and mode?

2. Is it possible for a set of data to have two means?

3. Is it possible for a set of data to have two modes?

4. The mode is sometimes a poor measure of central tendency. Explain why.

PRACTISE

A 1. Find the mean, median, and mode of each set of data.
a) 24, 90, 10, 78, 50, 35, 90, 33, 46
b) 5, 9, 1, 6, 4, 8, 3, 5, 6, 9, 7, 4, 8, 5, 5, 1
c) 100, 98, 155, 110, 99, 123, 140, 145

2. The annual family income for people living on one street is recorded.

| $61 000 | $113 000 | $43 000 | $101 000 |
| $77 000 | $105 000 | $106 000 | $66 000 |

Calculate the mean, median, and mode of the family incomes.

3. The daily high temperature, in degrees Celsius, was recorded for a two-week period.

26 27 28 30 34 35 36 37 38 37 35 36 34 29

a) Calculate the mean, median, and mode.
b) Which measure of central tendency gives the best idea of what the temperature was like these two weeks?

4. Sunita received the following marks on tests in geography.

60 60 53 96 91 92 72 60 60 67 74 60 85 84

Calculate her mean, median, and mode mark.

APPLY THE CONCEPTS

5. A high school basketball team collects the statistics of each of their games. Last season, their numbers of shots taken per game were as follows.

82 115 150 99 89 125 115 126 140 97 124 119 130 114

a) What was the mean, median, and mode for shots taken per game?
b) Which statistic is most useful in this context?

6. Mohan recently moved to a new town and wanted to find his average commute time. He recorded the time, in minutes, to travel from his apartment to the office for fifteen days.

65 80 58 56 67 60 75 85 55 60 52 70 63 59 61

a) Determine and compare the three measures of central tendency.

b) Which measure gives the most appropriate "average" commuting time? Explain why.

B 7. Twenty different mutual funds reported the following average annual returns.

30.76%	27.29%	13.95%	24.95%	−9.21%	−26.73%	5.63%
24.95%	9.1%	−17.71%	52.48%	5.63%	9.1%	9.1%
30.76%	24.95%	30.76%	−7.12%	−23.12%	25.49%	

a) Calculate the mean, median, and mode rate of return.

b) Which measure best shows the "average" rate of return? Explain why.

8. A salesperson records the number of items sold by price from a vending machine each week.

Number Sold	Price ($)
85	1.00
96	1.25
83	1.50
64	1.75
55	2.00

a) How many items were sold this week?

b) Determine the mean, median, and mode price.

9. Thinking/Inquiry/Problem Solving, Communication

a) Refer to the **Data Bank**, page 457. Determine the mean, median and mode length of Action Movies.

b) Use the data for Family Movies given in Example 3 on page 171. Determine the mean, median, and mode length.

c) Compare and contrast the measures of central tendency for the two types of movies. Give possible reasons for any similarities and differences.

10. Data Bank Use the Passenger Aircraft Operating Statistics on page 462.

a) Determine the mean, median, and mode for each distance.

   i) takeoff runway length

   ii) landing runway length

b) Which length needs to be longer? Is there a relation between takeoff runway length and landing runway length? Explain.

4.5 | Measures of Dispersion

In analysing data, it is often important to know whether it is widely spread out, or whether it is clustered around the mean.

DISCOVER

Spread of Data

The number line is a simple pictograph, showing the number of e-mail messages that Sanjay received each day during one week.

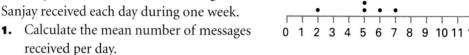

Number of E-mail Messages per Day

1. Calculate the mean number of messages received per day.
2. On how many days did Sanjay receive more than the mean number?
3. On how many days did Sanjay receive less than the mean number?
4. Are the data equally distributed on either side of the mean? Describe the distribution of the points.

Measures of dispersion are used to quantify the spread of the data. The range, the variance, and the standard deviation are known as the measures of spread, or dispersion.

The range is the simplest measure of dispersion. The **range** is calculated by finding the difference between the greatest and the least values in the data. The range is useful since it is easy to understand. However, if there are extreme data, the range is affected.

For example, the range of the values 1, 2, 4, 6, 9, 11, 15, 25 is $25 - 1$, or 24.

For each value in the set of data, its difference from the mean is called its **deviation from the mean**. The **variance** can be found by calculating the average squared difference (or deviation) of each value from the mean.

For example, consider the values 1, 3, and 8. The mean is 4. The variance, *Var*, is calculated as follows.

$$Var = \frac{(1-4)^2 + (3-4)^2 + (8-4)^2}{3}$$

$$= \frac{9 + 1 + 16}{3}$$

$$= \frac{26}{3}$$

$$\doteq 8.67$$

The **standard deviation**, usually denoted by the greek letter σ, is the most commonly used measure of dispersion. It is found by taking the positive square root of the variance.

Standard deviation = $\sqrt{\text{variance}}$

$$\sigma = \sqrt{\frac{\Sigma(x-\bar{x})^2}{n}}$$

The standard deviation is small if most data are close to the mean, but it is larger if the data are spread out further from the mean.

EXAMPLE

Finding Measures of Dispersion

Rachelle works part-time at a gas station. Her gross earnings for the past eight weeks are shown.

$55 $68 $83 $59 $68 $95 $75 $65

Calculate the range, variance, and standard deviation of her weekly earnings.

Solution

Method 1: Using a Calculator

The range is the difference between the greatest value and the least value.

$95 − $55 = $40

The range of Rachelle's earnings is $40.

To find the variance, first find the mean.

$$\text{Mean} = \frac{55 + 68 + 83 + 59 + 68 + 95 + 75 + 65}{8}$$

$$= \frac{568}{8}$$

$$= 71$$

The variance is calculated by summing the squares of the difference of each individual piece of data and the mean, and dividing by the total number of data.

$$Var = \frac{(55-71)^2+(68-71)^2+(83-71)^2+(59-71)^2+(68-71)^2+(95-71)^2+(75-71)^2+(65-71)^2}{8}$$

$$= \frac{(-16)^2 + (-3)^2 + 12^2 + (-12)^2 + (-3)^2 + 24^2 + 4^2 + (-6)^2}{8}$$

$$= \frac{1190}{8}$$

$$= 148.75$$

The variance is $148.75.

Standard deviation = $\sqrt{variance}$

$$\sigma = \sqrt{Var}$$
$$\sigma = \sqrt{148.75}$$
$$\sigma \doteq 12.20$$

A scientific calculator can be used to find the standard deviation of the data.

55 $\boxed{\Sigma+}$ 68 $\boxed{\Sigma+}$ 83 $\boxed{\Sigma+}$ 59 $\boxed{\Sigma+}$ 68 $\boxed{\Sigma+}$ 95 $\boxed{\Sigma+}$ 75 $\boxed{\Sigma+}$ 65 $\boxed{\Sigma+}$ $\boxed{\sigma xn}$ 12.19631092

The standard deviation is approximately $12.20.

Method 2: Using a Graphing Calculator
First, clear any previous lists.
Press $\boxed{STAT}$ **5** to select **5:SetUpEditor.**
Press $\boxed{ENTER}$.

See Appendix B for more information on entering data (page 427) and on calculating single-variable statistics (page 429) with a graphing calculator.

SetUp Editor is pasted to the home screen.
Press $\boxed{ENTER}$.

 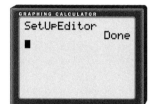

Enter the data in L1.

To perform statistical calculations, press $\boxed{STAT}$, then select **CALC**.

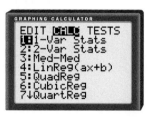

Select **1: 1-Var Stats.**
Press $\boxed{ENTER}$.

The mean, $\bar{x}$, is $71.
The standard deviation, σx, is approximately $12.20.

For a normal distribution, 68% of the data are within one standard deviation of the mean. For Rachelle's situation, this means that for 68% of the past eight weeks, her earnings were $71 plus or minus $12.20. You will learn more about normal distributions in the next section.

Another statistical calculation, Sx on a graphing calculator or σ_{n-1} on a scientific calculator, is found by dividing the sum of the squares of the difference of each value from the mean by $n - 1$, instead of by n, and taking the positive square root.

This value, $S_x = \sqrt{\dfrac{\Sigma(x - \bar{x})^2}{n-1}}$, is used to infer the standard deviation of the whole population from which a sample is chosen. For the data in the example, S_x is $13.04. This means that, provided other factors about her employment situation remain the same, Rachelle can expect that 68% of the time her weekly earnings will be $71 plus or minus $13.04, or between $57.96 and $84.04.

KEY CONCEPTS

Measures of dispersion are used to analyse the spread of data.

- The range is the difference between the least value and the greatest value.

- The variance, Var, is the sum of the squares of the difference of each value from the mean, divided by the total number of data.
$Var = \dfrac{\Sigma(x - \bar{x})^2}{n-1}$

- The standard deviation, σ, is the positive square root of the variance.

DISCUSS THE CONCEPTS

Use examples to support your answers.

1. Can the range of a set of data ever be negative?

2. Can the range of a set of data ever be zero? If so, what are the variance and the standard deviation?

3. Is the standard deviation ever greater than the range?

4. If the range of data is a large number, but the standard deviation is small, what can you infer about how the data are spread?

A 1. The following scores were recorded in Olympic archery trials. Calculate the range, the mean, the variance, and the standard deviation of this set of data.

800 799 900 952 909 650 850 844

2. On a dairy farm, the masses, in kilograms, of new-born calves are recorded. Calculate the range, the mean, and the standard deviation of this set of data.

58 67 93 52 79 63 72 66 84 78 85 70 59 65 74

B 3. A realtor recorded the values of the homes that she sold in one month.

$74 300	$97 700	$115 600	$173 400	$110 700	$119 700	$103 700
$100 200	$95 400	$100 500	$66 700	$109 800	$68 900	$123 000
$150 200	$199 600	$108 500	$184 000	$109 400	$66 700	

a) Calculate the three measures of dispersion for this set of data.

b) Describe the spread of the data.

4. Communication The business section of a newspaper records activities of the stock market.

Stock	52-Week High ($)	52-Week Low ($)	Dividend ($)	Close ($)
LKP	219.33	143.03	7.59	151.82
CMP	203.55	58.63	5.19	103.79
PDP	137.71	47.61	0	101.86
MEP	202.99	170.25	17.35	173.52
TMP	257.70	56.73	18.29	182.91
MIP	284.38	134.47	0	213.21
DBD	129.44	2.49	0.51	5.71
VIP	228.49	68.34	12.09	151.11

a) Find the range and the standard deviation of the 52-week high prices for the given stocks.

b) Find the range and the standard deviation of the closing prices for the given stocks.

c) Compare and contrast the statistical measures between the two sets of data. Which prices have a greater spread?

5. Data Bank Men's Olympic Triple Jump records have been kept since the first modern games in 1896. Refer to page 461.

a) Calculate the mean and the standard deviation of the winning distances.

b) Calculate the mean and the standard deviation of the distances that set new Olympic records.

c) Compare the mean and the standard deviation of the Olympic records and the same measures for the winning distances. How are they similar? Give reasons for any differences.

6. **Communication** The provinces of Canada are diverse in many ways. The population of each province and territory, on July 31, 2001, is listed. The numbers are in thousands.

Province/Territory	Population (thousands)
Newfoundland and Labrador	533.8
Prince Edward Island	138.5
Nova Scotia	942.7
New Brunswick	757.1
Québec	7 410.5
Ontario	11 874.4
Manitoba	1 150.0
Saskatchewan	1 015.8
Alberta	3 064.2
British Columbia	4 095.9
Yukon	29.9
Northwest Territories	40.9
Nunavut	28.2

a) Calculate the mean, the range, and the standard deviation.

b) Write a paragraph describing and interpreting the meaning of each of these statistics in the context.

7. Find the mean, median, mode, range, and standard deviation for the data you have collected. Are these measures appropriate for the data you have collected? Why or why not? Explain.

8. Go to *www.mcgrawhill.ca/links/MPCA12* and follow the links to research data on the numbers of households with at least one regular Internet user, by province. Calculate and comment on the mean and standard deviation of the data.

9. Collect data on the pulse, in beats per minute, of all the males in your class.

a) Find the mean, median, mode, range, and standard deviation for the data you have collected.

b) Determine the standard deviation of the pulse of all grade 12 males, when they are in some sedentary activity.

Properties of Common Distributions

Medical researchers have determined a normal interval for a person's blood pressure. Blood pressure is usually recorded as a fraction, for example, 120/80. When the heart contracts, the pressure is higher; this is the systolic measure. The normal range of systolic blood pressure is 110 to 140.

To determine what is normal, collect data and create a histogram. For example, if a researcher were to select a random sample of 100 adult women, measure their blood pressure, and construct a histogram, the researcher would obtain a graph like the one shown.

To improve the reliability of the data, the researcher would increase the sample size (as large as possible) and decrease the width of the classes. The revised histogram is shown.

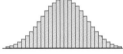

A smooth curve is drawn through the middle of each bar creating the characteristic bell shape of a normal distribution curve.

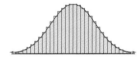

DISCOVER

Characteristics of a Normal Distribution

The data are the maximum daily temperatures recorded during the months of July and August. (They have been ordered for you.)

11, 12,
13, 13, 14,
15, 16, 16, 16
17, 17, 17, 18, 18
19, 19, 19, 19, 20, 20
21, 21, 22, 22, 22, 22, 22
23, 23, 23, 23, 23, 24, 24, 24
25, 25, 26, 26, 26, 26, 26,
27, 27, 27, 28, 28, 28
29, 29, 29, 29, 30
31, 31, 32, 32
33, 34, 34
35, 36

1. Display the data in a histogram.
2. Mark the midpoint of each bar in the histogram. Draw a smooth curve passing through the midpoint of each bar.

3. Determine the mean, median, and mode of the data. Mark these values on your graph using a vertical line for each measure.

4. The temperature data have a normal distribution. Describe the shape of the normal distribution curve.

5. Describe the relationship of the mean, median, and mode for a normal distribution.

6. Calculate the standard deviation. Draw vertical lines on your graph to mark one standard deviation on either side of the mean.

Data that are collected from a large sample of people or about naturally occurring phenomena usually have a normal distribution. For example, data for a large sample of men's shoe sizes, masses of infants, people's blood pressure, and heights of trees will produce a normal distribution curve.

The normal distribution curve shows the standard deviation marks on either side of the mean. For a normal distribution, 68% of the data lie within one standard deviation of the mean, and 95% of the data lie within two standard deviations of the mean.

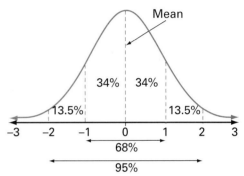

A normal distribution can be compared to the shape of a bell. If the shape of the distribution is symmetrical, then the mean, median, and modal class interval will all be similar. Sometimes the data collected do not have a normal distribution. Other distributions with distinct characteristics are examined below.

If the shape is lopsided, or asymmetrical, the distribution is said to be **skewed**. The data are **positively skewed** if the curve has a long right-hand tail or **negatively skewed** if the curve has a long left-hand tail.

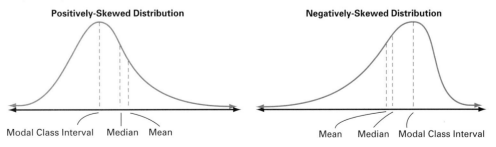

When a distribution is positively skewed, the mean is greater than the modal class interval and median (which remains in the middle). When a distribution is negatively skewed, the mean is less than the modal class interval and median.

EXAMPLE 1

Skewing

Study the three distributions shown.

a) Compare distributions B and C to the normal distribution, A.
Use the terms positively skewed and negatively skewed, giving reasons.

b) Suggest a set of data that might be modelled by each distribution.

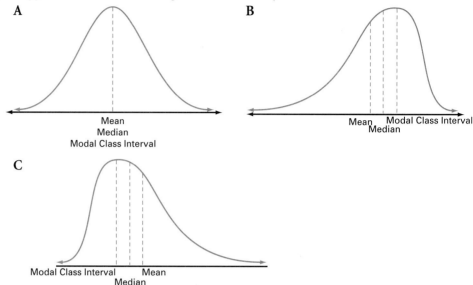

A

B

Mean
Median
Modal Class Interval

Mean   Modal Class Interval
Median

C

Modal Class Interval   Mean
Median

Solution

a) Distribution B is shifted to the right of the median. The mean has moved to
the left and the modal class interval has moved to the right. Distribution B
can be described as negatively skewed.

Distribution C is shifted to the left of the median. The mean has moved to
the right and the modal class interval has moved to the left. Distribution C
can be described as positively skewed.

b) Distribution B could be a model of the heights of basketball players in the
NBA or WNBA. Distribution C could be a model for the heights of students
in a brand new high school that has more grade nine students than any
other grade.

EXAMPLE 2	**Bimodal Distributions**

Some data sets do not follow the normal pattern. Consider the two sets of data.

A: The number of high school graduates in 1999, by province

   6715, 1643, 10 161, 8778, 82 200, 116 913, 11 829, 11 838, 27 386, 39 331

B: One-year closed mortgage rates of 14 financial institutions

   5.75, 5.7, 5.65, 6.95, 6.19, 6.19, 6.59, 6.45, 6.7, 6.75, 6.75, 6.85, 7.52, 5.6

a) How many modes does each data set have? Use the term bimodal, if possible.

b) Discuss whether either data set has a normal distribution.

Solution

a) In A, each value occurs only once; therefore, there is no mode. In B, there are two modes. The rates 6.19 and 6.75 both occur twice. This set of data is bimodal.

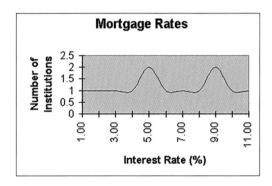

b) The data in A do not have a normal distribution because the populations of the provinces are very different. When graphed, the distribution for the data in B has two peaks. This is not a normal distribution either.

EXAMPLE 3	**Exponential Distribution**

Aircraft components must meet stringent safety regulations. It is essential to know the number of hours of use that can be expected before a part might fail due to stress (constant use in regular conditions). Tests are used to determine and record the lifetime of each component.

In a test situation, the failure times, in hours, for 40 particular components under stress are recorded.

21	36	42	54	59	77	95	100	117	159	184	256	332	428
442	482	596	646	652	738	892	908	918	1088	1220	1348	1384	1433
1744	1893	2341	2516	2671	3000	3044	3229	3989	4297	5390	6531		

a) Using technology, prepare a histogram for these data and draw a curve to represent the distribution.

b) Identify the type of distribution curve.

c) Calculate the mean, median, and mode.

d) On your graph, label the mean, median, and mode.

e) Discuss the position of the mean, median, and mode for this type of distribution.

f) Discuss the safety concerns based on these data.

Solution

a) The data are graphed using Fathom™.

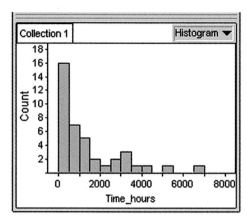

b) This distribution has a decreasing exponential curve.

c) Mean = 1383.8   Median = $\dfrac{738 + 892}{2}$ = 815

The mean lifetime of the component data is 1383.8 h, and the median is 815 h. There is no mode, since there are no repeated values.

d)

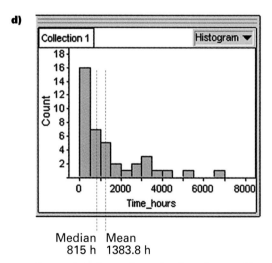

Median   Mean
815 h   1383.8 h

e) One might have thought that the median and the mean lines would be closer to the left side, or peak, of this exponential distribution curve, but this is not the case.

f) This aircraft component has a very high failure rate, under the median of 815 h. In fact, the number of failures in the first 100 h is very high. This particular component would probably not be classified as safe. Another type of component, or manufacturer, would probably be used.

KEY CONCEPTS

- A normal distribution is characterised by the following:

  i) the curve is symmetrical; the mean, median, and mode are the same or very close

  ii) the majority of the scores are clustered close to the centre of the distribution (between −1 and 1 standard deviations)

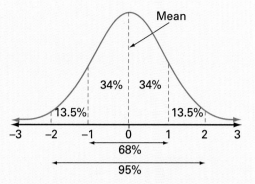

- The distribution of data may be positively or negatively skewed. In this case, the mean and mode are shifted to the left or right, depending on the direction of skewing.

- The distribution of data may be exponential if the majority of points are clustered toward one end of the curve.

DISCUSS THE CONCEPTS

1. In a normal distribution, a positively skewed distribution, and a negatively skewed distribution, a vertical line showing the position of the mode always passes through the maximum point of the curve. Explain why.

2. Provide two examples of data that would probably have a normal distribution.

3. Provide two examples of data that would probably not have a normal distribution. Explain what the distribution might look like and why.

PRACTISE

A 1. Copy this chart into your notes and describe the characteristics of each type of distribution.

Distribution	Characteristics
Normal	
Positively skewed	
Negatively skewed	
Bimodal	
Exponential	

2. The graphs represent the distribution of marks in mathematics, language arts, and chemistry courses at a large high school.

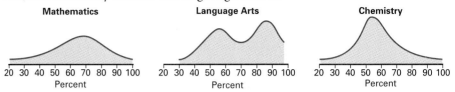

a) Describe the differences between the three distributions.

b) What term can be used to describe each distribution?

B 3. **Communication** The graphs represent the distribution of incomes in three different cities. Describe each distribution and what you can infer about pay scales in the cities.

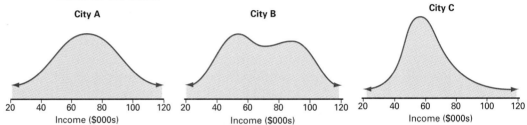

4. An advertisement for an antacid indicates that 68% of people who took the antacid experienced relief in 1 h or less. Draw a normal distribution curve and shade the area that represents the advertiser's results.

5. **Communication** A manufacturer of women's clothing has noted that the most common size for women 25 years ago was a size 8. Today, the most common size for women is a size 12.

a) Sketch a distribution for each situation, marking the mode on each curve.

b) Why is it important in clothing manufacturing to know the mode size?

c) Discuss whether these data mean that women are 50% larger than they were 25 years ago.

6. A pain reliever advertisement indicates that 19 out of 20 patients experience relief in 20 min or less. Sketch a normal distribution curve, and mark the position of the mean and one and two standard deviations on either side of the mean. Shade the section under the curve represented by the advertiser's claim.

7. Application The final marks for all 182 students who took grade 9 mathematics were recorded. A frequency table was used to organize the data.

Class Interval	Frequency
[30, 40)	11
[40, 50)	18
[50, 60)	25
[60, 70)	34
[70, 80)	42
[80, 90)	32
[90, 100)	20

a) Display the data in a histogram.

b) Join the middle of each bar to create a distribution curve. Describe the type of distribution.

8. Thinking/Inquiry/Problem Solving An automobile manufacturing company tested their new compact car. They recorded the stopping distance, in metres, from a speed of 100 km/h.
95, 98, 98, 98, 99, 101, 102, 102, 102, 102, 103, 105

a) Prepare a graph and identify the type of distribution.

b) Find the mean and the median. Mark these measures on your graph.

c) Calculate the standard deviation.

d) What range of stopping distances is within one standard deviation of the mean? If this is a normal distribution, what percent of the data should lie within this range?

e) How many of the measures recorded do lie within one standard deviation of the mean? What percent of the data recorded is this? Does this agree with a normal distribution pattern? If not, what can you conclude?

9. Data Bank Use the data on Length of Action Movies, on page 457.

a) Display the data in a histogram. Describe the type of distribution.

b) Mark the mean and median on your graph. (You may already have calculated these measures, see Section 4.4, question 9.)

c) Calculate the standard deviation.

d) If this is a normal distribution, what percent of the data should lie within one standard deviation of the mean? How many of the movie lengths do lie within this range?

e) If this is a normal distribution, what percent of the data should lie within two standard deviations of the mean? How many of the movie lengths do lie within this range?

f) Do the data seem to have a normal distribution or are they skewed? Explain your reasoning.

CAREER PROFILE

Food and Drug Technologist

Companies that produce food items and drugs follow strict guidelines to ensure that their products are safe for consumers. Food and drug technologists are involved in all aspects of manufacturing, from research and quality control, to sales and customer service.

Becoming a food and drug technologist involves a three-year diploma program offered at most colleges. Students study a variety of courses including biology, chemistry, interpersonal communications, and statistical quality control. The statistical quality control courses involve applications of central tendency and dispersion. The questions below are typical ones that food and drug technologists can answer.

1. In an experiment, students gradually increased the dosage of a particular drug to determine the minimum dosage required to produce the desired effect. The results are displayed in the graph.

   a) Describe the shape of the distribution.

   b) Estimate the mean and the median dose.

   c) If an acceptable response falls within two standard deviations, estimate the minimum dose required for the desired affect.

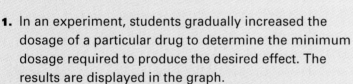

2. Rats were given acetaminophen and the concentration of the drug in their blood was measured over a 5-h period. The results are shown in the graph.

   a) Describe the shape of this distribution and discuss how it compares to the graph in question 1.

   b) Estimate how long, after the drug has been administered, the concentration of acetaminophen in the rat is the greatest.

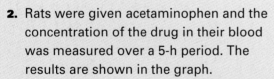

3. Visit the college of your choice to learn more about the prerequisites of the Food and Drug Technologist program.

CHAPTER PROBLEM WRAP-UP

Prepare a report, that could be presented either for a class presentation or wall display, describing the steps of your investigation and the conclusions based on your analysis. Include appropriate visuals to illustrate your work.
The following are the parts of the process that you should use to prepare your presentation.

1. Describe the type of sample chosen to represent the population you are studying. Justify your choice.

2. Present the survey that you used. Explain why you asked the questions that you did, and describe any revision that you needed to make to refine the process.

3. Organize the data you collected, and decide whether a histogram or bar graph would be better to represent the data. Explain your choice and draw the graph.

4. Find the mean, median, mode, and standard deviation for the data you have collected. Are these measures appropriate for the data you have collected? Why or why not? What do these measures tell you about the data collected and the research done? Do you need to revise any parts of your data-gathering process to obtain better results? Explain.

5. Provide a one-paragraph summary of the conclusions that you can make from your research.

Review

4.1 Collecting Data: Sampling Techniques, pages 158–162

1. a) Describe what a random sample is. Give two examples in which a random sample is the best method of selecting a sample.

b) Describe the difference between a stratified sample and a random sample.

2. A large multi-national company is preparing a report on pay equity by collecting data from a large sample of its employees. The data are organized into male and female categories. How would you classify this type of sample?

3. A random number generator is a popular way to select a sample without bias. ABC Cola wants to check its product as it comes off the production line. The company wants to conduct a taste test of the cola, but does not want to test every bottle. They decide to test 10% of the bottles in each batch.

a) Could they choose every 10th bottle? Why or why not?

b) How could they use random numbers to select which bottles to test?

4.2 Methods of Collecting Data, pages 163–167

4. Describe what is meant by each statistical term.

a) population   **b)** sample   **c)** primary source of data

5. How is validity of a study compromised if a survey is biased? Give an example to illustrate.

4.3 Representing Data, pages 168–177

6. Describe the similarities and differences between histograms and bar graphs.

7. The data show players' scores at a bowling alley one evening. Create a histogram to display the results.

Class Interval	Number of Players
[70, 80)	2
[80, 90)	6
[90, 100)	18
[100, 110)	29
[110, 120)	22
[120, 130)	4
[130, 140)	3

4.4 Measures of Central Tendency, pages 178–183

8. A class of grade 12 students has the following numbers of absences.

2 8 11 0 13 1 5 4 6 3 7 12 4 11 5 18 9 2 10 7

a) Determine the mean, median, and mode.

b) Which measure of central tendency best represents the data? Why?

9. The numbers of employees of ten large companies are as follows.

6400 8620 7650 9225 9800 7650 8230 8910 6870 6100

a) Determine the mean, median, and mode.

b) Which measure of central tendency best represents the data? Why?

4.5 Measures of Dispersion, pages 184–189

10. Find the range and the standard deviation of the data listed in question 11.

11. The daily high temperature was recorded, in degrees Celsius, in a Canadian city during the month of November.

0 3 7 13 3 4 10 −3 −5 0 9 10 7 1 5
4 4 −1 1 4 8 6 5 11 9 6 3 2 −2 −1

a) Find the range.

b) Determine the mean and the standard deviation.

4.6 Properties of Common Distributions, pages 190–197

12. The three distributions show the population by age group for three Canadian cities.

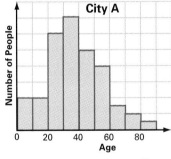 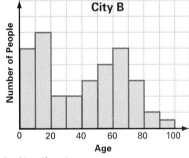 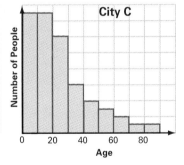

a) Describe each distribution.

b) What can you infer about the population of each city?

13. Hannah got a summer job working in a restaurant. She kept track of the money she earned in tips each day during the first four weeks.

$8 $15 $12 $25 $38 $23 $22 $58 $18 $3
$24 $16 $30 $29 $27 $19 $33 $35 $46 $35

a) Display the data in a histogram. Describe the distribution.

b) Find the mean, median, and mode. Show the mean and median on your histogram.

c) Find the standard deviation.

d) How much can Hannah expect to earn in tips 95% of the time?

ractice Test

Category	Knowledge/Understanding	Thinking/Inquiry/Problem Solving	Communication	Application
Questions	1–6	1, 2, 5, 6	1, 2, 4, 5	2, 4, 6

ACHIEVEMENT CHART CATEGORY CONNECTIONS

1. A parts manufacturer wants to test the accuracy of a robotic cutting arm. Each part must be within 0.3 mm of the standard size. What type of sampling selection would you recommend? How would they carry out the sampling?

2. The graph shows the number of condominium sales for two salespeople over a one-year period.

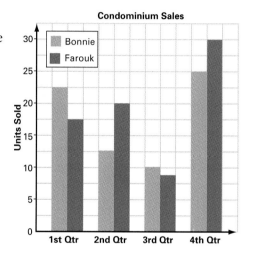

a) What type of graph is this?

b) In which quarters did Bonnie outsell Farouk?

c) Over the year, how many units did Farouk sell?

d) What was the median number of units sold per quarter by Bonnie?

3. The temperatures were recorded, in degrees Celsius, one day in August.

City	High (°C)	Low (°C)	City	High (°C)	Low (°C)
Barrie	33	20	Ottawa	34	21
Bracebridge	34	19	Owen Sound	34	19
Brantford	35	23	Parry Sound	34	19
Cornwall	34	21	Pembroke	32	18
Dryden	20	11	Peterborough	34	21
Hamilton	34	24	St. Catharines	36	24
Huntsville	34	19	Sudbury	29	16
Kenora	20	11	Sault Ste. Marie	26	15
Kingston	34	11	Thunder Bay	26	11
Kitchener	34	21	Timmins	22	13
London	35	23	Toronto	35	24
Niagara Falls	36	24	Windsor	36	25
North Bay	29	16			

a) Determine the mean, median, and mode high temperature.

b) Determine the mean, median, and mode low temperature.

c) Which is greater, the range of the high temperatures or the range of the low temperatures?

4. A manufacturer of handheld electronic games would like to find the lifetime of its product. Thirty game units are tested until they fail. The time to failure, in days, is recorded.

40.3	27.1	35.6	13.7	4.2	35.8	15.5	5.9	31.6	28.6	24.3
18.1	25.5	11.4	49.1	9.9	10.6	8.9	12.2	19.7	7.2	8.9
19.4	29.2	37.3	35.6	28.4	18.0	32.9	25.6			

a) What is the range of the values?

b) What was the mean number of days the units failed after?

c) Create and complete a frequency chart.

d) Draw a histogram based on the data in the chart.

e) Calculate the variance and the standard deviation.

f) What can you conclude about this model game unit based on the data?

5. Changes in climate or environment can affect the normal growth patterns of plants and animals. The distribution curves for data gathered in these situations would be skewed. For each situation below, sketch a distribution curve.

a) Data on the increase in the height of trees were recorded. Record amounts of precipitation this summer has resulted in unusual rapid growth of trees.

b) Data on a large sample of young children in a Middle Eastern country were collected. The lower than normal mass of the children is a result of the lack of food during many years of political unrest.

ACHIEVEMENT CHECK Knowledge/Understanding Thinking/Inquiry/Problem Solving Communication Application

6. There are nine branches of a national clothing store in the local area. Each day the manager takes a "midday read" of store sales in the area. The amounts reported for the stores one Tuesday are as follows.

$1256 $2726 $1224 $2588 $3294 $1893 $2537 $3177 $2460

a) Determine the range, mean, median, and mode.

b) Calculate the standard deviation.

c) Represent the data in a histogram. Show the measures of central tendency. Describe the type of distribution.

d) Defend the use of a histogram to represent this data rather than some other type of graph.

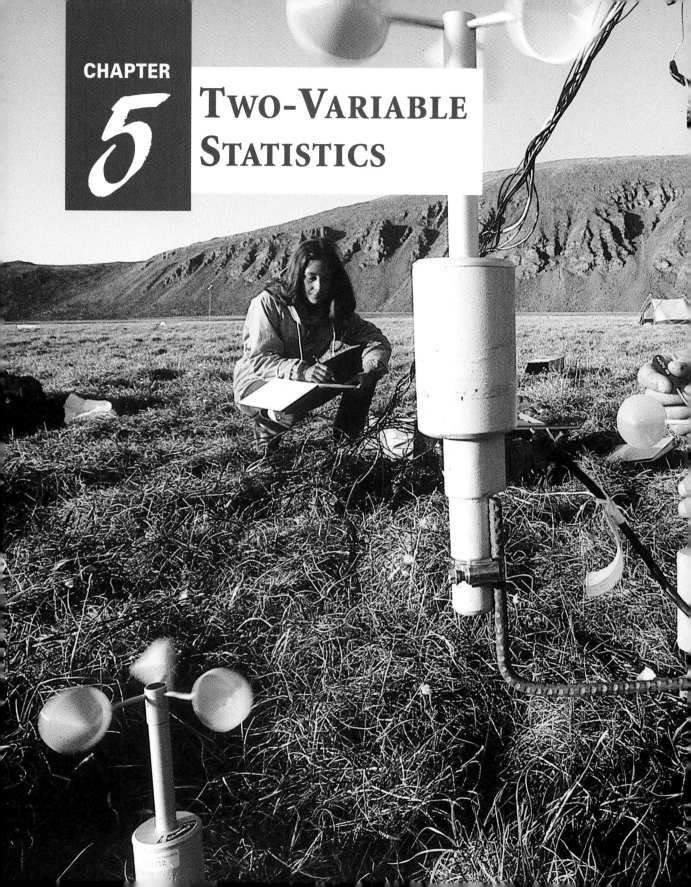

Specific Expectations	Sections
• Determine appropriate methods for collecting, storing, and retrieving, from primary or secondary sources, data involving two variables.	5.1, 5.2
• Construct a scatter plot to represent data using technology.	5.3
• Determine an equation of a line of best fit, using the regression capabilities of graphing technology.	5.4
• Calculate and interpret the correlation coefficient, using appropriate technology.	5.4
• Describe the relationship between two variables suggested by a scatter plot.	5.4
• Describe possible misuses of regressions.	5.5
• Make and justify statements about a population on the basis of sample data.	5.5
• Collect, organize, and analyse data to address problems or issues, and calculate relevant statistical measures.	5.1, 5.2, 5.3, 5.4, 5.5
• Formulate a summary conclusion to a problem or an issue, by synthesizing interpretations of individual statistical measures.	5.1, 5.2, 5.3, 5.4, 5.5
• Formulate extending questions related to the conclusion reached in the investigation of a problem or an issue.	5.1, 5.2, 5.3, 5.4, 5.5
• Communicate the process used and the conclusions reached in the investigation of a problem or an issue, using appropriate mathematical forms.	5.1, 5.2, 5.3, 5.4, 5.5

Get Ready

1. **Linear graphs** For each line, indicate whether the slope is positive, negative, or zero.

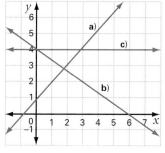

For help with the graphing calculator, see Appendix B, page 425.

2. **Graphing using technology** Using technology, graph each line.

   a) $y = 4x - 2$ **b)** $y = -0.5x + 1$ **c)** $y = 0.248x - 12$

3. Setting the window on a graphing calculator to accommodate a graph is sometimes difficult. The diagram indicates the position of each of the variables found in $\boxed{\text{WINDOW}}$. Discuss how you determine these four values.

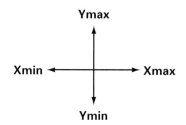

4. Identify some trouble-shooting strategies for a graphing calculator. How might you fix each of the following problems?

   a) ERR: INVALID DIM
       1:Quit

   b) ERR:SYNTAX
       1:Quit
       2:Goto

   c) No graph appears on the screen when $\boxed{\text{GRAPH}}$ is pressed.

   d) ERR:DIM MISMATCH
       1:Quit

5. **Linear graphs** This chapter focuses on using technology to create and analyse graphs. A good understanding of graphing by hand will help when trouble-shooting. Graph each line using pencil and paper.

   a) $y = 2x + 1$ **b)** $y = \dfrac{2}{3}x - 4$ **c)** $y = -\dfrac{3}{5}x + 2$

6. **a)** Describe why a sample is used when gathering data.

   b) List three ways to collect sample data.

7. There are various different methods of choosing a sample in data collection.

   a) Discuss the difference between *random* and *stratified* sampling techniques.

   b) What is meant by the term *biased sampling*?

8. The daily maximum temperature and the number of bottles of spring water sold were recorded.

Temperature (°C)	Number of Bottles of Spring Water Sold
5	4
10	25
14	38
17	54
22	50
25	82
30	103
33	126

   a) Using paper and pencil, display the data on a scatter plot. Use a ruler to draw a line of best fit.

   b) Is the slope of the line of best fit positive or negative?

   c) Describe how the points are scattered about the line of best fit.

   d) Discuss whether your line of best can be used to interpolate the number of bottles of spring water sold at any temperature.

   e) Using your line of best fit, how many bottles of water would be sold on a day when the maximum temperature is –5°C? 38°C?

CHAPTER PROBLEM

Do you get frustrated when you see a problem and think you have a good solution to it, but nobody will listen to you? What you need is good data that is well organized and analysed to support your idea. Of course, you will also need solid mathematical reasoning. Put all this together in an effective presentation and people will listen.

This chapter will provide you with the skills to make a valid case to support your call for change. For this Chapter Problem, you are to identify a school issue that concerns you and a possible solution. In a small group, discuss issues that affect you, such as cafeteria food, school hours, attendance rules, or dress code.

5.1 | Discover E-STAT

E-STAT is Statistics Canada's interactive learning tool designed with the needs and interest of the education community in mind. E-STAT offers an enormous warehouse of reliable statistics about Canada and its ever-changing people.

E-STAT will be an on-going tool used in your statistics work. Follow the steps below to learn more about E-STAT.

DISCOVER

Part I Beginning to Use E-STAT

1. Go to the E-STAT site as follows:
   Go to Statistics Canada's home page.
   Click **English**.
   Select **Learning Resources** and you should see the following form.
   Alternatively, you can access E-STAT by going to *www.mcgrawhill.ca/links/ MPCA* and using the link.

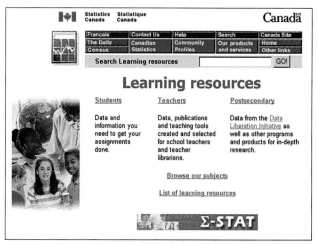

Source: Statistics Canada's Internet Site, http://www.statcan.ca/english/edu/index.htm, November 7, 2001.

2. Click on the E-STAT logo at the bottom.

3. At the bottom of the page, click **Accept**.

4. You will need to enter your identification number and password. Your teacher will be able to supply this for you. Select **OK** when finished.

Source: Statistics Canada's Internet Site, http://www.statcan.ca/english/edu/index.htm, November 7, 2001.

5. You are now presented with the following window, outlining the table of contents for E-STAT.

6. Identify how E-STAT categorizes data.

7. Under **People**, select **Travel and Tourism**.

8. Prepare a summary of the overview for Travel and Tourism.

9. Scroll down to the graph at the bottom of the page to obtain the data to answer the following questions.
   a) Where do most of the visitors to Canada come from?
   b) During what period was there consistent growth in the number of visitors?
   c) The period 1990 to 1993 shows a decline for most visitors. What could be responsible for this decline?
   d) Make up two more questions related to this graph, and then, exchange your questions with another person or group.

10. Scroll down to the very bottom and select **Articles**. Select an article and prepare a brief summary of the topic.

11. Select **Back** to return to the Travel and Tourism overview page.

12. Select **Data**, located at the bottom of the page.

13. Prepare your own report, based on the amount of money spent on various methods of travel. Use the following steps.
   a) Select **Tourism** from the Travel and Tourism main menu.
   b) You are now presented with a list of **Active tables**. Tables are used to organize the data in Statistics Canada's database. Each table is identified by a seven-digit code. How many tables are available under **Tourism**?
   c) Identify the number linked to the matrix title "Tourism demand in Canada, quarterly." Click on that number.
   d) Now you need to select some items from the lists provided.
      Canada has already been chosen in the **Geography** list.
      Choose 1992 constant prices from the **Prices** list.
      Choose **Seasonally adjusted** from the **Seasonal adjustment** list.
      Choose **Tourism demand** from the **Categories** list.

E-Stat Link

For help with using E-STAT data, click on **Getting you started**, for a quick guide.

In the **Expenditures** list, hold down **Ctrl** on your keyboard and click on the following:

>**Tourism expenditures**
>>**Transportation**
>>**Accommodation**
>>**Food and beverage services**

Choose **Mar 1986** in the **From:** field, but do not change the **To:** field.

e) Select **Continue**. On the next screen, select **Time Series**, then select **Continue** again.

f) At this point, you can accept all the default values (values indicated by the computer).

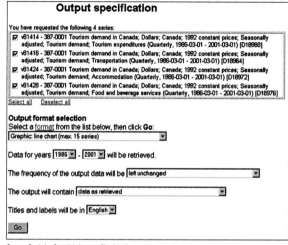

Source: Statistics Canada's Internet Site, http://www.statcan.ca, November 7, 2001.

14. Select **Go** to view your graph.

a) What type of graph is presented?

b) Write an explanation to interpret the data presented in your graph.

15. Now, you can use the same data to compare the percent of tourism expenditures in each of the three areas selected.

a) Select **Back** to return to **Output format selection**.

Change "The Output will contain" option (the second last option) to "each series expressed as a percentage of the first one." Select **Go**.

b) Analyse the graph, and explain any similarities and differences between this graph and the one in question 14.

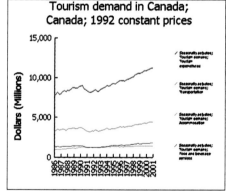

Source: Statistics Canada's Internet Site, http://www.statcan.ca, February 4, 2002.

Part II Exploring Environmental Issues

The navigational frame below will guide your search for articles or data to answer the related questions. Where there are arrows, you need to make a selection; otherwise, follow the path by clicking on the link indicated.

▶ **E-STAT - Table of Contents** ◀
Land and Resources
  Environment

1. Read the Environment Overview article
   a) Briefly describe the main concepts presented.
   b) Identify the format, such as table, line graph, or map, of the visual representation at the bottom of the screen of this overview.
   c) Discuss whether it would be suitable to represent this data in another format.

From the overview screen, you can go directly to the articles by clicking **Articles** at the bottom of the form.

▶ Articles ◀
Land and Resources
  Environment

2. a) Identify the title of the **Canada Year Book** article.
   b) Read the article and describe the current trend presented in the article, as well as what might be expected in the future. Then, return to the previous screen. Select **Data** and the following path.

   ▶ Data ◀
   Land and Resources
     Environment
       Environmental Impact

3. Select a table that is of interest to you and prepare a graph to represent the data. Include a paragraph to summarize the data.

Part III Using E-STAT, Prepare a One-Page Report

Select an area that is of interest to you from the table of contents. Read the overview and related article. Summarize the information. Illustrate the key points in your summary with graphs.

Methods for Collecting, Storing, and Retrieving Data

Information is all around you. Every day you use data to communicate, analyse, and interpret information. We are all numbered and surveyed. Our habits are stored in databases and analysed to learn about who we are, what products we like, and how we spend our money.

Statistics involves collecting, organizing, summarizing, analysing, and presenting data. The outcomes of statistical research can be used to describe a situation, **descriptive statistics**, or to make generalizations about a population based on the sample data, **inferential statistics**.

EXAMPLE 1

Identifying Types of Data Sources

A Rhonda's Pizza

Rhonda's Pizza is a family-owned restaurant. To decide how many employees are needed for night shifts, Rhonda keeps a record of the number of pizzas sold over a three-month period. The average number sold each night is shown in the table.

Day of the Week	Number of Pizzas Sold
Monday	52
Tuesday	55
Wednesday	60
Thursday	72
Friday	145
Saturday	123
Sunday	46

B Byberg Engineering

Steve Byberg manages his own engineering company. He relies on other companies for some data. A water testing company measured the turbidity of water samples for Steve. (Turbidity indicates the presence of particles in the water due to problems in the water treatment process.)

Sample Number	Turbidity
1	0.15
2	0.21
3	0.23
4	0.11
5	0.12
6	0.18
7	0.08
8	0.29

a) Identify the two variables in each situation.

b) Is the source of data in each example primary or secondary? Discuss which is more reliable.

Solution

a) In A, Rhonda's Pizza, the two variables are Day of the Week and Number of Pizzas Sold.

In B, Byberg Engineering, the two variables are Sample Number and Turbidity.

b) A is an example of a primary source of data—data that were collected first-hand. Rhonda collected her own data.

B is an example of a secondary source of data—data that were obtained by someone else. A primary source of data is often more reliable because the person conducting the research is in control of the data collection and, hopefully, can ensure that no errors have occurred.

Secondary sources of data may not be as reliable as primary sources of data because errors might have occurred in the data collection or communication. However, a secondary source is often the only method available for obtaining data. In this case, you should use reliable sources of data collected by official or reputable agencies, such as Statistics Canada, or certified agencies specializing in data collection, such as consumer research companies.

Researching data and preparing a report is a multi-step process and can be divided into four steps. This chapter will focus on the first three steps.

i) Collecting Data

ii) Storing and Organizing Data

iii) Retrieving Information—presenting the facts

iv) Analysing Data—interpreting the facts

The data you gather must be accurate; otherwise, your results will be very misleading. There are basically three ways to obtain data:

- Data made available by others (secondary data)
- Data resulting from an experiment (primary data)
- Data collected in an observational study (primary data)

Data can be organized or stored in many different ways, such as by age, year, gender, geographic region, and income. Organizing data into groups often helps identify patterns in the data.

Organizing data is not an afterthought. If a survey is used, consideration needs to be given to how the data will be organized. Similarly, if variables are measured or manipulated, thought needs to be given to how the data will be organized.

EXAMPLE 2

Understanding the Data Research Process

Research has shown that there is a reasonable correlation between self-assessed knowledge (what people think they can do) and the actual performance of the same task (what people actually can do).

A college surveyed a sample of its students from different areas of study on their self-assessed basic numeracy skills. Students were asked how much difficulty they would have with each of the following tasks.

A Determine how much change you should get from a $10-bill if you purchase a pen that costs $2.75.

B Determine the percent of Liberal support, if 15 people vote Conservative, 10 people vote NDP, and 20 people vote Liberal.

C Explain what *square root* means.

The results of the survey are shown in the double-bar graphs.

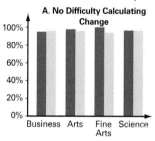 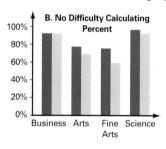 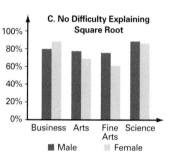

a) Discuss how the data in this survey were collected. Give one other example of this type of source of data.

b) Describe the other ways to collect data, and give an example of where each could be used.

c) Discuss how these data were organized.

d) List three other ways in which data can be organized.

e) Summarize the results as presented in each graph.

f) Discuss the inferences you might make based on this research.

Solution

a) The data in this example model were collected by observation, since the researcher is not influencing the participants' responses. Another example of observational data collection is in the study of animal migration patterns. The researcher observes the animal population at regular intervals. Tagging animals may be done to help with the recording or tracking of the animals. When using data collected by others, be sure to understand how both variables (year and number of animals) are defined.

b) Data can be obtained from others, that is, from secondary sources. Sometimes you do not have the capability to collect the data yourself. In this case, you need to use data that has already been gathered. Newspaper journalists rely on data collected by others, such as Statistics Canada. Journalists must be careful to cite the source of their information.

Data can also be collected from an experiment. It is important when collecting data by this method that you thoroughly define the scope of the experiment and both variables. Researching the volume of cafeteria garbage each day of the week is an example of an experiment. This is a primary source of data.

c) The data in this example were organized and displayed in a double-bar graph by college departments and by gender.

d) Data can be organized in many ways, depending on the nature of the data. For example: engine type: 4, 6, or 8 cylinder versus fuel consumption
students: grade level and number of credits
provinces: area and population

e) Graph A: Overall, approximately 90% of students perceive that they have no difficulty with this type of question. The differences between males and females are minor.

Graph B: Most of the groups surveyed perceived calculating percent to be more challenging. Students in business and science are more self-confident with this skill than students in arts and fine arts.

Graph C: Males in business perceive that they have more difficulty explaining square roots than females do. In the other departments, females perceive that they have slightly more difficulty than males do.

f) Overall, students in business and science have fairly high levels of self-assessed basic numeracy. In contrast to other departments, females in business and science have the same or higher levels of self-assessed basic numeracy.

Part f) in the preceding example models **inferential statistics,** since a sample of college students was used to make some generalizations about all the students (the population). Other examples of inferential statistics include the following:
• Political polls
• Marketing surveys
• TV surveys of popular shows

If all, or almost all, students completed the survey, this would model **descriptive statistics**—the data describe all the students of the college. Examples of descriptive statistics include the following:
- Surveying all students in your class to obtain data relating the circumference of the wrist and height
- Counting the number of deer in a provincial park each year
- Measuring the rainfall each day

KEY CONCEPTS

- Primary sources of data involve collection of data first-hand, such as conducting a survey yourself. Secondary sources of data are prepared by somebody else, such as Statistics Canada.

- There are four steps involved in statistical research:
  - **i)** Collecting
  - **ii)** Storing and organizing
  - **iii)** Retrieving
  - **iv)** Analysing

- Data can be collected in three ways:
  - **i)** Data made available by others
  - **ii)** Data resulting from an experiment
  - **iii)** Data collected in an observational study

- When collecting data, you have to decide whether to use inferential or descriptive statistics. Inferential statistics is when a sample of the population is used to make a generalization about the entire population. Descriptive statistics involves surveying all, or almost all, of the population.

DISCUSS THE CONCEPTS

1. Identify and describe the four steps of statistical research.

2. Data can be collected in three ways. For each, identify whether the data are from a primary or secondary source. Explain why.

3. Is the mark that you will receive at the end of this course a descriptive statistic or an inferential statistic?

4. Discuss how samples are used in inferential statistics.

A **1.** Classify the following as either a primary or a secondary source of data.

a) You count the number of cars passing through an intersection during rush hour.

b) Your group measures the height and records the shoe size of each person in your class.

c) You obtain a table from E-Stat, showing the population growth over the past 50 years in Ontario.

d) You survey students at your school, asking their age and the number of soft drinks they consume in a week.

e) You use the Internet to find the number of pandas in the world each year for the past 50 years.

2. When researching, it is important to identify the variables you are measuring. For each example, identify which is the independent variable (*x*-axis) and which is the dependent variable (*y*-axis).

a) The time of day and the level of hunger a student feels

b) The number of humpback whales and the year

c) The population of various cities and the consumption of electricity

d) A person's net worth and the person's age

3. Identify each example as either a primary or secondary source of data.

a) Data found in an almanac   **b)** Data resulting from an experiment

c) Data you collect in an observational (survey) study

d) Data presented in this textbook

4. Identify each example as either inferential or descriptive statistics.

a) Census data   **b)** Political polls

c) Survey of 10 randomly selected grade 12 students

5. Most often, data are gathered from a sample of a population. Where or when would data be collected from an entire population?

B **6.** Discuss why insurance companies are interested in the data shown in this graph.

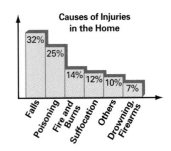

Causes of Injuries in the Home

7. For each statement, decide whether descriptive or inferential statistics have been used.
   a) In 2000, Canada consumed 15% more energy per person than the United States.
   b) Two out of ten illegal immigrants in Canada go to Vancouver.
   c) Expenditures for the cable industry were $5.66 billion in 1996.
   d) In 2000, the median household income for Canadians aged 25–34 was $42 896.
   e) The national average annual expenditure on medicine per person is $1052.

8. Refer to the school issues that you discussed in the introduction to the Chapter Problem, on page 207.
   a) Choose one that interests you and identify the variables.
   b) Discuss what you want to change.
   c) Identify how you would collect supporting data.

9. Communication Milk marketers use sampling to determine the quality of a batch of milk. If a sample of the milk passes the test, then the entire batch is transported to a dairy processing plant where the milk is processed. Discuss with a partner other uses of sampling in everyday life.

10. A.C. Nielsen is an American consumer research company best known for monitoring television viewing habits. The primary tool is an electronic gadget, called a people meter, that works like an ordinary remote control, but keeps a minute-by-minute record of the stations being watched and allows the viewer to record which family members are watching the television. Five thousand households throughout the United States are paid a small fee for using people meters and allowing their televisions to be connected to their telephone lines. Each day, a computer automatically processes the viewing habits recorded during the preceding 24 h. The company then computes a rating (an estimate of the fraction of all homes in the nation that watched a particular program) and an audience share (an estimate of the fraction of the television sets that were turned on and tuned to that program).
   a) Identify the sample data and population in this research.
   b) What type of statistics is this?
   c) Discuss how the results of this statistical analysis affect the most popular shows and the shows at the bottom of the viewer preference list.

11. Before Nielsen ratings were introduced, television networks gauged the popularity of their shows by analysing comments from fan mail. Explain why this procedure was unreliable.

12. Nielsen Media Research has expanded into Canada. As of September 2001, 3150 homes in Canada have people meters. These homes are located as follows:

450 Montréal   625 Vancouver   450 Calgary   700 Toronto

Does this placement of people meters provide a good sample of the Canadian population? Why or why not?

13. Thinking/Inquiry, Problem Solving, Communication
   a) Inferential statistics uses probability, relying on the chance of an event occurring. Using an example, discuss the chances of the event occurring.
   b) Discuss why it is more practical to obtain data from a sample than from the entire population.

14. Communication, Application Survey your class to determine their self-assessed basic numeracy skills. Use Example 2 as your guide or create your own numeracy questions. Since high school students are not streamed into study areas, use students' intended college study area or design your own strategy to differentiate between groups of students.
   a) At the beginning of your report, include a brief explanation of the purpose of the study. Identify your sample and population groups.
   b) For each question, prepare an individual summary and graph.
   c) Prepare an overall conclusion and recommendation for each area.
   d) Include your actual data in an appendix at the back of your report.

ACHIEVEMENT CHECK   Knowledge/Understanding   Thinking/Inquiry/Problem Solving   Communication   Application

15. Honest Auto has branches in Toronto, Montreal, and Vancouver, with a total of 316 salespeople. The national sales manager wants to obtain a random sample of 40 of his staff to determine their average gross sales per month. One of the three branches is randomly selected and 40 salespeople are selected from this branch.
   a) Identify the type of sample chosen.
   b) Describe other ways of choosing the sample.

---

16. Communication Insurance companies use inferential statistics to determine their vehicle insurance rates. Go to *www.mcgrawhill.ca/links/MPCA12* and follow the links for some sources of information.
   a) What type of data are insurance companies interested in? Identify two possible variables insurance companies would use in their analysis.
   b) Describe how this analysis affects insurance premiums.

5.3 Scatter Plots

When two variables are graphed, they often establish a pattern or relationship, such as a line or curve. Scatter plots are created to determine whether a relationship exists. If the scatter plot suggests a possible relationship between the variables in a clustered pattern that resembles a line, parabola, or exponential graph, then further analysis is necessary.

Before creating a scatter plot, you need to identify which variable or data will be placed along the x-axis (independent variable) and which will be placed along the y-axis (dependent variable). Incorrect placement of these variables may result in an error in interpreting the relationship.

EXAMPLE 1

Identifying Independent and Dependent Variables

For each relationship, sketch the axes and label the independent and dependent variables.

a) The age of a child and the child's height

b) The number of hours a student studies per course and his or her average mark in the course

c) The time of day and the number of cups of coffee sold

d) The amount of snowfall and the number of collisions on major highways

e) A person's income and the amount she or he spends on clothing

Solution

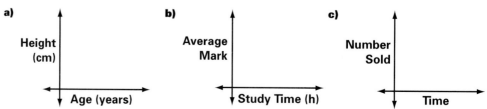

a) Height (cm) / Age (years)   b) Average Mark / Study Time (h)   c) Number Sold / Time

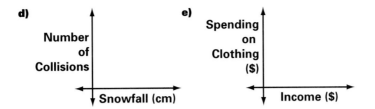

d)

Number
of
Collisions

Snowfall (cm)

e)

Spending
on
Clothing
($)

Income ($)

In each, the independent variable is on the horizontal axis and the dependent variable is on the vertical axis.

EXAMPLE 2

Interpreting Scatter Plots

A number of scatter plots are shown. Create a table to classify the different types of scatter plots. Use the following terms:
- Type of correlation (linear or exponential)
- Strength of correlation (strong—points are close to the pattern, or weak—points are spread further apart)
- Direction (positive—moving up toward the right, or negative—moving up toward the left)
- Outlier(s) (A point does not conform to the pattern of the other points. This point needs careful consideration; it may or may not have been a data entry error or an error in data collection.)

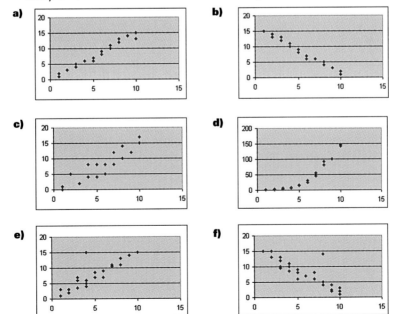

Solution

Scatter Plot	Type of Correlation	Strength	Direction	Outliers
a	line	strong	positive	no
b	line	strong	negative	no
c	line	weak	positive	no
d	exponential	strong	positive	no
e	line	weak	positive	yes
f	line	weak	negative	yes

Scatter plots can be drawn and analysed most efficiently using technology—a graphing calculator, or Microsoft® Excel or Fathom™ on a computer. Each Discover that follows explores one of these applications.

DISCOVER

Creating and Analysing Scatter Plots Using a Graphing Calculator

The Boston Marathon has been run annually since 1897. The table shows the men's record times from 1927 to 2001. Note that, at the time of going to print, the 1994 record set by Cosmas Ndeti of Kenya remained unbeaten.

Year	Time (min)
1927	160.37
1928	157.12
1929	153.13
1933	151.02
1939	148.85
1940	148.47
1942	146.85
1947	145.65
1957	140.08
1963	138.97
1965	136.55
1967	135.75
1969	133.82
1970	130.50
1975	129.92
1979	129.45
1981	129.43
1982	128.87
1986	127.85
1994	127.25

Sports Link

Many Canadians run the Boston Marathon. Sometimes they win. The second running, in 1898, was won by Ronald J. MacDonald in a time of 2:54:38. In 1980, Jacqueline Gareau had the best women's time, 2:34:28.

1. Enter the data.
To open the **Stat List Editor**, use (STAT) 1.
Enter the year in **L1** and the times in **L2**.
To make it easier, you can enter just the last two digits of the year.

See page 427 for an introduction to the statistical functions of the graphing calculator.

2. Create a scatter plot.
   Press (2nd) (Y=) to access the **STATPLOT** menu.
   Press **1** or (ENTER), then (ENTER) again to turn Plot 1 on.
   The graph type should be the scatter plot, the first graph shown.
   Check that the Xlist is L1 and the Ylist is L2.
   Select a Mark—the square or cross are easiest to see.

3. Set the WINDOW.
4. To view the scatter plot, press (GRAPH).
5. Describe the shape of the pattern of points (linear, exponential, quadratic, no relation).
6. If the data are approximated by a line, does the line have a positive or a negative slope?

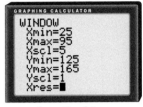

7. Considering the data set, discuss whether the relationship you described is appropriate. If not, discuss why and include another possible relationship.

DISCOVER

Creating and Analysing Scatter Plots Using Spreadsheet Software

These instructions are for Microsoft® Excel. You may need to modify them if you are using other spreadsheet software.

For help with entering and graphing data in spreadsheets, refer to page 432 (Microsoft® Excel) or page 436 (Quattro® Pro).

1. Enter the data in the spreadsheet as shown.

	A	B	C
1	Advertising Expenditures	Toy Sales	
2	(thousands of dollars)	(thousands of dollars)	
3	1	100	
4	5	300	
5	8	400	
6	6	200	
7	3	100	
8	10	400	
9	2	175	
10	2.5	200	
11	4	250	
12	7.5	325	
13	6.75	300	
14			

2. Select both columns, including the titles.
3. Click the graph icon to start the chart wizard.
4. Select **XY Scatter** graph and accept the first chart type.
5. Click **Next** to move to the next screen.
6. Accept the default range. Click **Next**.
7. Add the following titles:
   Chart Title: Toy Sales vs. Advertising Costs
   Value (X) Axis: Advertising Costs (thousands of dollars)
   Value (Y) Axis: Toy Sales (thousands of dollars)

8. Click **Next** to move to the final window.

9. Select either option for the chart location.

If you place your scatter plot as an object in the spreadsheet, you may have to re-size (click and drag a corner square) the scatter plot to show all the data.

10. Your graph should look like the screen shown.

11. Describe the relation (linear, exponential, quadratic, no relation) that seems to approximate this scatter plot.

12. Describe the strength of the relation, weak or strong.

13. State the direction, positive or negative, of the relation.

DISCOVER

See page 438 for help creating a case table in Fathom™.

Creating Scatter Plots Using Fathom™

The data in **Collection 1** represents the number of community college diplomas in social sciences and services over a seven-year period.

1. Enter the data into a case table as shown.

	Year	Graduates
1	'90-'91	10214
2	'91-'92	11057
3	'92-'93	12513
4	'93-'94	13526
5	'94-'95	14304
6	'95-'96	15803
7	'96-'97	16779

2. Click on the graph icon to create a scatter plot.

3. In your case table, click on the Year attribute (column header) and drag it to the x-axis on your graph.

Click on the Graduates attribute and drag it to the y-axis. Your graph should look similar to the one shown.

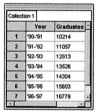

4. Complete a table, with headings as shown, to describe the characteristics of this scatter plot.

Scatter Plot Type	Strength	Direction	Outliers

KEY CONCEPTS

- Independent variables exist on their own and are not influenced by other factors. Examples include time, year, and age. When graphing, the independent variable is placed along the horizontal or *x*-axis.

- Dependent variables rely on another factor. For example, the amount of your paycheque from a part-time job depends on the number of hours you worked. When graphing, the dependent variable is placed along the vertical or *y*-axis.

- A scatter plot is the graph of a set of ordered pairs (*x*, *y*) used to describe a relationship between independent and dependent variables. If a relationship exists, further analysis may be done.

- Examples of some types of scatter plots are shown.

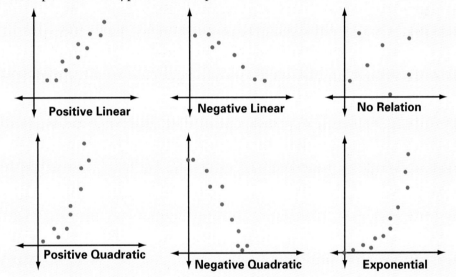

- You can create a scatter plot using technology. Three tools that can be used to create scatter plots are a graphing calculator, Microsoft® Excel, and Fathom™.

DISCUSS THE CONCEPTS

1. Discuss the difference between dependent and independent variables.

2. Describe an example of a relation with two variables. Identify the dependent and independent variables, and make a sketch of the axes, including labels.

3. Discuss the purpose of creating a scatter plot, and describe the information that may be obtained from it.

4. Select a tool for creating a scatter plot and, in point form, give the steps used to create the scatter plot.

PRACTISE

A 1. Identify the independent and dependent variables for each relation.
  a) The area of a house and the cost of building materials
  b) The population of a town, recorded every 4 years over a 40-year period
  c) A study to determine if a child's toy rocket launcher meets the advertised height of 25 m. The toy was tested 50 times.
  d) The level of a person's physical fitness, determined in part by measuring age and heart rate

2. For each scatter plot, identify the following:
  i) The type of relationship (linear or non-linear)
  ii) The strength of the relationship (strong or weak)
  iii) The direction of the relationship (positive or negative)
  iv) The presence of an outlier

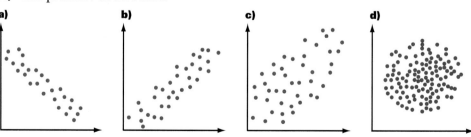

a)   **b)**   **c)**   **d)**

B 3. Communication When you are using a graphing calculator, you sometimes get unexpected screen messages.
  a) Describe when a **Mismatch Error** will occur when using a graphing calculator.
  b) When creating a scatter plot with a graphing calculator, describe what you should do when no graph appears after pressing (GRAPH).

4. Application The data were collected from Mr. Li's grade 11 Mathematics of Personal Finance course.

Student	Final Mark	Number of Days Absent
A	35	6
B	42	5
C	63	3
D	70	1
E	71	1
F	70	2
G	60	2
H	50	4
I	87	0
J	68	2
K	60	3
L	88	1
M	87	0
N	55	4
O	63	3

a) Display the data in a scatter plot.

b) Identify the type of relationship.

c) What possible explanation might there be for the outliers in this data set?

d) What message does the relationship in this data seem to convey?

5. Problem Solving, Application A store manager wants to find out whether there is a relationship between the age of an employee and the number of sick days taken each year.

Age	18	26	39	48	53	58
Number of Days Absent	16	12	9	5	6	2

a) Use the data to decide whether a relationship exists and if so, the type of relationship.

b) Describe how you can strengthen your analysis of this data set.

6. Application Expresso Legatto is a small coffee house in Bowmanville that is open from 6 a.m. until 6 p.m. Merna recorded the number of cups of coffee sold throughout the day.

Hour	1	2	3	4	5	6	7	8	9	10	11	12
Number of Cups	35	45	52	31	44	46	31	23	46	21	19	15

a) Using technology, create a scatter plot.

b) Analyse your scatter plot to determine whether a pattern exists.

c) If the owner needs one employee for every 25 cups of coffee sold per hour, is it possible to determine how many employees are required for the morning shift from 6 a.m. to 1 p.m.?

7. Two students were concerned about the volume of garbage in the cafeteria, and decided to prepare a report to present to the school council. After the data were gathered, they each created their own version of a scatter plot using the same data set. Their graphs are shown.

Graph A   Graph B

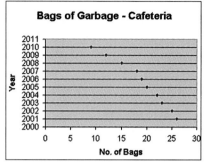

a) Compare the two graphs and discuss how they differ.

b) Which graph is better? Why?

8. **Application** The data below represent the number of reported cases for all ages for chicken pox in Canada.

Year	Number of Cases of Chicken Pox
1986	37 599
1987	52 957
1988	42 312
1989	41 560
1990	20 254
1991	13 687
1992	4 514
1993	3 509
1994	3 117
1995	1 932

Source: Statistics Canada, #H510821, matrix 1089

a) Create a scatter plot for this data.

b) Use the scatter plot to identify the relation in the data.

 9. Go to *www.mcgrawhill.ca/links/MPCA12* and follow the links to find data on the Boston Marathon Women's winning times from 1966 to date. Create a scatter plot of the data. What pattern emerges?

Line of Best Fit

The term *regression* comes from an investigation conducted by Sir Francis Galton. He measured the heights of very tall and very short parents and their children. Galton found that the heights of the children regressed toward the mean population height. Now the term **regression** is used to describe the process of identifying relationships between data.

You learned in the previous section that a scatter plot is useful in visually identifying relationships in data. In this section, you will use the regression capabilities of the graphing calculator and of computer software to identify the relationships found in data.

Drawing a line on a scatter plot that passes through or close to as many points as possible will identify the equation that represents the data. This line is called the **line of best fit**. Knowing the equation of the line of best fit allows you to predict future trends based on the current data by simply extending this line.

EXAMPLE 1

Using Linear Regression

a) The scatter plot shows the average annual exchange rate in Canada for U.S. dollars. Describe where you think the line of best fit should be drawn.

Year	Exchange Rate
1990	0.8541
1991	0.8728
1992	0.8273
1993	0.7751
1994	0.7322
1995	0.7286
1996	0.7334
1997	0.7222
1998	0.6741
1999	0.6731

b) When the linear regression function on the graphing calculator is used (you will learn more about this in the next Discover), the screen shown is displayed. State the meaning of each variable and write the equation of the line of best fit.

c) Enter the equation from part b) and graph it. Use this line to find the approximate exchange rate in 2000.

Solution

a) The line of best fit should be drawn as close as possible to most of the points. The line will slope down to the right, so it has a negative slope.

b) *a* is the slope of the line, and *b* is the *y*-intercept.
The equation of the line of best fit, with *a* and *b* rounded to four decimal places, is $y = -0.0222x + 0.8595$.

c) Use the TRACE key to find the point on the line of best fit that corresponds to 2000. The years are represented by $x = 0$, $x = 1$, and so on. Trace to $x = 10$. The approximate average exchange rate for U.S. dollars in 2000 was 0.6370.

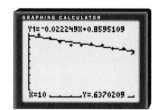

The **correlation coefficient** is a variable used to represent the strength of a relationship. This value is given by the variable *r*. The coefficient of determination, r^2, is another measure of the strength of a relationship. This section will focus on the correlation coefficient, *r*.

DISCOVER

Linear Regression With a Graphing Calculator

Statistics Canada collects data from across Canada to determine the average price of various foods and other consumer items. The table shows the average quarterly price of one dozen eggs over a five-year period.

Year	Quarter	Price of Eggs ($)
1996	1	1.65
	2	1.67
	3	1.71
	4	1.76
1997	1	1.76
	2	1.77
	3	1.78
	4	1.79
1998	1	1.78
	2	1.81
	3	1.78
	4	1.79
1999	1	1.79
	2	1.77
	3	1.75
	4	1.78
2000	1	1.82
	2	1.83
	3	1.88
	4	1.90

Source: Statistics Canada's Internet Site,
http://estat.statcan.ca/cgi~win/CNSMCGI.EXE, table 326-0012, November 7, 2001.

For more information on entering data and creating scatter plots using a graphing calculator, see page 430 of Appendix B.

1. Using a graphing calculator, prepare a scatter plot of the data. One possible way of entering the data and the related window settings are shown. Explain the way this was done.

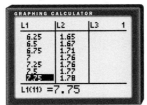

2. To find the line of best fit, press **STAT** and cursor right to **CALC**.
Select **4:LinReg(ax + b)**.
Press **ENTER**.
To view the correlation coefficient, r, you need to activate **DiagnosticOn**.
Press **2nd 0** to access the **CATALOG**.
Scroll to **DiagnosticOn** and press **ENTER** twice.
Repeat the steps, if necessary, to view the correlation coefficient, r.

3. **a)** Identify the correlation coefficient and discuss how well the data fall into a linear pattern.
   b) Compare the signs of the correlation coefficient, r, and the slope.
   c) Describe how the variables a and b relate to the equation of a line in the form $y = mx + b$.
   d) Write the linear equation in your notebook, rounding the values of a and b to two decimal places. Then, enter this equation into the graphing calculator, using **Y=**. Press **GRAPH**.

4. Describe the position of the line of best fit in relation to the points in the scatter plot, and discuss the relationship between the correlation value and the position of the line of best fit.

5. Use your regression model to extrapolate the average price of one dozen eggs in the first quarter of 2001. Use E-STAT to check the result against the real average cost.

Linear Regression Using Microsoft® Excel

The amount of rainfall influences growing patterns in nature. Farmers know very well how too much rainfall or too little rainfall can affect their crops. In the summer of 2000, a high volume of rainfall early in the season reduced the harvest that year. One year later, Canada experienced the least amount of rainfall in 50 years, which also reduced crop levels.

The relationship between rainfall during the growing season and the average mass of a carrot grown in a field is shown in the table.

Rainfall in Growing Season (cm)	Average Mass of Carrot (g)
15	87
18	89
20	93
22	101
23	102
25	108

For help with graphing data using Quattro® Pro, see page 436 in Appendix B.

1. Enter the data into a Microsoft® Excel spreadsheet.

2. Create a scatter plot by selecting the graph icon, followed by the scatter plot icon, **XY(Scatter)**.

3. Add the appropriate titles and place the graph on the worksheet.

4. To draw the line of best fit, complete the following steps:

i) Select **Chart** from the main menu. (The chart menu bar may need to be inserted if the word Chart does not appear in your menu bar.)

ii) Select **Add Trendline**.

iii) Select **Linear**.

Your scatter plot and trend line should look similar to the one shown.

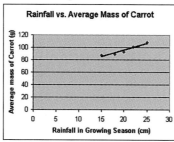

5. Discuss whether extending the line of best fit to determine the amount of rainfall needed to grow larger carrots is realistic.

Linear Regression Using Fathom™

Mechanical engineering technicians were randomly surveyed to collect data on their number of years experience and their current salary.

Years of Experience	Salary ($)
1	27 800
2	44 500
2	34 300
5	37 800
6	51 175
6	53 500
8	58 200

See page 438 of Appendix B for help with entering data, graphing, and linear regression in Fathom™.

1. Create a case table of the data.
2. Create a graph (scatter plot), with Years of Experience along the *x*-axis and Salary along the *y*-axis.
3. Describe the strength of the relationship.
4. From the **Graph** menu, select **Least-Squares Line**. The equation of the line of best fit will be shown automatically below the graph.

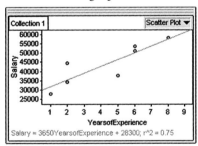

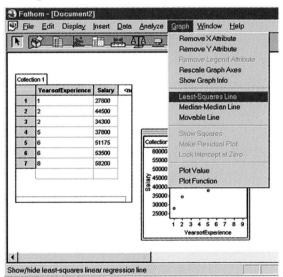

5. Summarize the relationship between the years of experience and the salary a person earns.
6. Discuss whether this relationship would apply to other occupations.
7. Discuss whether this relationship continues until a person retires, or whether there is a point where the relationship plateaus.

As you saw in the previous section, some scatter plots have stronger relationships than others. It is important to know how strong or weak a relationship is to determine how relevant the line of best is. If a relation is very weak, then the line of best fit may be of little use, since its accuracy in predicting future trends will not be reliable.

EXAMPLE 2

Interpreting the Correlation Coefficient

An *r*-value of ±1 represents the strongest relationship possible; the closer the *r*-value is to 0, the weaker the relationship. Rank the relationships shown, in order of strength.

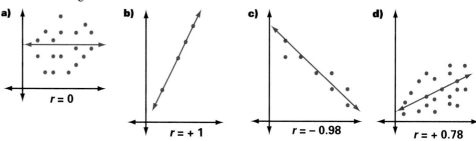

Solution

Using the correlation coefficients, the graphs ranked from strongest to weakest relationship are b, c, d, a.

EXAMPLE 3

Interpreting the Sign of the Correlation Coefficient

The graphs below all represent fairly strong relationships. Discuss the connection between the correlation coefficient and the direction of the graph.

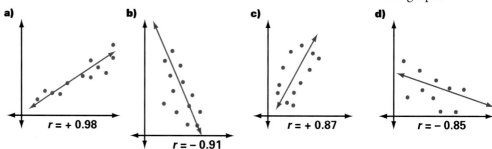

Solution

The graphs with a negative slope also have a negative correlation coefficient. Similarly, the graphs with a positive slope have a positive correlation coefficient.

KEY CONCEPTS

- Linear regression is the process of creating a line of best fit.

- The line of best fit is drawn close to most of the data points.

- The line of best fit allows for further analysis, such as extrapolating or interpolating data.

- When the line of best fit passes through all the points, the correlation coefficient, r, is +1 or −1. The closer the correlation coefficient is to 0, the further away the points are from the line of best fit.

- A negative correlation coefficient corresponds to a negative slope and a positive correlation coefficient corresponds to a positive slope.

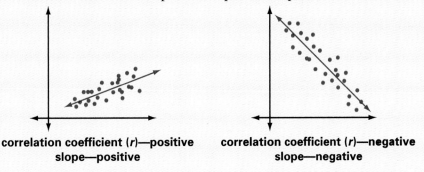

correlation coefficient (r)—positive   correlation coefficient (r)—negative
slope—positive   slope—negative

DISCUSS THE CONCEPTS

1. Describe how the correlation coefficient relates to the scatter plot and the line of best fit.

2. Describe the information that can be obtained from a line of best fit.

3. Describe the appearance of a scatter plot with a correlation coefficient of 1, −1, 0.5, and 0.

PRACTISE

A 1. Some possible values for the correlation coefficient are −0.85, 0.6, 0.01, and 0.94. Which of these values do you associate with each of the following?
   a) Strong, positive relation   **b)** Strong, negative relation
   c) Very weak relation

2. Sketch a scatter plot to model each of the following correlation coefficients.
   a) −0.9   **b)** 1   **c)** 0.8   **d)** 0.002

For help with linear
regression, see page
232 (Microsoft® Excel),
or page 231 (graphing
calculator), or page 233
(Fathom™).

3. **Advertising** An advertising consultant wants to find out whether there is a relationship between the number of radio advertisements aired per week and the amount of sales, in thousands of dollars, of the product. The data for the sample are shown below.

Number of Advertisements per Week	2	5	8	8	10	12
Sales ($000s)	2	4	7	6	9	10

a) Use technology to create a scatter plot.
b) Identify the correlation coefficient.
c) Describe the type of relation and the strength of the relation.

B **4.** **Communication, Application** An insurance company wants to determine the strength of the relationship between the number of hours a person works and the number of injuries or accidents that person has over a period of one week. An injury can be anything from a paper cut to a lower back injury from lifting incorrectly. Accidents range from spilling coffee to misreading a label, resulting in an imbalance in a mixture. Use the data shown.

Hours Worked	40	32	36	44	41	50	45	20	60
Number of Accidents/Injuries	2	0	1	3	2	8	5	0	10

a) Using technology, create a scatter plot of the data.
b) Identify the correlation coefficient.
c) Prepare a summary to be distributed to companies describing the type of relation and the strength of the relation. Indicate the impact that working longer hours has on the number of injuries and accidents in the workplace.

5. **Thinking/Inquiry/Problem Solving** Shawn is the manager of a fitness club. He wants to determine if a person's age is related to the number of hours of exercise per week. This information will allow him to focus his advertising best. He collected data from some randomly selected current members.

Age	22	26	32	38	49	52	44	25	43	51	52	35	22	20	48
Hours	7.5	5	2	3	8	6	2.5	6	3.5	5	5	2	20	8	4

a) Using technology, create a scatter plot of the data.
b) Identify the correlation coefficient.
c) Describe the type of relation and the strength of the relation.
d) If this is a good model for the relation, identify the ages that Shawn should focus his marketing on. If it is not, what advice can you give to Shawn to help identify a focus group?
e) Discuss the suitability of using this model to extrapolate or interpolate other ages and estimated amount of exercise time.

6. **Application** The airline prices shown here were the advertised summer specials from Toronto. Using linear regression, describe the relationship between airline ticket prices and the distance travelled. Include the correlation value and the regression model. Discuss the appropriateness of this model to determine ticket prices. Give possible reasons for any outliers

Destination	Distance (km)	Ticket Price ($)
Montreal, PQ	554	198
Ottawa, ON	399	158
Fredericton, NB	1373	238
Halifax, NS	1929	258
Charlottetown, PE	1738	358
Sydney, NS	1788	398
St. John's, NF	3141	398
Winnipeg, MB	2084	258
Thunder Bay, ON	1397	190
Saskatoon, SK	2874	268
Regina, SK	2669	268
Calgary, AB	3491	288
Edmonton, AB	3390	312
Victoria, BC	4721	430
Vancouver, BC	4537	398

7. **Data Bank** Use the data on Carbon Dioxide Emissions in Canada, on page 458. Carbon dioxide (CO_2) is a by-product of burning fossil fuels, such as the gasoline we use in our vehicles and factories. Carbon dioxide from fossil fuels is harmful to humans, as well as the environment.
   a) Explain why Statistics Canada gives carbon dioxide emissions in both total amounts and per capita.
   b) Using technology, create scatter plots and identify the line of best fit or the curve of best fit for both total amount of carbon dioxide emissions and carbon dioxide emissions per capita.
   c) Compare the results for the two relations.
   d) An environmentalist states that carbon dioxide emissions are still increasing, resulting in damage to the ozone layer. A spokesperson from a large factory that uses fossil fuel claims that carbon dioxide emissions are actually decreasing. Using your analysis of the data here, and other data that you can find, prepare a report to support one side of this debate. Include at least one graph and a written argument to support your position.

8. Application The coach of the Clarington Tiger Cats pee-wee team wants to determine whether there is a relationship between how fast a player can run 40 yards and how far the player can throw a football. The table lists the sprint time and distance thrown for each player.

Player	Sprint Time (s)	Distance Thrown (yd)
Milos C.	7.93	18
Dale R.	7.65	18
Nick F.	7.20	21
Connor J.	8.13	13
Kevin H.	7.62	16
Troy J.	7.14	18
Raj M.	8.22	20
Nick J.	8.84	17
Brandon K.	7.82	13
Brandon B.	7.90	13
Tyler C.	6.94	26
Josh M.	6.56	18

a) Using technology, create a scatter plot comparing sprint time and distance.

b) Review the data for any outliers and give possible reasons for them.

c) Discuss whether the coach can use this information to select next year's team.

9. Data Bank Use the data on The Solar System, page 465, to determine the relationship between the orbital period of each planet and its relative distance from the sun. (AU is the mean distance between the sun and Earth, and is equivalent to 149 600 gigametres.)

a) Describe the relationship.

b) How could an astronomer use this data to identify the orbital period of a newly discovered planet if the planet's distance from the sun is known?

ACHIEVEMENT CHECK Knowledge/Understanding Thinking/Inquiry/Problem Solving Communication Application

10. The manager of Honest Auto wants to find the relationship between the number of radio advertisements that are broadcast each week and the number of cars sold. A random sample of the records for eight weeks produces the following data.

Number of radio advertisements: 3, 16, 7, 4, 15, 7, 8, 5

Number of cars sold: 4, 40, 16, 9, 38, 16, 17, 10

a) Using technology, construct a scatter plot for the data.

b) Draw the line of best fit.

c) Give the correlation coefficient.

d) Prepare a short summary, based on the above analysis, that the manager could present to the board of the company.

Uses and Possible Misuses of Sample Data

It is important to recognize the limitations and restrictions of sample data. You have already examined some of the possible misuses of sample data by considering
- the nature of any outliers
- weak correlation coefficients

Further examination of misuses of regression and other misuses will provide a better analysis of the data you are working with.

EXAMPLE 1

Misuse of Regression

Identify the anomaly in each of the scatter plots.

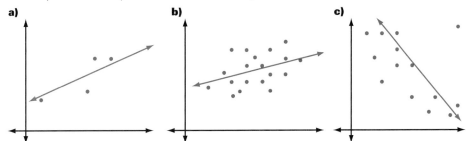

Solution

a) There is not enough data to analyse. Although there is no minimum number of data, typically, the more the better.

b) The correlation value is so low that the regression line is not reliable. It would be best to conclude that there is no correlation for this set of data.

c) The point in the upper right corner does not fit the pattern of the other points. This point is an outlier. The line of best fit for this data set shifts a bit to the upper right to accommodate the outlier. The outlier could be a data entry error or an error in data collection, or it may be a valid point. If an error is confirmed and can be corrected, then the line of best fit can be redrawn with the revised point. If an error is definitely confirmed, but cannot be corrected, then the outlier can be removed and the line of best fit drawn without that point. However, outliers may very well represent true data, and should not be discarded without good reason.

Extrapolation involves extending the line of best fit. Sometimes this may not be appropriate. Factors that can influence and alter an established trend include the following:
- Environment
- Society
- Politics
- Economics

There are numerous situations involving inappropriate extrapolation due to these influencing factors. The question below will highlight the misuse of extrapolation and heighten your awareness to the possibility of this misuse of sample data in other situations.

EXAMPLE 2

Inappropriate Extrapolation

The scatter plot shows the percent of people who favoured Party A over the past six months. Discuss whether this data should be used to estimate the number of votes Party A will receive in an election three months later.

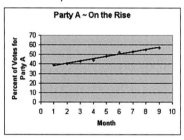

Solution

This data set is very sensitive to people's feelings or preferences. It is not based on concrete numbers and, therefore, can change unexpectedly.

Changes in party preference are influenced by economical and political situations, such as announcements of cut backs in funding for health care or education. Announcements such as these, by the governing party, might lower this party's rating. Alternatively, announcements for increased spending on health care, or lower taxes, might increase this party's popularity. Political parties have a good understanding of what affects their voters and strategically place these announcements.

In these situations, extending the line of best fit is risky because this data set can be influenced by other factors.

KEY CONCEPTS

- Outliers will cause the line of best fit to shift in the direction of the outliers and may result in a misleading line of best fit. Check the outlier for a data entry or data collection error. If an error is identified, it should be corrected or the outlier should be removed. If no error can be confirmed, the outlier should be kept.

- Insufficient data may result in misleading conclusions.

- Before extrapolating data, be sure the correlation coefficient for the line of best fit is reasonably strong. If the relationship is weak, your analysis should indicate this, and caution should be used when identifying trends.

- Be sure to examine external factors, such as environment, society, politics, or economics, that may influence future trends before extrapolating.

DISCUSS THE CONCEPTS

1. Describe two factors that might lead to a weak correlation coefficient for the line of best fit. Explain why.

2. Even if the correlation coefficient is strong, why should you be cautious about extrapolating using the line of best fit?

PRACTISE

Ⓐ 1. Describe the misuse of regression in each graph.

a)
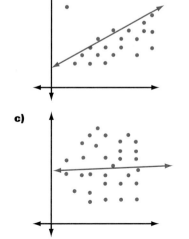

b)
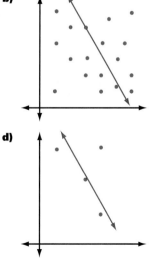

c)

d)

2. Give an example of a data set that may be influenced by each factor.
   a) environment
   b) society
   c) economics
   d) politics

APPLY THE CONCEPTS

3. **Communication, Application** The amount of money spent in Ontario on pollution control is shown.

Year	Thousands of Dollars
1992–93	251 741
1993–94	230 537
1994–95	170 680
1995–96	144 699
1996–97	137 772

Source: Statistics Canada, Canada Year Book 2001, p. 64, #160200-XKE

   a) Analyse the data to determine whether a trend exists and, if so, the strength of the relationship.
   b) Discuss whether it is possible to extrapolate the amount of money that will be spent on pollution control in future years.

B 4. **Data Bank** Use the Passenger Aircraft Operating Statistics, on page 462. Determine whether there is a relationship between the number of seats on an aircraft and its speed.

For help with linear regression, see section 5.4 page 232 (Microsoft® Excel), or page 231 (graphing calculator), or page 233 (Fathom™).

5. **Problem Solving, Application** Angela is practising with her dog Scruffy for an upcoming competition. Scruffy is timed on ten trials of the obstacle course. His times are recorded in the table.

Trial Number	1	2	3	4	5	6	7	8	9	10
Time (s)	52	62	51	58	57	135	54	54	52	51

   a) Determine a linear model for this data.
   b) Discuss any possible outliers in the data, and describe how they could be corrected.
   c) The cost to enter a competition is approximately $25. Scruffy will have a chance at winning if his time is below 60 s. Using your model, discuss whether it is worth spending the $25 to enter Scruffy in this competition.

6. **Application** Canadian forests are monitored annually by the Ministry of Natural Resources. The table shows the number of hectares of Canadian forests harvested each year.

Year	Area Harvested (ha)
1991	851 781
1992	908 459
1993	962 216
1994	1 000 185
1995	1 017 657
1996	1 043 918
1997	1 022 539

Source: Statistics Canada, Canada Year Book 2001, p. 69

a) Create a scatter plot of the data, and use linear regression to find the equation of the line of best fit.

b) Discuss whether this linear model can be used to determine the area of Canadian forests that will be harvested in the next five years.

7. The Ministry of Natural Resources also monitors reforestation in Canada. The table shows the area reforested by seed and seedlings over a seven-year period.

Year	Reforestation (ha)
1991	509 675
1992	465 547
1993	453 701
1994	488 927
1995	459 429
1996	450 173
1997	457 780

Source: Statistics Canada, Canada Year Book 2001, p. 69

a) Use linear regression to describe the relationship between time and area reforested.

b) How strong is the linear model? Can it be used to estimate the number of hectares that will be reforested in the next five years?

c) Compare this set of data with the data in the previous question. How does the area reforested compare with the area harvested?

8. **Thinking/Inquiry/Problem Solving** Fires in Canadian forests are monitored weekly during the fire season. The total number of fire starts for each week from May 1 to August 13 in the summer of 2000, and the ten-year averages for those weeks, are shown.

Week	Number of Fire Starts in 2000	Ten-Year Average, 1990-99
1	200	380
2	105	185
3	190	275
4	160	280
5	180	340
6	100	315
7	105	450
8	115	445
9	410	415
10	155	325
11	195	370
12	290	320
13	470	405
14	425	605
15	400	610

For help with creating scatterplots, see page 231 (graphing calculator), page 232 (Microsoft® Excel), page 436 (Quattro® Pro), or page 233 (Fathom™).

a) Use technology to create a scatter plot for the fire starts by week in 2000. Which data point looks as though it might be an outlier? Remove this point and draw the line of best fit for the remaining data. How strong is the correlation? Is it reasonable to extrapolate the number of fire starts in week 18 (August 28–September 3) from this line of best fit? Explain.

b) On the same screen, create a scatter plot of the ten-year averages by week. Draw the line of best fit for this data. How strong is the correlation?

c) Compare and contrast the 2000 data with the ten-year average. Was the year 2000 unusual? Should the outlier in part a) have been removed?

9. **Communication, Application** The sales for a new business totalled $2500 on the first day of opening, $2800 on the second day, and $3200 on the third day. Feeling confident, the owner uses linear regression to model her success. She claims, "At this rate, I will sell $10 883.33 in just 25 days."

a) Using technology, create a scatter plot and use linear regression to check the accuracy of the owner's claim.

b) Discuss whether it is realistic for the owner to think she will sell $10 883.33 of merchandise on day 25.

c) Advise the owner on what she needs to do to substantiate a claim like this.

10. Application The population in Canada was recorded every ten years from 1851 to 1951. After this period, the census was taken every five years. The population at the end of each census period is shown.

Year	Population of Canada (millions)
1861	3.230
1871	3.689
1881	4.325
1891	4.833
1901	5.371
1911	7.207
1921	8.788
1931	10.377
1941	11.567
1951	13.648
1956	16.081
1961	18.238
1966	20.015
1971	21.568
1976	23.450
1981	24.820
1986	26.101
1991	28.031
1996	29.672

Source: Statistics Canada, Canada Year Book 2001, Population data by year 1861-1996

a) Create a scatter plot using technology. Use linear regression to find the equation of the line of best fit.

b) Use your model to determine the population in 2001. Use Statistics Canada's website to check how close this is.

c) Use the model to predict the population in 2011 and 2021. Compare your answers with the predictions made by Statistics Canada, as listed in the **Data Bank**, page 454. Give possible reasons for any differences. Make a list of the factors that influence population change in Canada. How have these factors changed with time?

CAREER PROFILE

Mechanical Engineering Technician—Non-Destructive Evaluation

Non-destructive evaluation (NDE) technicians are employed in a number of different industries, such as power supply, pipeline, shipbuilding, and aerospace. They inspect parts and joints for damage that is invisible to the naked eye, but can be seen using radiographic, ultrasonic, or magnetic inspections.

Students studying non-destructive evaluation technology learn to use the equipment to conduct these inspections in the classroom session, as well as during their field-place term. One of the courses that students take is called Statistical Methods in Quality Assurance. The topics covered in this course (scatter plots, correlation coefficients, regression analysis) are introduced in this chapter. Students further their understanding of these statistical methods by using the concepts in industry-related applications.

1. Durham College in Oshawa offers the Mechanical Engineering Technician (NDE) course.
   a) Research to determine if the course is offered at other locations.
   b) Find the cost of the course.
   c) At Durham College, the NDE program involves a paid work term. Discuss the advantages of this type of learning.

2. A non-destructive test of concrete was conducted using ultrasonic pulse velocity. The results are displayed in the graph.
   a) Identify the three types of surfaces tested.
   b) A line of best fit is shown for the data for each type of surface. Of the three, discuss which has the highest correlation. Explain your answer.
   c) Based on the outcome of this test, what surface is best for this type of testing?

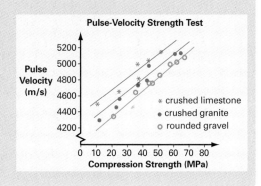

CHAPTER PROBLEM WRAP-UP

Do you ever get frustrated when you think you have the solution to a problem, but nobody will listen to you? What you need is good data that is well organized and analysed to support your concern. Of course, you will also need solid mathematical reasoning. Put all this together in an effective presentation and people will listen.

Throughout this chapter, you have seen some school-related issues raised. Now it is your turn to make a difference.

1. In a small group, discuss school issues that affect you and your colleagues, such as cafeteria food, school hours, attendance rules, dress code, or the school recycling policy.
   a) List at least three issues that interest you.
   b) For each of these issues, identify all concerns related to the issue.
   c) Identify the variables.

2. Select one issue that you could feasibly change, if you had the data, statistical analysis, and presentation to support your idea.
   Prepare a report on your issue.
   Your report should include a visual aspect of your research, including a written explanation and appropriate tables, graphs, or formulas, as well as an oral presentation (preferably to the audience for whom the issue is directed).

*R*eview

5.1 Discover E-STAT, pages 208–211

1. **a)** Outline the steps you need to use to access data stored in E-STAT.

  b) Identify the six main organizational categories for data in E-STAT.

2. The microwave oven has become a common appliance in Canadian homes. Using E-STAT, prepare a marketing report, showing the growth in popularity of microwave ovens in Canadian households.

5.2 Methods for Collecting, Storing, and Retrieving Data, pages 212–219

3. When starting a statistical research project, the first step involves data collection. Describe the three ways you can collect data.

4. Describe each data collection method below as inferential or descriptive.

  a) One thousand people are randomly selected and contacted by telephone to learn more about the spending habits of people in the 705 area code.

  b) Students are surveyed in the school cafeteria to determine their views on school issues.

  c) A pharmaceutical company attached a self-addressed survey to their product to learn what people like and do not like about their product.

5. State whether each of the following is an example of a primary or secondary source of data.

  a) A survey you did of 50 people in the subway to determine travel time to work or school

  b) The Statistics Canada data base

  c) A newspaper report giving the number of accidents and time of day

  d) A record of measurements that you made of the height and mass of ten cats in your neighbourhood as part of an experiment to determine whether a relationship exists

6. Home insurance companies use inferential statistics to determine their insurance rates. Discuss the type of data home insurance companies are interested in, and identify two possible variables insurance companies would use in their analysis.

5.3 Scatter Plots, pages 220–228

7. Use one term from each descriptor set to describe each of the following scatter plots.
Strength of a relation: strong, moderate, weak, or none
Direction of relation: positive or negative

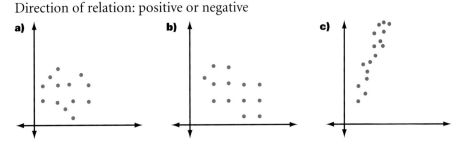

8. For each situation, identify the independent and dependent variables.
  a) A person's age and his or her heart rate
  b) The amount of pollution in a lake and the number of fish
  c) The age of a child and the number of words known

9. a) Describe how you can tell whether a variable is independent or dependent.
  b) Why is it important to know which variable is independent and which is dependent?

10. A six-month exercise program recorded the total mass lost for three people since the beginning of the program.

Total Mass Lost (kg)			
Month	Person 1	Person 2	Person 3
1	5	4	4
2	11	7	8
3	16	11	10
4	20	14	13
5	24	16	14
6	26	17	14

  a) Using technology, create a scatter plot. Enter the month followed by the mass lost for each person.
  b) Describe the type of relation.
  c) Using the data given, create a new table that records the mass lost by each person each month. Draw three scatter plots to show each person's loss per month. Show the line of best fit for each person's record. You may choose to show all three on one graph by using different colours or symbols. What type of relation do these graphs seem to model?

5.4 Line of Best Fit, pages 229–238

11. Select one tool (graphing calculator, Fathom™, or spreadsheet software) and list the steps you would use to create a line of best fit. Start from entering the data and end at displaying the line.

12. For each correlation coefficient, describe the strength of the relation.

a) 1   b) −0.78

c) 0   d) 0.98

13. Television commercials Have you ever noticed how commercials are more frequent toward the end of a televised movie? Determine the relationship between television commercials in a two-hour movie and their frequency by recording the start time of each commercial. Use a table similar to the one shown to record your data.

Minutes Since Start of Movie	Commercial Number
0	0
	1
	2
	3

a) Describe the pattern.

b) Use regression analysis to identify an equation to model this relation.

c) Discuss the appropriateness and strategy in the frequency of commercial intervals.

14. Health care The total amount of money spent on health care by the federal, provincial, and local governments is shown in the table.

Year	Expenditures (billions of dollars)
1975	12.3
1980	22.4
1990	61.3
1991	66.5
1992	70.1
1993	72.9
1994	73.6
1995	74.6
1996	75.6
1997	78.0

Source: Statistics Canada, Canada Year Book 2001, p. 130

a) Use technology to create a scatter plot.

b) Using linear regression, show the line of best fit. Describe the strength of this relation.

c) The amount spent on health care is increasing. Does this mean that Canadians need more health care than in the past?

d) Estimate the healthcare expenditures in 2005 and 2010.

5.5 Uses and Possible Misuses of Sample Data, pages 239–245

15. Outliers Sometimes elements in a set of data are referred to as outliers.

a) What is an outlier?

b) Describe how outliers can affect the line of best fit.

c) When can you omit an outlier?

16. Linear regression The growth of pumpkins in a field depends on the amount of water received. An agricultural research station collected the following data.

Daily Amount of Water (cm)	Average Mass of Pumpkin (kg)
0.25	0.9
0.50	1.8
0.75	2.2
1.00	3.1
1.25	3.4
1.50	3.7
1.75	4.1
2.00	4.3
2.25	4.8
2.50	5.0
2.75	5.3
3.00	5.6

a) Using technology, create a scatter plot.

b) Use linear regression to draw the line of best fit.

c) Discuss the strength of this relationship.

d) Using your line of best fit, estimate the average mass of a pumpkin that receives 10 cm of water daily.

e) Discuss whether this is a realistic application of your linear model.

Practice Test

ACHIEVEMENT CHART CATEGORY CONNECTIONS			
Category	Knowledge/Understanding	Thinking/Inquiry/Problem Solving	Communication	Application
Questions	1–8	6–8	1–4, 7, 8	3, 7, 8

1. Discuss the use of samples in statistics. Give two examples of questions for which a sample would be used. Give at least one example of a question for which you would need to poll the whole population.

2. The three methods of data collection are listed below.
   a) Copy them into your notebook, and indicate whether the method can be classified as a primary or secondary source of data.
   b) For each collection method, describe an example where it would be necessary to use that method.
   i) Data made available by others
   ii) Data resulting from an experiment
   iii) Data you collect in an observational study

3. Statistical research can be divided into several steps. Identify the steps and describe what is involved in each step.

4. Compare inferential and descriptive statistics, and include an example of each.

5. For each statement below, indicate whether inferential or descriptive statistics have been used.
   a) During 2001, Canadian forests grew by 4%.
   b) The recent Canadian census indicates a noticeable decrease in the number of marriages.

6. For the scatter plots shown, identify the following:
   i) The independent and dependent variables
   ii) The type of relationship (positive/negative, linear or non-linear, strong or weak)
   iii) Any data points that you would consider outliers

a)

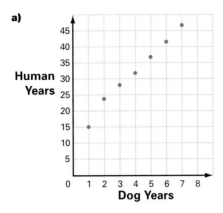

b)

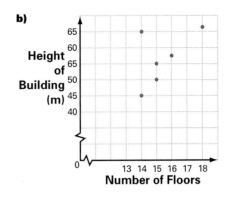

7. Life Expectancy People are more conscientious about leading a healthy lifestyle than in the past. Advances in medicine have made it possible for many illnesses to be cured. These and other factors have resulted in an increase in life expectancy. The table shows the life expectancy of males and females in Canada.

Year Born	Males (Years)	Females (Years)
1970	67.1	74.7
1975	68.8	76.6
1980	70.4	77.8
1985	71.1	78.2
1990	71.8	78.8
1995	72.8	79.7
2000	77.2	80.2

Source: Statistics Canada, Canada Year Book 2001, p. 139

a) Using technology, prepare a scatter plot for males.

b) Find the line of best fit.

c) Referring to the correlation value, discuss whether this equation is a good model for the data.

d) Use the equation to estimate the life expectancy for males born in 2010.

e) Repeat parts a) to c) for females.

f) Use the equation to estimate the life expectancy for females born in 2005.

g) Discuss whether these equations can be used to model the life expectancy for males and females into the next century.

h) Discuss why insurance companies are interested in this type of model related to life expectancy.

ACHIEVEMENT CHECK | Knowledge/Understanding Thinking/Inquiry/Problem Solving Communication Application

8. GiftsforU has just opened a new store. On the first six days, the daily sales, to the nearest dollar, were as follows:

Day	1	2	3	4	5	6
Sales ($)	1500	1650	1850	2000	2050	2080

The manager has asked you to predict the total sales for the month (30 days).

a) Use technology to analyse the data given and make an informed prediction. Identify factors that may affect the sales as time goes on.

b) Write a summary of your conclusions, including appropriate mathematical calculations and forms.

INTERPRETING AND ANALYSING DATA

Specific Expectations	Sections
• Retrieve information from various sources.	6.1
• Identify options that meet certain criteria, using more than one chart, spreadsheet, or schedule.	6.1
• Make informed decisions, using data provided in chart, spreadsheet, or schedule format and taking into account personal needs and preferences.	6.1
• Explain the meaning, and the use in the media, of indices based on surveys.	6.2
• Enter data or a formula into a graphing calculator and retrieve other forms of the model.	6.3
• Assess the validity of conclusions made on the basis of statistical studies, by analysing possible sources of bias in studies.	6.4
• Explain the use and misuse in the media of graphs and commonly used statistical terms and expressions.	6.4
• Collect, organize, and analyse data to address problems or issues, and calculate relevant statistical measures.	6.1, 6.2, 6.3, 6.4
• Formulate a summary conclusion to a problem or an issue, by synthesizing interpretations of individual statistical measures.	6.1, 6.2, 6.3, 6.4
• Formulate extending questions related to the conclusion reached in the investigation of a problem or an issue.	6.1, 6.2, 6.3, 6.4
• Communicate the process used and the conclusions reached in the investigation of a problem or an issue, using appropriate mathematical forms.	6.1, 6.2, 6.3, 6.4

et Ready

1. **Graphing data** Describe the type of data you would use each type of graph to display.
   a) a line graph   **b)** a circle graph, or pie chart   **c)** a histogram

2. **Interpreting Graphs** Refer to the graph and write a statement using the following terms.
   a) extrapolate
   b) interpolate

3. **Interpreting Graphs** Identify the independent and dependent variables in question 2.

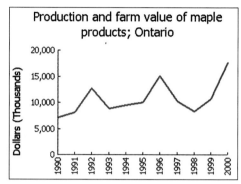

Source: Statistics Canada's Internet Site, http://estat.statcan.ca/cgi~win/CNSMCGI.EXE, table 001-0008, November 7, 2001.

For help with graphing data in spreadsheets, see page 432 (Microsoft®Excel or page 436 Quattro ®Pro).

4. **Graphing and interpreting data** The table gives information on the area planted and seeded in Canada over a ten-year period.

Year	Area Planted and Seeded (hectares)
1987	411 586
1989	492 341
1991	508 579
1993	436 425
1995	449 923
1997	458 638

   a) What type of graph is most suitable for displaying this data? Explain why you chose this type and not others.
   b) Using spreadsheet software, or otherwise, prepare a graph of the data.

5. The average number of days absent from work varies depending on gender and occupation. The table shows the average number of days women in Canada were absent from work.

Year	Number of Days
1990	12.6
1991	12.9
1992	12.8
1993	13.4
1994	13.4
1995	13.4
1996	12.2
1997	13.8
1998	13.8
1999	14.4

   a) Using technology, create a scatter plot of this data.
   b) Draw the line of best fit on your graph. Discuss the strength of this relation.
   c) As an employer, how many days might you expect a woman to be absent from work, on average, in 2005? 2010?
   d) Women are absent from work slightly more often than men. Give possible explanations why this is so.

6. **Graphing and interpreting data** A school timetable is shown.

Period	Time
P1	8:15 – 9:28
TAP	9:33 – 9:39
P2	9:44 – 10:57
Lunch	10:57 – 11:47
P3	11:47 – 1:00
P4	1:05 – 2:18

a) How long is each class?
b) How much time is allocated to move between classes?
c) How much time is allocated for lunch?
d) If a fast-food restaurant is 18 km away and you drive there and back at an average speed of 60 km/h, how much time do you have at the restaurant to eat lunch?

7. **Percents** One year, the cost of a lube, oil, and filter for a standard car was $24.98. Three years later, the cost is $32.98. Calculate the percent increase.

8. **Graphing data** Describe the inappropriate technique that has been used in drawing each graph so that it conveys a poor image of the data.

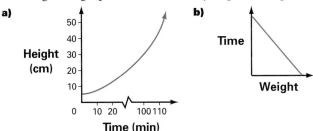

CHAPTER PROBLEM

The number and type of sporting injuries have been increasing and have prompted safety regulators to re-examine the use of safety equipment. Statistics are being kept to analyse injuries in various sporting activities so that recommendations for future safety regulations can be made.

Research a sporting activity that interests you. Identify the number and types of injuries related to your sporting activity. Throughout this chapter, you will see various ways to interpret and analyse the statistics that you gather.

There are many web sites that may be helpful in your research. For some starting points, go to *www.mcgrawhill.ca/links/MPCA12* and follow the links.

Retrieving Information

Data you find can be static or dynamic. Static data are facts obtained from a medium that cannot be changed; for example, the information presented within this textbook. Dynamic data are facts that are continually changing or are updated regularly; for example, the data found in Statistic Canada's database, which is updated every four years, or a hospital patient's chart on which temperature and blood pressure are updated every six hours. The more current your data are, the better your analysis of the situation and estimates of future trends will be.

Information is found in many different places, such as textbooks, almanacs, newspapers, the Internet, and electronic databases. When retrieving information or data, you need to ensure it is from a reliable source. Obtaining data from a newspaper that obtained its data from Statistics Canada is considered a secondary source. It is preferrable to go directly to the source to ensure no bias is present in the data. Alternatively, you can collect your own data by conducting a survey—this is a primary source.

Information can be represented in the form of graphs, spreadsheets, schedules, or charts. The following examples demonstrate how information is retrieved using each source.

EXAMPLE 1

Retrieving Information From a Graph

The graph represents the ratio of women's to men's incomes, in Canada, at four different times.

a) On average, are women paid less than men are for the same type of work? Explain.

b) Using the data provided in the graph, discuss whether women's and men's incomes are getting closer or further apart.

c) In 1988, if the average men's income was $50 000, what was the average women's income?

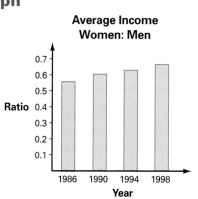

Adapted from: Statistics Canada's Internet Site, http://estat.statcan.ca/cgi~win/CNSMCGI.EXE, table 202-0101, November 7, 2001.

d) Give some reasons that might explain why men and women do not earn the same average income.

Solution

a) The data in this graph were obtained by dividing the average women's income by the average men's income for each year. Any value below 1 indicates women have a lesser average than men. A value of 1 would indicate men and women have the same average income.

For each of the four years shown, the ratio $\dfrac{\text{women's income}}{\text{men's income}}$ is less than 1.

Each year, on average, women are earning less than men. However, the graph shows average incomes. The graph gives no information about types of work, so you cannot tell from this graph whether women are paid less than men for the same type of work.

b) The ratios shown are 0.56, 0.60, 0.63, and 0.66. This indicates that the gap between average income for men and women is getting smaller.

c) The ratio of incomes for women to men in 1988 was 0.59.
If the average man earned $50 000 in 1988, then
$$\frac{\text{women's income}}{\text{men's income}} = 0.59$$
$$\frac{\text{women's income}}{50\ 000} = 0.59$$
$$\text{women's income} = 0.59 \times 50\ 000$$
$$= 29\ 500$$
The average woman's income in 1988 was $29 500.

d) Since the graph is comparing the average income for all men and women, the data are not taking into consideration the proportion of women and men in different occupations. Traditionally, more women have worked in lower-paying occupations, such as retail and hospitality service. A more accurate comparison of women's to men's income would be to analyse the data further and compare by occupation, and then, compare occupations based on education required, physical or mental demands, and stress and responsibilities.

EXAMPLE 2

Retrieving Information From a Spreadsheet

The spreadsheet shows enrolment figures for Ontario colleges in 2000.

	A	B	C	D	E	F	G	H
1		Tuition	Number of Students		Where They Come From (%)			
2	College	($)	Full Time	Part Time	High School	Workforce	Other	
3	Algonquin	2076	10 447	19 636	37	40	23	
4	Boreal	2303	1 332	74	45	55	0	
5	Cambrian	2165	3 992	2 592	33	0	67	
6	Canadore	2103	2 680	1 442	40	55	5	
7	Centennial	2114	10 225	14 734	39	54	7	
8	La Cite	2000	3 528	1 235	36	35	29	
9	Conestoga	1929	4 467	11 771	60	25	15	
10	Confederation	2137	3 197	10 000	41	49	10	
11	Durham	2165	4 700	7 773	60	0	40	
12	Fanshawe	2100	8 854	15 879	48	0	52	
13	George Brown	2100	8 863	40 000	34	66	0	
14	Georgian	2305	5 284	8 776	50	0	50	
15	Grand Lacs	1654	154	11	10	11	79	
16	Humber	2033	11 588	19 055	31	0	69	
17	Kemptville	2024	334	11	81	11	8	
18	Lambton	2240	2 366	1 335	44	56	0	
19	Loyalist	2144	2 887	5 342	47	28	25	
20	Michener Institute	1700	582	5 330	16	8	76	
21	Mohawk	2033	7 426	18 984	no data available			
22	Niagara	1716	5 020	2 466	48	22	30	
23	Northern	2040	1 378	5 105	45	55	0	
24	Ridgetown	2504	342	39	70	20	10	
25	St. Clair	2103	5 373	7 746	no data available			
26	St. Lawrence	2118	4 562	4 000	55	40	5	
27	Sault	1698	2 476	1 793	33	20	47	
28	Seneca	2135	13 428	26 353	49	30	21	
29	Sheridan	2488	10 306	15 029	57	21	22	
30	Sir Sanford Fleming	2186	5 096	2 820	no data available			
31								
32								
33								
34								
35								

Maclean's Guide to Canadian Universities and Colleges 2001.

a) How many students were in full-time attendance at an Ontario college in 2000?

b) What is the average tuition fee for Ontario colleges?

c) Attending a smaller college is important for some students. Reorder the data so that the colleges are listed from smallest to largest, with respect to the number of full-time students. Name the first five colleges in this order.

d) Which college had the greatest number of full-time students? Which had the greatest number of part-time students?

e) What is the range of tuition fees? Discuss why tuition fees for the same program at different colleges can vary by more than $800.

Solution

a) To calculate the number of students in full-time attendance at an Ontario College,
in cell **C31** enter the formula **=SUM(C3:C30)**.
In 2000, 140 787 students were in full-time attendance at an Ontario College.

For more information on spreadsheets, see Appendix B, page 432 (Microsoft®Excel) or page 436 (Quattro ®Pro).

b) To calculate the average tuition fee,
in cell **B31** enter the formula **=AVERAGE(B3:B30)**.
In 2000, the average tuition fee for an Ontario College was $2082.60.

c) Select cells: **A3** to **G30**.
Click on **Data**, **Sort**, select **Ascending**.
Click **OK** .
The five smallest colleges, with respect to the number of full-time students, are Grand Lacs, Kemptville, Ridgetown, Michener Institute, and Boreal.
(Note: Location and college specialization may be factors influencing student enrolment. In this example, Grand Lacs and Boreal are francophone colleges; Kemptville and Ridgetown are agriculture colleges, and Michener Institute is a medical technology institution.)

d) Scanning the data in column C, or by using the **MAX** function of the spreadsheet, it can be seen that Seneca had the greatest number of full-time students, 13 428. The data in column D show that George Brown had the greatest number of part-time students, 40 000.

e) By applying the **MAX** and **MIN** functions of the spreadsheet to column B, or by visual scanning, it is seen that Ridgetown had the greatest tuition fee, $2504. Grand Lacs had the least fee, $1654. The range is $2504 − $1654, or $850. Tuition fees vary considerably. It is necessary to look closely at what exactly is included in the tuition fee. Some colleges include mandatory administrative fees, such as Athletics, Student Council, and Technology, as part of their tuition fee, while other colleges do not. The tuition fee for some programs is higher to cover additional supplies or equipment rental fees.

EXAMPLE 3

Retrieving Information From Transportation Schedules

Gary wants to travel from Toronto to Nanaimo, British Columbia. The map shows the routes he is considering. If he takes the ferry across to Victoria, he will need to make arrangements with his friend Pat to pick him up at the bus depot in Victoria and drive him to Nanaimo, a one-hour drive.

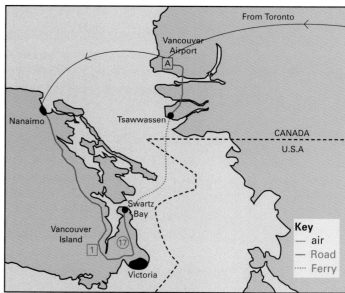

Flights: Toronto (ON) – YYZ to Vancouver (BC) – YVR				
Flight #	Departure	Arrival	Travel Time	Cost ($)*
105	08:45 T2	10:50	5 h 05 min	649.49
3105	09:15 T2	11:20	5 h 05 min	649.49
3981	10:15 T2	12:06	4 h 51 min	649.49
101	11:15 T2	13:06	4 h 51 min	649.49
3983	12:15 T2	14:06	4 h 51 min	649.49
3979	13:15 T2	15:06	4 h 51 min	649.49
3989	14:15 T2	16:06	4 h 51 min	649.49

Flights: Vancouver (BC) – YVR to Nanaimo (BC) – YCD				
Flight #	Departure	Arrival	Travel Time	Cost ($) *
1835	10:50	11:15	0 h 25 min	185.11
1837	12:20	12:45	0 h 25 min	185.11
1839	14:30	14:55	0 h 25 min	185.11
8350	16:05	16:30	0 h 25 min	207.56
1843	18:30	18:55	0 h 25 min	185.11

Ground Transportation: Bus to Tsawwassen Ferry, Ferry to Swartz Bay, and Bus to Victoria Bus Depot		
Departure Vancouver Airport	Arrival Victoria Bus Depot	Cost ($) *
11:45	15:20	32.50
13:45	17:20	32.50
15:45	19:20	32.50
17:45	21:20	32.50

* Prices include all applicable taxes.

a) Study the schedules shown to determine the two possible travel itineraries: one by air for both parts of the journey, and one by air to Vancouver and then, ground transportation. Gary would like to arrive in Nanaimo by 6:00 p.m., with minimum delays. Explain your reasoning for selecting the schedule of each itinerary.

b) Calculate the total travelling time for each itinerary.

c) Calculate the cost for each itinerary.

d) Explain why the arrival time in Vancouver makes it appear as though Toronto to Vancouver is a short flight.

Solution

a) Itinerary 1: A possible travel itinerary for flying from Toronto to Vancouver, and then, to Nanaimo is presented in the chart.

Flight #	From	Departure	To	Arrival	Cost ($)
3979	Toronto	13:15 T2	Vancouver	15:06	649.49
8350	Vancouver	16:05	Nanaimo	16:30	185.11

This itinerary will allow Gary to arrive in Nanaimo by 4:30 p.m., with time to spare. Almost an hour is allowed for connecting time.

Itinerary 2: A possible travel itinerary for flying from Toronto to Vancouver, and then, taking the ground transportation (bus, ferry, bus) is presented in the chart.

Flight #	From	Departure	To	Arrival	Cost ($)
105	Toronto	08:45 T2	Vancouver	10:50	649.49

From	Departure	To	Arrival	Cost ($)
Vancouver Airport	11:45	Victoria Bus Terminal	15:20	32.50

This route takes longer, but is less expensive. Sufficient time (55 min) is allowed for Gary to pick up his baggage and walk to the bus loading area.

b) Total travelling time:
Itinerary 1: 4 h 51 min + 25 min = 5 h 16 min
Itinerary 2: 5 h 05 min + 3 h 35 min + 1 h (drive to Nanaimo) = 9 h 45 min

c) Cost:
Itinerary 1: $649.49 + $185.11 = $834.60
Itinerary 2: $649.49 + $32.50 = $681.99

d) The arrival time in Vancouver is Pacific Standard Time. Vancouver is three hours behind Eastern Standard Time. For example, when it is noon in Toronto, it is 9:00 a.m. in Vancouver.

EXAMPLE 4

Retrieving Information From a Chart or Table

Data are frequently presented in charts. You can easily obtain charts from newspapers, the Internet, or E-STAT. The following chart was obtained from E-STAT. To obtain this data yourself, follow the path below.

Area Name:	Population percentage change, 1991-1996
Alberta	5.9
British Columbia	13.5
Manitoba	2
New Brunswick	2
Newfoundland	-2.9
Northwest Territories	11.7
Nova Scotia	1
Ontario	6.6
Prince Edward Island	3.7
Quebec	3.5
Saskatchewan	0.1
Yukon Territory	10.7

Source: Statistics Canada's Internet Site, http://estat.statcan.ca/cgi-win/CNSMCGI.EXE, Population Percent Change 1991-1996, by Provinces and Territories, November 7, 2001.

► DATA ◄

People
   Population and demography
      Census Databases
         Population Characteristics
            1996 Census of Population all tables
            Geography
               1996-Canada, Provinces & Territories
            Characteristics
               Population Percent Change 1991–1996
            Table

a) Generally, how has the population changed from 1991 to 1996?

b) The change for the province of Newfoundland is −2.9%. Explain what this means and give a possible explanation for this change in population.

Solution

a) Generally, the population increased from 1991 to 1996; although, the rate of growth varied from province to province.

b) The change for the province of Newfoundland, −2.9%, indicates that the population was decreasing. The main factor creating a decrease in population is a slow economy. People in Newfoundland left to look for employment in other provinces, resulting in a decrease in population.

KEY CONCEPTS

- Data can be retrieved from electronic databases (such as E-STAT), reference books, newspapers, and the Internet. Sources of data can be static or dynamic.

- Data is more reliable if it comes from a primary source as opposed to a secondary source. The more current (dynamic) the data are, the more up to date the information is.

- Information is presented in many forms, such as charts or spreadsheets, graphs, tables, and schedules.

- Information retrieval can involve reading values presented in graphs or charts, and making observations and conclusions based on these values. Information can also be obtained by combining sources or by performing calculations based on the information presented.

DISCUSS THE CONCEPTS

1. Describe sources you can use to retrieve the following information.
   a) population of Canada
   b) current hockey statistics
   c) bus schedule for your area

2. Data are often presented in the form of a table or a graph. In this section, you studied tables involving airline and bus schedules and a graph comparing men's and women's average incomes. Describe other types of data that can be found in tables and graphs.

3. Give two examples of situations in which you could obtain data from a primary source. Give two examples of situations in which you would have to obtain data from secondary sources.

APPLY THE CONCEPTS

A 1. For each source of information, identify whether it is primary or secondary, as well as dynamic or static.
   a) a newspaper report on economic growth in Canada
   b) population by age data tabulated by Statistics Canada
   c) a survey conducted by your group
   d) a table from a textbook showing the impact of forest fires on the area of Canadian forests

2. Study the graph, which shows the number of students enrolled at colleges in Ontario.

   a) Over the nine-year period, how has full-time enrolment in Ontario Colleges changed?

   b) Estimating the number of students at colleges is vital in the planning of accommodation and courses. How could the enrolment pattern over the next five years be predicted using this graph?

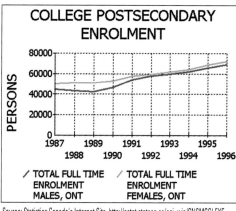

Source: Statistics Canada's Internet Site, http://estat.statcan.ca/cgi~win/CNSMCGI.EXE, table 477-0006, November 7, 2001.

B 3. The graph compares crimes of violence and property crimes committed by young offenders in Canada.

   a) Write a paragraph, describing the information presented in the graph.

   b) To determine whether Ontario youths are committing less or more crimes than in the past, what other information is needed?

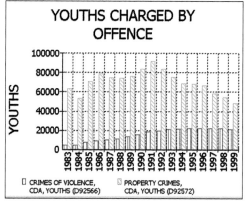

Source: Statistics Canada's Internet Site, http://estat.statcan.ca/cgi~win/CNSMCGI.EXE, table 252-0002, November 7, 2001.

4. This graph, showing gasoline prices, appeared in the *Toronto Star* newspaper on May 26, 2001.

   a) What makes this graph more interesting than other traditional graphs?

   b) What information would a reader retrieve from this graph?

   c) Study the scale of each axis. Describe the units of measure and any other scale techniques used.

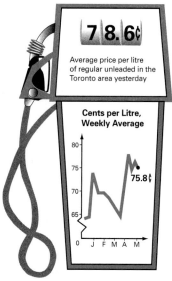

Toronto Star Staff, Ministry of Energy, Science and Technology

5. **Communication** Study these graphs.

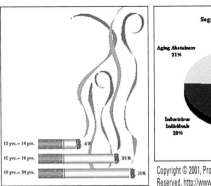

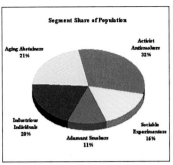

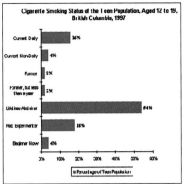

Graphic artists try to create visually appealing graphs to attract the reader's attention. Describe the visual techniques used in each graph.

6. **Application** The top ten movies, by gross revenue up to the end of 1999, are listed.

	Movie	Year	**Gross Revenue in U.S. Dollars Since Opening**
1	Titanic	1997	600 787 052
2	Star Wars	1977	460 935 655
3	Star Wars: The Phantom Menace	1999	431 065 444
4	E.T.	1982	399 804 539
5	Jurassic Park	1993	356 839 725
6	Forrest Gump	1994	329 690 974
7	The Lion King	1994	312 855 561
8	Return of the Jedi	1983	309 125 409
9	Independence Day	1996	306 052 958
10	The Sixth Sense	1999	293 476 912

Work with a partner to answer the following.

a) Is there any relationship between the year the movie was produced and the gross revenue?

b) Classify the type of each movie, such as comedy, science fiction, adventure, or children's movie.

c) Based on the information presented in the chart, what type of movies seems to appeal to more people?

d) From the information presented in the chart, can you determine which movie was seen by the most people? What other indicator could be used to determine the movie seen by most people?

e) Go to *www.mcgrawhill.ca/links/MPCA12* and follow the links to obtain data on the American film industry's top 50 movies. Repeat parts b) and c) for these movies.

7. **Inquiry/Problem Solving** Study the train schedules shown.

TORONTO • LONDON • SARNIA/WINDSOR									
TRAIN	**km**	**71**	**73**	**75**	**83**	**683**	**79**	**47**	**67**
DAYS/JOURS				x6,7	6,7			x6	x6
Toronto	0	0750	1200	1600	1715	1740	1845	2055	2115
Oakville	34	0811	1222	1623			1907	2115	2136
Aldershot	56	0825	1235		1746		1920		
Brantford	97	0853	1305	1708	1815		1950		
Woodstock	140	0918	1334		1842		2017		
Ingersoll	154	0928					2028		
London	185	0955	1405	1805	1913	2040	2055		
Strathroy					1934	2100			
Watford					1948	2114			
Wyoming					2001	2127			
Sarnia					2019	2145			
Glencoe	235	1022					2124		
Chatham	290	1054	1506	1909			2157		
Windsor	359	1134	1546	1956			2238		

SARNIA/WINDSOR • LONDON • TORONTO										
TRAIN	**km**	**40**	**640**	**82**	**682**	**70**	**670**	**72**	**76**	**78**
DAYS/JOURS		x6,7	6	x6,7	6,7	x7	7			
Windsor	0					0600	0625	0955	1405	1730
Chatham	69					0650	0715	1037	1447	1812
Glencoe	124						0748	1109		1845
Sarnia				0525	0610					
Wyoming				0545	0630					
Watford				0558	0643					
Strathroy				0612	0657					
London	174			0635	0720	0758	0823	1141	1548	1918
Ingersoll	205					0819				1937
Woodstock	219			0701		0832	0855	1208		1949
Brantford	262			0728		0859	0923	1235	1648	2016
Aldershot	303					0928	0953	1302	1705	2043
Oakville	325	0615	0710			0943	1008	1316	1718	2057
Toronto	359	0635	0730	0825	1010	1004	1030	1337	1738	2118

Key: x6,7 means this train does not run on Saturday or Sunday   Via Rail Canada

a) If a passenger boards the noon train from Toronto to Windsor, what time should she arrive in Windsor?

b) For a one-day trip from Toronto to Windsor, what is the maximum length of time you could have in Windsor?

c) A Brantford resident would like to travel to Sarnia. At what times can he expect a train leaving Brantford for Sarnia?

d) Is it possible to travel from Woodstock and arrive in Oakville by 9:00 a.m.?

e) How far is it from Chatham to Glencoe?

f) Make a challenging question using this table and share strategies for solving it with your group.

8. Select an out-of-town college (possibly the one you plan to attend). Plan your first trip home for the holidays. Assume you will be using public transportation. Include an itinerary, including times and locations of bus and/or train stations, as well as the cost for the trip.

9. The average monthly precipitation is recorded, in millimetres, at many weather stations across the country. The data for Kenora and Sault Ste. Marie are shown for four months.

	January	April	July	October
Kenora	27.7	36.7	89.9	46.2
Sault Ste. Marie	74.4	65.2	65.6	83.2

a) Which of these two places has the most precipitation, based on the four months shown?

b) From the data, can you tell which is the wettest month of the year in Kenora?

 c) Which of these places do you think has the greater annual precipitation? Go to *www.mcgrawhill.ca/links/MPCA12* and follow the links to obtain data that will check your prediction.

10. The Trans-Canada Trail, started in 1992, is a project to make a hiking or biking path that forms a continuous trail through all provinces and territories. The table shows the progress by the end of 2001.

TRANS-CANADA TRAIL PROGRESS			
Province/Territory	**Objective (km)**	**Total Registered (km)**	**Percent Complete**
Newfoundland	900	885	
Nova Scotia	700	538	
New Brunswick	820	630	
Prince Edward Island	350	350	
Quebec	1400	1069	
Ontario	3400	1896	
Manitoba	1188	1091	
Saskatchewan	1400	489	
Alberta	2200	867	
British Columbia	2700	1029	
Northwest Territories	586	586	
Yukon	1600	653	
Nunavat	to be determined		

a) Which provinces have reached more than 75% of their objective?

b) Which provinces have completed less than half of their objective?

c) Describe how you would calculate the percent of the full length of the trail that was completed in 2001.

d) Go to *www.mcgrawhill.ca/links/MPCA12* and follow the links to obtain current data on the progress of the trail. Make a list of the provinces and territories in descending order by the percent of their trail objective completed.

11. The graph represents the number of wrist injuries at a snowboard park north of Toronto.

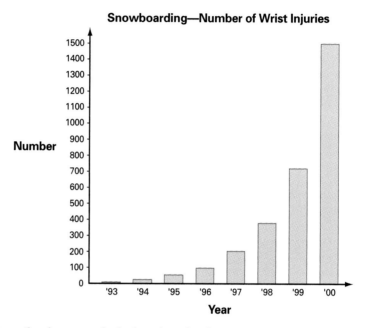

a) Describe the type of relation that the data seem to fit.

b) There appears to be an increase in wrist injuries since 1997. With a partner, discuss why this has happened.

c) What other information is needed to determine whether there will be more wrist injuries in snowboarding this year than in the past?

Discover Information From Indices

The department of labour monitors the change in prices for goods and services in Canadian cities. Data are obtained by comparing how the average prices of products and services change over time. Prices fluctuate due to political and economical activities. The graph shows the percent change in the Consumer Price Index (CPI) for a five-year period.

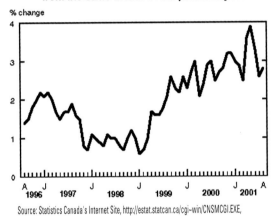

Percentage change in the Consumer Price Index from the same month of the previous year

Source: Statistics Canada's Internet Site, http://estat.statcan.ca/cgi~win/CNSMCGI.EXE, table 326-0002, November 7, 2001.

1. During what one-year period did the CPI increase most rapidly?

2. During what one-year period did the CPI remain relatively stable, with moderate growth?

The CPI is an effective tool used to measure inflation (increase in prices) or deflation (decrease in prices). The CPI might affect you in the following areas: rental increases, cost of living pay raises, and amounts paid out by government-sponsored social programs.

The CPI is a measure of relative price levels. The costs of about 600 items, representing typical household expenses including food, shelter, clothing, transportation, and recreation, are recorded. Prices in 1992 are used as the base for comparison; the price of an item in 1992 is deemed to be 100. For example, the CPI for September 2000 listed these two fruits.

Apples: 118.0   Oranges: 116.2

From these numbers, we can tell apples increased in price by 18.0% and oranges increased by 16.2% since 1992.

3. Study the CPI graph obtained from Statistics Canada.

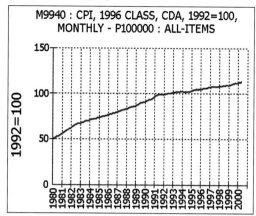

M9940 : CPI, 1996 CLASS, CDA, 1992=100, MONTHLY - P100000 : ALL-ITEMS

Source: Statistics Canada's Internet Site, http://estat.statcan.ca/cgi~win/CNSMCGI.EXE, table 326-0002, November 7, 2001.

a) Along the vertical axis you see 1992 = 100. Describe how to use this statement to interpret the data.

b) What was the percent increase in the CPI from 1980 to 2000?

c) Is the following statement true or false? Explain why or why not.
If 1 L of milk cost $1.00 in 1992, then the graph tells us that 1 L of milk cost 50¢ in 1980.

4. If the average of all the items in the CPI for September 2000 was 114.0 for Newfoundland, 114.9 for Ontario, and 114.3 for British Columbia, what might you conclude from this information?

5. The CPI for all items for January of each year is shown in the table. The base for comparison is 1992 = 100.

a) Use a graphing calculator to plot this data and to obtain the equation of the line of best fit.

b) Use your equation to predict the CPI for 2001.

c) Use Statistics Canada's Internet site to check your prediction.

Year	CPI
1980	49.8
1981	55.8
1982	62.1
1983	67.3
1984	70.9
1985	73.5
1986	79.7
1987	79.7
1988	83.0
1989	86.6
1990	91.3
1991	97.6
1993	101.2
1994	102.5
1995	103.1
1996	104.8
1997	107.0
1998	108.2
1999	108.9
2000	111.4

Adapted from: Statistics Canada's Internet Site, http://estat.statcan.ca/cgi~win/CNSMCGI.EXE, Statscan CPI for January by year 1980-2000, November 7, 2001.

6. You can use E-STAT to find lists of actual products and their prices. Over 600 products ranging from dairy products to spark plugs are listed.

   a) Select a food item or product and follow the path to study the price changes.

   ▶ Data ◀
   Economy
     Prices and price indexes
       Prices
         [9935] Average retail prices for food and other selected items
           select the product or service you are looking for

   b) Prepare a summary of how the product you selected compares with the same product in another province. Include a graph and table.

Remember, the CPI is not a cost of living index. Calculating the change in your cost of living involves more items, and factors such as where you live will influence it. Many people rely only on the CPI to measure price changes. Other indices are available and should be used to obtain an overall picture. The chart below highlights some of the other indices used to determine inflation or deflation, and includes a brief description of each index.

Industrial Product Price Index	Measures changes in prices received by Canadian manufacturers for goods as they leave the factory.
Raw Materials Price Index	Measures price changes for the purchase of raw material by Canadian industry.
New Housing Price Index	Measures changes in the builders' selling prices of new residential houses.
Farm Product Price Index	Measures the change in prices received for agricultural commodities at the first transaction point.

7. **Problem Solving, Communication** Choose one index other than the CPI and research to find out more about it. Prepare a graph to show how the index changed from 1980 to 2000. Compare your graph to the first graph in this section. Do the percent changes follow similar trends? If not, try to explain any differences.

Representing Data in a Variety of Ways

The data you collect can be represented in many ways. To analyse data, you may find it helpful to represent it in any of the following forms:
- an equation
- a table
- a graph (bar chart, histogram, circle graph or pie chart, broken-line graph, scatter plot)

Many tools or application packages are available to help you display data in various ways. The tool you use will depend on the availability of that tool, as well as the relevance of using that tool for the particular data. Try to choose the most suitable method of representing the data. The technology you use should also be the most appropriate, but as in real-life applications, you may not always have access to the technology of your choice and may have to use what is available.

Deciding which method to use to present your data will depend on the type of data you are working with. In the following Discover, you will investigate a variety of ways to represent data related to the number of visits, or hits, to an Internet page.

DISCOVER

Choosing Ways of Presenting Data

The number of hits to a web page is recorded in the table.

Week	1	2	3	4	5	6	7	8	9
Hits	50	75	155	315	640	1200	2475	4821	9864

For help with the statistical functions of the graphing calculator see page 429.

1. Identify the independent and dependent variables.
2. Using a graphing calculator, make a scatter plot of this data.
3. Identify the growth pattern: linear, quadratic, exponential, or other.
4. Describe the advantages of displaying data in a scatter plot.
5. Using the most appropriate form of regression, identify the equation to represent this data.
6. In the first 10 weeks, what is the potential total number of hits for this web page?

7. What other types of tools and graphs would be appropriate to represent this data and to predict the future trend? Explain why you selected the type of graph you did.

8. Experiment with other tools to represent this data.

Before data are represented in the form of a graph, they are typically collected and stored in a table. Statistics Canada's database stores all its information in tables. In some reports, tables appear in the appendix (at the end of the report), and graphs generated from the tables are inserted in the actual report. This is particularly useful when someone needs to extract an exact number that is not easily visible in the graph.

Using technology to represent your data will allow you to move from one format to another with ease. The following examples demonstrate how easy it can be to represent data in a variety of ways.

EXAMPLE 1

Equation to Table of Values Using a Graphing Calculator

The cost of renting a moving van is given by $C = 49.95 + 25.95t$, where C is the cost, in dollars, and t is the time the van is kept for, in hours.

a) Enter the equation in a graphing calculator.

b) Display a table of values for times from 6 h to 12 h.

Solution

For help with graphing see Appendix B, page 425.

a) Enter the equation $y = 49.95 + 25.95x$.

Choose appropriate WINDOW settings.

b) To obtain a table of values for $x = 6$ to $x = 12$ in increments of 1, press [2nd] [WINDOW] to access **TBLSET**.

Press [2nd] [GRAPH] to access **TABLE**.

 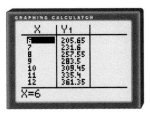

EXAMPLE 2

Table of Values to Graph Using E-STAT

When consumers experience steep gasoline price increases, they tend to be more concerned about the fuel efficiency of their vehicles. Reviewing gasoline prices over the past 20 years, you will see that the price has increased significantly.

a) Use E-STAT to obtain a table showing the average retail price for unleaded gasoline at a self-service filling station in Ottawa.

b) By how much did the price increase from 1990 to 2000?

c) A company pays their employees an amount per kilometre when it is necessary for them to travel, for business purposes, using their own vehicles. The rate per kilometre has not been adjusted since 1990. Display a table and graph showing how gasoline prices changed over the next 10 years.

d) What other information could you use to convince the company to increase the rate per kilometre?

Solution

a) Follow the path to access the data using E-STAT.

▶ Data ◀

Economy
  Prices and price indexes
    Data
      Cansim II
        Prices
          [326 0009] Average retail prices for gasoline at self-serve filling stations
            Ottawa
            Regular unleaded gasoline at self-serve filling stations
            From: Jan 1990   To: Jan 2001

Click: Continue
Click: Table
Click: Continue
  HTML Table, time in rows
Click: Go

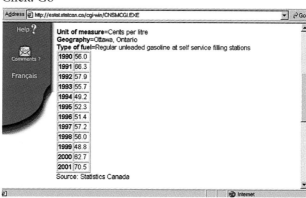

Source: Statistics Canada's Internet Site, http://estat.statcan.ca/cgi-win/CNSMCGI.EXE, table 326-0009, November 7, 2001.

b) To calculate the increase, subtract the 1990 price in the table from the 2000 price.

62.7 − 56.0 = 6.7

The gasoline price increased by 6.7¢ from January 1990 to January 2000.

c) To display the data in a graph, go back three windows.

Then, select **Time series** graph and move forward as in part a).

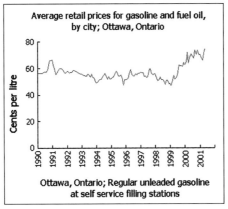

Source: Statistics Canada's Internet Site,
http://estat.statcan.ca/cgi~win/CNSMCGI.EXE, table 326-0002, November 7, 2001.

d) Print your graph. With a pencil, draw a line of best fit. Extend the line beyond 2000 to extrapolate the cost of gasoline in the future. The main concept is that gasoline prices will continue to increase. There will be some increases or decreases, but overall, the linear pattern indicates an increase in gasoline prices. This graph could be used to convince the company that the rate per kilometre should be reviewed regularly.

A computer spreadsheet is another tool that you can use to represent data. Data can easily be entered, manipulated, and represented in many forms. The next example demonstrates the process involved in creating a bar graph.

EXAMPLE 3

Table of Values to Graph Using Spreadsheet Software
The investment period and interest rates for Canada Savings Bonds are shown in the table.

Term (years)	0.25	0.5	1	2	3	4	5	7	10	20	30
Yield (%/year)	4.4	4.4	4.6	5.0	5.1	5.4	5.6	5.9	5.9	6.1	6.1

Present this data as a bar graph so that readers can visualize changes more easily.

For more
information on
graphing data in
spreadsheets,
see page 432
(Microsoft®
Excel) or page 436
(Quattro® Pro).

Solution

Follow these steps. (These steps are for Microsoft® Excel.
You may need to modify the steps for another spreadsheet.)

1. Enter the data into a spreadsheet, as shown.
2. Select the right column of data by clicking and dragging the mouse.
3. Select the **Chart Wizard** and accept the default histogram.
4. Select **Next**.
5. Select **Series** (tab near the top left of form).
6. Click the symbol for Category (**X**) axis labels—this will take you
    back to the spreadsheet.
7. Select the first column, not including the header.
8. Click the symbol to return to the spreadsheet—this will use the first
    column (the term) as the *x*-axis.
9. Select **Next**.
10. Enter the title: Canada Bond Rates June 2001.
11. Enter the Category (**X**) axis: Term (in years).
12. Enter the Value (**Y**) axis: Rate (percent).
13. Remove the legend by clicking the legend tab, and deselect the show
    legend box.
14. Select **Next**.
15. The next window asks where to Place Chart. Select **as object in sheet 1**.
16. Select **Finish**.

The resulting bar graph is shown.

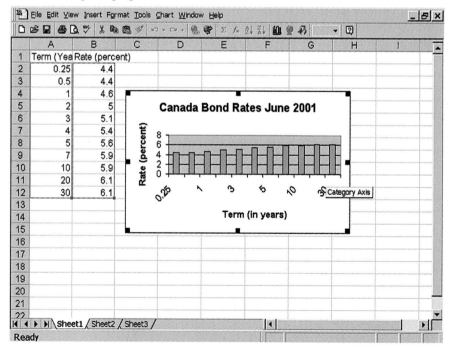

KEY CONCEPTS

- Data can be represented in many forms, such as an equation, a table, and a variety of graphs.

- The tools you can use include a graphing calculator, E-STAT, spreadsheet software, Fathom™, or any other graphing package you have access to.

- The method in which the data are represented will depend on the type of data, as well as the objective of the data display.

DISCUSS THE CONCEPTS

1. For each method of representing data, give an example where it would be suitable to use that method.

   a) a table   b) an equation

   c) a bar graph   d) a scatter plot

2. Describe how to represent the same data in the form of an equation, table, or graph, using a graphing calculator.

3. Discuss the advantages and disadvantages of using technology to represent data.

APPLY THE CONCEPTS

Σ-STAT

B 1. Using E-STAT, prepare a presentation on population growth in Ontario. Use the best format to indicate how the population is changing and predict the population for the next 10 and 20 years. This path will lead you to the data.

   ▶ Data ◀

    People

      Population and demography

        Cansim II

          Population characteristics

            [051-0001] Estimates of Population

          Ontario

Source: Statistics Canada's Internet Site, http://estat.statcan.ca/cgi~win/CNSMCGI.EXE, Instructions lead to E-STAT 051-0001, population data, November 7, 2001.

For help with linear regression, see page 431 (graphing calculator) or page 441 (Fathom™).

2. Using E-STAT, locate the population of your city or area over the past 50 years.

   a) Enter the data in your graphing calculator or import it into Fathom.

   b) Using linear regression, obtain the line of best fit equation to represent the growth of your area.

c) Using the equation you obtained, find the predicted population in the next 10, 20, and 30 years.

d) Discuss why it is important for urban planners to be able to predict the population growth for communities.

3. Using E-STAT, prepare a graph comparing the populations in three towns or cities in your area with the total population of the province. Include labels and a title.

4. Survey your classmates to determine the number of days they were absent from school last year. Record your data in a table similar to the one started.

Number of Days Absent	Number of Students
1–4	
5–8	
9–12	

a) Prepare a histogram to represent your data.

b) Give a brief analysis of the data presented.

5. Use the data from question 4, and add a third column to show the students' average marks.

Number of Days Absent	Number of Students	Average (for all students in this category)
1–4		
5–8		
9–12		

a) Review the data. What patterns, if any, exist? Prepare a report on the relationship between the number of days a student is absent and average mark. Include a graph and/or table.

b) Extend your report to include students from different grades. Analyse the data to determine if there is any relationship between days absent, grade, and subject.

6. Access Statistics Canada's database and complete the following.

a) Select one health-related issue, such as physical fitness, stress, diabetes. Prepare an advertisement to provide information about your topic. Include the following:
   • two formats (graph, equation, or chart)
   • a title to capture the reader's attention

b) In a small group, discuss which of the two formats is better.

7. The value of the Canadian dollar compared to the U.S. dollar fluctuates daily and is dependent in part on the Canadian and American economies. Use E-STAT or other Internet sites to research the value of the Canadian dollar compared to the U.S. dollar over the past 5 to 10 years. Prepare a graph showing the data.

8. Communication The table shows how the average weekly television viewing time for Canadians has changed.

Year	Time (h)
1991	23.3
1992	23.0
1993	22.8
1994	22.7
1995	23.2
1996	22.8
1997	22.7
1998	22.3
1999	21.6
2000	21.5

Source: Statistics Canada's Internet Site, http://estat.statcan.ca/cgi-win/CNSMCGI.EXE, Average weekly TV viewing time Canada, 1991-2000, from The Daily, October 23, 2001.

a) Prepare a graph to show the change in average weekly viewing hours over time.

b) Describe any patterns you see in the data. Suggest possible factors that may influence the pattern.

c) In a small group, discuss how the data might be different if you were looking at a segment of the population. Do you think males watch more television than females or the reverse? Which age groups do you think watch more than the national average? Which province do you think has the greatest weekly average? The least?

d) Refer to page 456 in the **Data Bank**. Does the data confirm your answers to the previous question? Do teens in Ontario watch more than the Canadian average for their age group? How do young males (18–24) compare to the national average for that age group?

9. Application

a) Obtain data for the population of Canada from 1871 to 2001, in increments of 10 years.

b) What is the best type of format to represent this data?

c) Identify the type of growth.

d) Present this data in the most suitable format.

10. **Application** Use the **Data Bank**, page 455, to obtain data on the number of secondary school graduates, each year from 1995 to 1999.
    a) Draw a graph to show the number of secondary school graduates by province in 1999.
    b) Draw a graph to show how the number of secondary school graduates in Ontario has changed over this time period.

11. **Communication** When displaying data in the form of an equation derived from a scatter plot, it is important to indicate the correlation value. Discuss the significance of including the correlation value.

12. **Whales** Northern Right Whales are an endangered species. The table gives the total number of Northern Right Whales and the total number of calves born each year.

Year	Total Number of Northern Right Whales	Total Number of Calves Born
1991	293	17
1992	301	12
1993	309	8
1994	307	8
1995	295	7
1996	299	21
1997	317	19
1998	307	5
1999	303	4
2000	299	1

See the Technology Appendix for help with graphing data using a graphing calculator (page 430), in Microsoft® Excel (page 432), in Quattro® Pro (page 436), or in Fathom™ (page 440).

   a) Prepare a graphic representation of this data, to show how the population of Northern Right Whales is changing. Predict the future number of the species, if the established pattern continues.
   b) Update your report by using the Internet to research more recent data.

13. **Communication** Obtain data on numbers of an endangered species that interests you. Prepare appealing visuals that you could use if you were asked to make a presentation about the species to a younger audience.

14. Thinking/Inquiry/Problem Solving A table and graph highlighting how mortgage payments are allocated at the end of each year for the first 10 years is shown. The initial mortgage was $175 000 at 7% per year, compounded semiannually, with monthly payments of $1236.86.

Year	Principal ($)	Interest ($)	Balance ($)
1	2 677.17	12 165.19	172 322.83
2	5 547.88	24 136.85	169 452.12
3	8 626.11	35 900.98	166 373.89
4	11 926.87	47 442.59	163 073.13
5	15 466.23	58 745.58	159 330.77
6	19 261.46	69 792.72	155 738.54
7	23 331.05	80 565.49	151 668.95
8	27 694.82	91 044.08	147 305.18
9	32 374.06	101 207.21	142 625.94
10	37 391.56	111 032.08	137 608.44

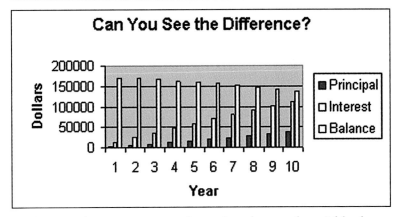

a) Prepare three separate graphs to show how each variable changes over the term of the mortgage.

b) Describe the type of growth for each of the three variables.

c) What if the monthly mortgage payments were increased by $100? Would the patterns you described in part b) change? If so, how? Would the graphs in part a) change? If so, how?

Validity and Bias

If you look carefully, you will often see statistics used in a misleading manner by the media. Sometimes the scales are distorted; sometimes the labels are misleading. In other cases, the samples used to obtain data are not truly representative of the whole population. People need to be critical and analytical when reading statistics. Consider the scenario in this Discover.

DISCOVER

Misleading Graphs

Profits for a newly established small company are lower than projected. However, management has reason to believe things will turn around in the next quarter. Not wanting to alarm their shareholders, they are considering presenting their quarterly report using one of the following graphs.

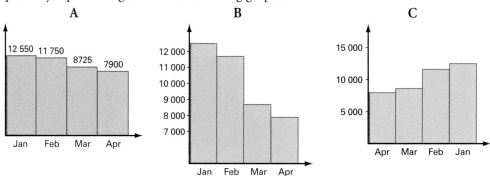

1. The data are shown in the table.

Month	Profit ($)
January	12 550
February	11 750
March	8 725
April	7 900

   Discuss how each graph has been drawn to present a misleading picture of the data.

2. Make a broken-line graph to represent the data. Compare your graph to the three graphs above.

3. Which graph would you recommend the company use in its report? Why?

Another thing that can be confusing about data analysis is the terms that are used to describe results. For instance, the results of tests such as course admission tests sometimes rank your score in terms of a **percentile**.

The words *percent* and *percentile* mean different things. Percentile is a term used in educational and health-related fields to indicate the position of an individual in a group, while percent is simply a ratio out of 100.

For example, if you scored 116 out of 150, your percent is $\frac{116}{150} \times 100$, or 77.3%.

To determine your percentile rank, this score needs to be compared with the scores of the other students who wrote the same test. If you scored better than three quarters of the class, then you scored in the 75th percentile.

The percentile in which a score of X falls is calculated using the following formula:

$$\text{percentile} = \frac{\text{number of scores below } X + 0.5 \times \text{number of scores equal to } X}{\text{total number of scores}} \times 100$$

EXAMPLE 1

Finding Percentiles

A test has a total score out of 150. The results achieved by a class of 17 students are listed.

112, 98, 72, 77, 86, 116, 59, 103, 116, 132, 146, 125, 119, 108, 96, 122, 134

Calculate the percentile rank of a score of 116. Round to the nearest unit.

Solution

See Appendix B, page 427 for information on entering and sorting data with a graphing calculator.

First, arrange the scores in ascending order.

59, 72, 77, 86, 96, 98, 103, 108, 112, 116, 116, 119, 122, 125, 132, 134, 146

Then, use the formula.

$$\text{percentile} = \frac{\text{number of scores below } X + 0.5 \times \text{number of scores equal to } X}{\text{total number of scores}} \times 100$$

$X = 116$

number of values below $X = 9$

total number of values $= 17$

$$\text{percentile} = \frac{9 + 0.5 \times 2}{17} \times 100$$

$$\doteq 59$$

A score of 116 for this class is in the 59th percentile. This is just above the median score of 112.

EXAMPLE 2

Identifying Sample Bias

A representative from a research department of a large store went to an up-scale mall and surveyed shoppers to determine how much they spent on Valentine's Day gifts. After spending the entire day at this mall, the surveyor concluded that, on average, Canadian men spend $75 and women spend $92.

a) What is wrong or missing in this survey?

b) Suggest a more accurate way of conducting this survey.

Solution

a) The researcher surveyed people in an up-scale mall. The people at the mall probably do not reflect all shoppers in the Canadian population, as suggested by the spending amounts in the conclusion.

b) The outcome of the survey could have read "shoppers in this mall spend ..." Alternatively, the representative should increase the sample to include a broader range of the Canadian population by visiting several shopping malls.

Consumers are continually bombarded with statistics supporting various products or services. At election time, the news often quotes statistics about polls, giving people's voting choice.

45% Party A
38.2% Party B
16.8% Undecided

These polls are often biased! We tend to believe what we see or hear when the data appear to be statistically based. The questions we need to ask ourselves are "Who conducted the survey?" and "How was the survey conducted? When? Where?" In statistical terms, "Are these statistics biased?" and "How valid are the data?"

You have probably heard or read statements such as the following in advertisements.
- 8 out of 10 patients had immediate success with…
- 9 out 10 dentists recommend…
- 70% success rate

Where do these numbers come from? Who is sponsoring these surveys? How was the sample obtained? How many people were in the sample? Ask yourself these questions before making any decisions based on what appear to be the facts!

EXAMPLE 3

Assessing the Validity of Data

Assess the validity of the data presented in each statistical expression.

a) 8 out of 10 patients experienced relief in 20 min.

b) 81% on a test is an outstanding score.

c) During a federal election campaign, a report in a newspaper included the following report based on a survey of readers.

If a federal election were held tomorrow, which party would you vote for?

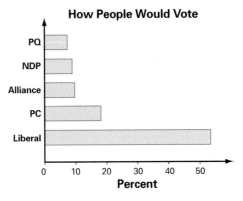

How People Would Vote

d) A manufacturer is thinking of changing the packaging for its product. A market research company asks people, at random, in a supermarket, "Do you prefer this new colourful package over the old one?"

Solution

a) The expression is not valid because
- the size of the sample group is not defined (Were 10 people involved? 20? 100?)
- the characteristics of the sample group are not defined (How many males and females? The drug may affect males and females differently. Was the weight of the individuals taken into consideration?)
- the term relief is not very descriptive (Does relief mean they felt a little better or moderately better or perfect?)

b) You cannot tell whether 81% was an outstanding score. Perhaps all other scores on the test were greater than 81%. The percentile is needed to clearly identify whether 81% is an outstanding score.

c) The sample is not well defined. Readers of one particular newspaper sometimes support or prefer one political party over another. It is important to identify any political bias that the newspaper might have, or to choose a random sample of all Canadian voters.

d) The question is somewhat biased. By describing the "new colourful package," it tends to sound better than the old one. An unbiased question would be "Which package do you prefer?"

 To review many interesting articles on the misuse of statistics, especially in the media, go to *www.mcgrawhill.ca/links/MPCA12* and follow the links.

KEY CONCEPTS

- Bias can occur when
  - **i)** the sample is too small
  - **ii)** the sample is not reflective of the whole population
  - **iii)** the survey questions elicit a desired response

- Data are misused in graphs when there is some type of distortion or poor labelling.

- Percent indicates how well an individual scored. Percentile is used to rank an individual score, compared to other scores in a group.

DISCUSS THE CONCEPTS

1. Give two examples where information is deliberately misrepresented to give a false impression.

2. Explain the difference between percent and percentile.

3. Describe how sample bias can occur.

4. The results of polls of voters for an upcoming election as reported in a national newspaper and in a local city newspaper may differ. Explain how this may occur.

A **1.** Select a brand of shampoo.

   a) Design two misleading questions you might ask of people in a store near the shampoo products.

   b) Trade questions with a partner and discuss why these questions are misleading.

2. During the Toronto Jazz Festival, people participating in the event were surveyed to determine their favourite type of music. The surveyor concluded that Canadians overwhelmingly prefer jazz music to any other music. Discuss the bias and describe the changes required to validate the statement.

3. In preparing a survey to compare two drinks, a question is drafted in three possible ways. Which question is most misleading? Explain why.

   A Drink 1 is very tasty. Don't you agree?

   B How would you describe the aftertaste of drink 3?

   C Which drink do you prefer?

4. In a survey conducted in a health food store, it was found that 97% prefer organic vegetables. Comment on the validity of the statement: "Organic vegetable farmers are enjoying huge successes since 97% of the people surveyed prefer their products."

5. Lotto 6/49 results are available from Western Canada Lottery Corporation. The following graphs show the number of times each of the first ten numbers were drawn during one year. Which graph presents the data more realistically? Why?

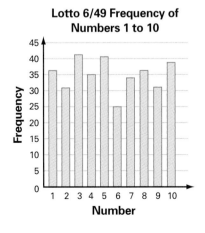

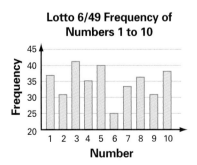

Data from The Western Canadian Lottery Corporation

6. Study the two graphs shown. Describe how each is misleading.

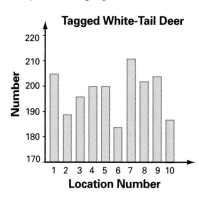

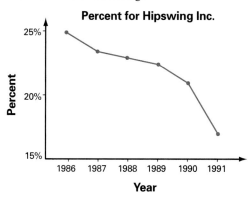

B 7. **Communication** For a one-week period, keep a list of misleading statistics that you see presented on television or in magazines. Select one of the products and prepare an oral presentation to explain how the statistics were used in a misleading way.

See page 427 for help with entering and sorting data using a graphing calculator.

8. Liam scored 78% on a history test. On the same test, the other students in his class scored 82, 89, 77, 92, 88, 98, 79, 81, 73, 80, 74, 77, 86, 79, 97, 68, 97, 74, 82, 87, 79, and 90.
a) Calculate Liam's percentile rank.
b) Stacey was the person who scored 86%. What percentile was her score in?

9. **Thinking/Inquiry/Problem Solving** Katie scored 72% on a test in Mrs. Hallett's period 1 math class. The scores for the other students in the same class were 63, 58, 79, 80, 55, 67, 69, 79, 91, 42, 67, 84, 75, 63, 85, 77, 50, 71, 86, and 60. Jenna also scored 72% on the same test in Mrs. Hallett's period 4 class. The scores for the other students in this class were 74, 88, 64, 85, 62, 58, 63, 45, 79, 81, 49, 87, 58, 79, 72, 97, 89, 64, 72, 67, 88, 76, and 70.
a) Both girls got the same mark. Does this mean that their percentile rank with their own math class is the same? What if the results for both classes were grouped together? Explain your reasoning.
b) Calculate the three percentiles to check your answer in part a).

10. **Communication** To determine how many people own their own vehicles (do not have a car loan), a researcher surveyed 200 people entering a Toronto Maple Leafs hockey game. The results of the survey indicated that the majority of people (71%) do not have a loan on their vehicles. Discuss the validity and bias of this survey.

11. A manufacturer of hockey equipment wants to run the following advertisement:

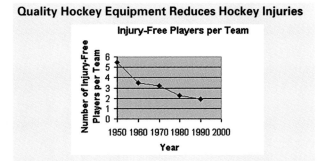

Quality Hockey Equipment Reduces Hockey Injuries

a) Discuss why someone might think that the number of injuries in hockey is decreasing, based on the data presented in this graph.

b) Safety equipment in hockey has not reduced the number of injuries. Discuss why the number of injuries in this data does not reflect this.

c) Prepare an alternative to present the data more realistically. Visit a hockey web site to obtain more data, if necessary. Go to *www.mcgrawhill.ca/links/MPCA12* and follow the links.

ACHIEVEMENT CHECK   Knowledge/Understanding   Thinking/Inquiry/Problem Solving   Communication   Application

12. You have been hired by a research company to help design and test sample questions for a new survey. The company has a new client who wishes to market children's backpacks in the most popular colours. Your boss wants to show this new client what the difference is between a poor survey and a good survey.

You are to describe the features of a survey in two ways:
• one that will produce reliable results
• one that will be misleading
Make sure you indicate which is which.
Your presentation should include the following:
• a description of the sample to be surveyed
• the actual survey question or questions
• examples of the types of graphs that will be used to display the resulting data

CAREER PROFILE

Machinist Apprenticeship

Machinists design, install, and operate machines primarily used in the manufacturing industry. They read blueprints and manuals, and interact with engineers, machine assemblers, and programmers to solve problems. Machinists need to have a good understanding of the characteristics of the materials they are working with.

Machinist apprentices attend classes to learn the theory, and spend a great deal of time in the field getting hands-on experience. The combination of class and field time varies, depending on the college. In some programs, students attend class one day a week and then, work as an apprentice the other days. Other programs have students attend class for several weeks, followed by several weeks in the field. Visit the college of your choice to learn more about the type of machinist apprentice program offered.

Steel is a very versatile material, and combined with other alloys and a modified heat treatment, it can be made stronger and more durable. A machinist working with steel can interpret graphs similar to the one shown.

1. Identify the dependent and independent variables.

2. Describe the two features of steel manufacturing that are being compared along the horizontal axis.

3. Describe the shape of the graph.

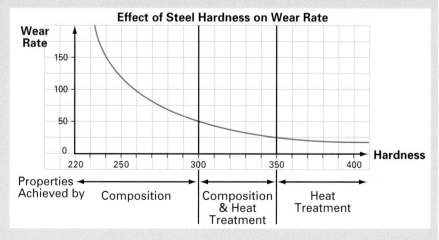

CHAPTER PROBLEM WRAP-UP

The increase in hockey-related injuries is staggering, as shown in the graph.

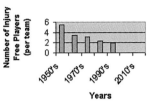

Hockey Injuries on the Rise

Will there be anyone without injuries available to play in 2010? Following the line of best fit, it is certain that there will be fewer injury-free players. The number of injury-free players in the 1950s was considerably higher than the other years. This might seem strange, as in those days, players did not wear helmets or other protective gear.

Continue the analysis of hockey injuries, or research a sporting activity that interests you. Identify the number of injuries and the types of injuries related to your sporting activity.

Prepare a report on the frequency of sporting injuries in your chosen sporting activity. Your report should include the following:
- a graph or chart identifying the pattern
- an analysis of injuries in your sporting activity, including possible reasons for changing trends
- a presentation of your data in an alternative format
- a discussion of whether your data contain any bias
- a discussion of how your data could be misrepresented to present a different view

6.1 Retrieving Information, pages 258–270

1. a) Explain the difference between static data and dynamic data.

b) Give two sources of each type of data.

2. Information is presented in many forms. The form of bar graph shown is sometimes called a ribbon graph.

a) Discuss how this unusual graph uses colour to represent the level of distraction.

b) What percent of people think cellular phones are less distracting than drinking or eating?

c) Identify the top two activities that are deemed more distracting than a cellular phone.

d) Discuss how easily information is obtained from this graph. Suggest how the data might have been better presented.

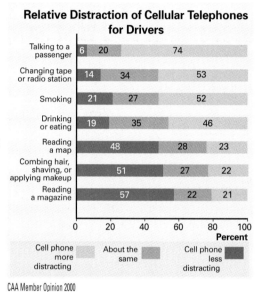

Relative Distraction of Cellular Telephones for Drivers

CAA Member Opinion 2000

3. Understanding how much (or how little) you earn on various investment options may influence your personal investment choices. Consider the table below, which gives the rates at some Canadian financial institutions in August 2001.

WHAT YOUR SAVINGS EARN 08.10.01

	Savings monthly	Accts daily	min inv	Term Deposits 30-59 days	60-89 days	90-119 days	120-179 days	min inv	Certificates/Debentures interest paid annually 1 yr	2yr	3 yr	4 yr	5yr
BCI Bank	0.35	0.15	5000	2.75	2.75	2.75	2.75	4000	3.10	3.50	3.70	4.00	4.30
Bank of Montreal		0.10	5000	2.75	2.75	2.75	2.75	1000	3.05	3.50	3.70	4.00	4.30
Bank of Nova Scotia		0.15	5000	2.75	2.75	2.75	2.75	500	3.05	3.50	3.70	4.00	4.30
CIBC			5000	2.75	2.75	2.75	2.75	1000	3.15	3.50	3.70	4.00	4.30
Citibank Canada		0.10	5000	2.75	2.75	2.75	2.75	5000	3.15	3.50	3.70	4.00	4.30
Ing Direct		4.00						0	4.05	4.70	4.85	5.10	5.35
NBG Bank	0.75	1.00	5000	2.75	2.75	2.75	2.75	1000	3.05	3.50	3.70	4.00	4.30
Royal Bank		0.15	5000	2.50	2.50	2.50	2.50	500	2.80	3.30	3.70	4.00	4.30
TD Canada Trust		0.15	5000	3.10	3.10	3.10	3.10	1000	3.15	3.50	3.70	4.00	4.30

a) Which of the institutions listed do not pay any interest on a monthly savings account?

b) On term deposits, what is the minimum investment amount for the majority of financial institutions?

c) If you had $5000 to invest for a minimum of three years, which investment option would earn the greatest interest rate?

6.2 Discover Information From Indices, pages 271–273

4. The CPI is an index used to measure inflation.

a) What do the letters CPI stand for?

b) Discuss how the CPI is used to give a measure of inflation.

c) List at least two other price indices that should be used in conjunction with the CPI to measure inflation or deflation.

d) CPI graphs include, beside the vertical axis, the statement 1992 = 100. Explain what this means and how you reference it.

5. Using E-STAT, select a food item or product to review the price changes over the past 50 years. Include the following.

a) a suitable visual presentation

b) a summary of your graph or chart

6.3 Representing Data in a Variety of Ways, pages 274–283

6. List four methods that you can use to represent data. For each method, indicate when it would be best to use that method to display data.

7. Using E-STAT, prepare a brief report on the change in population of one province other than Ontario over time. Select the most suitable format to display the data.

8. The residential property tax in Toronto is based on the assessed value of the home. Is sufficient data shown to indicate a pattern? If so, identify the tax for a $125 000 townhouse and for a $600 000 condominium.

Assessment ($)	Property Tax ($)
150 000	1610.32
200 000	2147.09
250 000	2683.86
300 000	3220.63
350 000	3757.40
400 000	4294.18
450 000	4830.95
500 000	5367.72

9. The table compares the percent of women holding parliamentary seats.

	1987	1995	1999
Canada	10	18	21
Denmark	32	34	37
Sweden	32	40	43
United States	5	11	13

  a) Discuss how Canada is changing, with respect to the percent of women occupying parliamentary seats.

  b) Discuss whether the trend line would, at some point in the future, show women occupying more seats than men in parliament.

  c) Discuss how Canada compares with the other countries listed, with respect to the number of women occupying parliamentary seats.

10. Use the **Data Bank**, page 454, or another source, to gather data on the projected population of Canadians aged 15–24 years, for the years 2011, 2021, 2031, and 2041.

  a) Display the data in a suitable graph.

  b) Use your graph to predict the population of females aged 15–24 in 2015.

  c) Use your graph to predict the population of males aged 15–24 in 2045.

  d) Which of the two preceding answers do you think is more likely to be reliable? Explain.

See the Technology Appendix for help with graphing data using a graphing calculator (page 427), in Microsoft® Excel (page 432), in Quattro® Pro (page 436), or in Fathom™ (page 440).

6.4 Validity and Bias, pages 284–291

11. Give an example of each type of bias.

  a) the sample size is too small

  b) the sample does not reflect the population well

  c) the survey question is misleading

12. Discuss how data have been misrepresented in this graph.

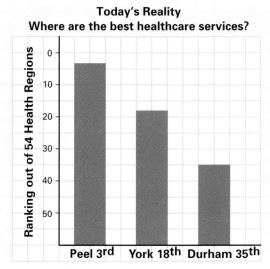

Today's Reality
Where are the best healthcare services?

This Week [Whitby], Tuesday Edition, June 26, 2001 and Maclean's Magazine, 3rd Annual Ranking of "Where We Get the Best Healthcare"

13. As an informed newspaper reader, discuss the validity of the following statements involving data.

a) Current reader polls indicate the following choices:
Party A: 37%
Party B: 32%
Party C: 17%
Undecided: 14%

b) Four out of five people preferred Champ Gum over all other gum.

c) 80% of migraine sufferers experienced relief in less than 15 min.

14. The graph shows Internet use, as of August 2001.

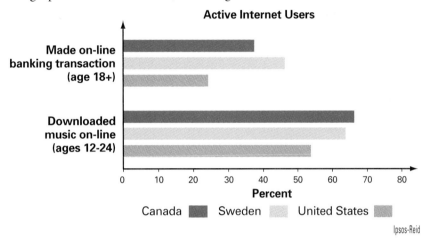

Discuss the validity of the following statements, based on the data presented in the graph.

a) Canadians are more like Swedes than Americans, when it comes to Internet use.

b) Fewer people use the Internet regularly to do their banking than to download music on-line.

c) More Swedes use the Internet for downloading music than Americans.

15. Freya got 78% on a recent geography exam. The results for her classmates were as follows:
55, 83, 72, 78, 66, 39, 60, 88, 75, 64, 68, 71, 57, 90, 59, 74, 70, 81, 58
Determine Freya's percentile rank.

Practice Test

ACHIEVEMENT CHART CATEGORY CONNECTIONS			
Category	Knowledge/Understanding	Thinking/Inquiry/Problem Solving	Communication	Application
Questions	1–8	5–8	1, 2, 4–8	5, 6, 8

1. Describe how the CPI is used to monitor how prices change.

2. Explain the difference between percent and percentile.

3. Rakan scored 91% on a science test. The other students in his class scored 45, 69, 75, 89, 53, 81, 69, 78, 76, 41, 64, 72, 90, 95, 88, 91, 58, 71, 76, 86, 74, and 66. Calculate Rakan's percentile rank.

4. Discuss the validity of the following statements.
   a) Data show that more than half the patients felt better in less than 1 h.
   b) Seven out of ten people preferred brand A over two other brands.

5. The data show the average local monthly cellular phone bill in Canada from 1990 to 1999.

Year	1990	1991	1992	1993	1994	1995	1996	1997	1998	1999
Average Local Monthly Bill ($ per user)	80.90	72.74	68.69	61.49	56.21	51.08	47.70	42.78	39.43	41.24

   a) Represent this data graphically.
   b) Describe the change and give reasons why this change in cost has occurred.

6. The table gives the full-time enrolment in post-secondary education in Ontario.
   a) Prepare a graphical representation to compare the change in enrollment for males and females over time.
   b) Describe any trends that you see.
   c) Use your graph to extrapolate the number of male and female students in 2000. How could you check the accuracy of your estimate?

Year	Males	Females
1987	44 949	50 080
1988	43 469	50 952
1989	42 401	50 936
1990	46 633	53 024
1991	54 055	57 307
1992	57 801	59 335
1993	60 026	61 660
1994	62 304	64 129
1995	65 747	68 756
1996	69 295	71 962

Adapted from: Statistics Canada's Internet Site, http://estat.statcan.ca/cgi~win/ CNSMCGI.EXE, Full-time enrolment in post-secondary education in Ontario, by year 1987-1996, November 7, 2001.

7. Use this train schedule.

Train	km	31	631	33	635	35	37	39
Days/Jours		x6,7	6		6,7	x6	x6,7	
Montreal	0	0645	0735	1000	1255	1510	1625	1800
Dorval	19	0702	0752	1017	1313	1527	1642	1824
Coteau	63							1849
Alexandria	100	0755	0844	1103	1359	1621	1729	1920
Maxville	117			1115				
Casselman	140	0818	0907	1126				
Ottawa	187	0849	0939	1156	1458	1720	1835	2012

MONTREAL • ALEXANDRIA • OTTAWA

Train	km	30	630	32	632	634	34	36	38
Days/Jours		x6,7	6	x6	6	6,7	x6	x6,7	
Ottawa	0	0645	0730	0915	0945	1305	1505	1645	1800
Casselman	47								1840
Maxville	69			0958	1028				
Alexandria	87	0737	0823	1011	1041	1359	1558		1902
Coteau	124	0808	0855				1626		
Dorval	168	0832	0935	1058	1135	1451	1708	1836	1956
Montreal	187	0848	0935	1114	1151	1508	1708	1852	2012

OTTAWA • ALEXANDRIA • MONTREAL

Key: x6,7 means this train does not run on Saturday or Sunday

Via Rail Canada

a) What time will the 6:45 a.m. train from Ottawa arrive in Montreal?

b) Create an itinerary for a one-day round trip on Monday to Montreal from Ottawa, spending the maximum possible time in Montreal. Include the departure and arrival times, as well as the journey time.

c) Repeat the previous question for a trip made on a Saturday.

Σ-STAT

ACHIEVEMENT CHECK	Knowledge/Understanding	Thinking/Inquiry/Problem Solving	Communication	Application

8. Use E-STAT to access the lists of actual products and their prices. Choose a food item or product and follow the path to study the price changes.

a) Prepare a report to summarize the data you found. Present your research in text form, in chart form, and in a graph.

b) Compare prices for the product in Ontario and one other province. Display the data in charts and graphs. Give possible reasons for the difference in pricing.

*C*umulative Review

Chapters 4–6

1. Identify each of the following as inferential or descriptive statistics. Which are samples and which are surveys of the whole population?
   a) Survey of randomly selected households
   b) Every employee at a factory is surveyed about their work environment
   c) Speeding tickets are tabulated at a regional police office to determine the average speed of the motorists who were caught speeding

2. List the three measures of central tendency and prepare a brief description of each.

3. Make a sketch of each type of distribution. On each sketch, mark the position of the mean, median, and mode.
   a) normal distribution
   b) positively skewed
   c) negatively skewed

4. Sometimes the distribution of data can be described as bimodal or exponential. Draw a sketch of each of these distributions.

5. Give an example of a survey for each type of sampling. Describe how you would choose the sample.
   a) random sampling
   b) stratified sampling

6. The durations of eruptions, in minutes, for the geyser Old Faithful, in Yellowstone National Park, on fifty consecutive occasions are given.

See Appendix B for help with calculating single variable statistics using a graphing calculator (page 429) or Fathom™ (page 439).

4.37	1.97	3.70	1.67	4.13	3.87	4.50	3.80	4.60	3.95
4.00	3.92	3.43	1.67	4.10	4.03	4.35	4.00	4.00	2.27
3.50	2.33	2.27	1.80	4.58	4.08	3.83	4.40	4.42	1.90
2.25	1.88	4.05	1.90	4.50	4.70	4.60	4.25	4.63	1.95
1.73	1.80	3.33	2.93	4.83	4.93	4.73	2.00	3.50	4.12

   a) Calculate the mean, median, and mode.
   b) Organize the data into a frequency table using half-minute class intervals: [1.50, 2.00), [2.00, 2.50), [2.50, 3.00), and so on.
   c) Display the data in a histogram. Mark the position of the mean, median, and mode.
   d) Is this a normal distribution? If not, what type of distribution is it?

e) Calculate the range and the standard deviation.

f) Go to *www.mcgrawhill.ca/links/MPCA12* and follow the links to obtain more data and to watch Old Faithful in action.

7. Identify the dependent and independent variables in each of the following.

a) Year and number of college applicants

b) Number of collisions and time of day

c) Total precipitation and month

8. When owners put their house up for sale, they specify how much they hope to sell the house for; this is the list price. The house may not actually sell for this price. The table shows the list price and the selling price of homes in Uxbridge, Ontario.

List Price ($) (What They Asked)	Selling Price ($) (What They Got)
185 000	184 500
139 900	138 000
244 500	242 500
149 000	149 000
209 900	210 000
234 500	231 900
178 500	178 000
210 000	210 000
162 900	16 000
155 000	144 500

a) Using technology, prepare a scatter plot.

b) Examine your scatter plot for the presence of outliers. Determine whether these outliers are errors or simply anomalies, and correct the data set, if necessary.

For help with linear regression, see section 5.4, page 232 (Microsoft® Excel), See Appendix B, page 431 (graphing calculator), or Appendix B, page 441 (Fathom™).

c) Use linear regression to find the equation to model the relation between list price and selling price.

d) State the correlation coefficient and describe the strength of the relation.

e) Use your linear model to find the selling price of a house that is listed for $182 500.

f) Prepare a statement about the relationship between the list price and selling price in the real estate market.

9. A large company has manufacturing plants in four different cities. At the plant in Ottawa, all the products were tested and 99.8% of the products passed the health and safety test. The company released this statement: "99.8% of all products manufactured by our company pass the health and safety test."

a) Discuss how the company has misused the data.

b) What would they have to do to make their claim valid?

10. The bus on Route 1 in Peterborough travels from downtown to the university and back. Part of the bus schedule is shown.

a) How long is the journey from the downtown terminal to Trent University?

b) If you wanted to meet a friend at the university at noon, at what time would you need to catch a bus from the downtown terminal?

c) Explain what Dial-a-Bus means.

Route #1	
Depart Downtown Terminal	**Depart Trent University**
Monday to Friday	
6:15 AM	6:45 AM
7:15 AM	7:45 AM
and every 30 min until	
6:15 PM	6:45 PM
7:15 PM	7:45 PM
Dial-a-Bus Zone 1	
8:15 PM	
9:15 PM	
10:15 PM	

11. Sometimes utility bills show customers their utility usage in the form of a graph. The amount of hydro used over a 12-month period is shown.

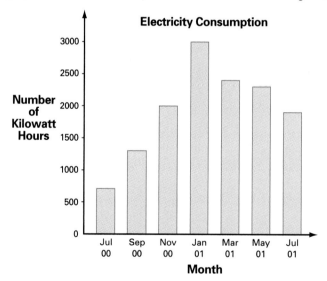

a) Each bar represents the amount of electricity used during that month and the previous month. During which months did the customer use the greatest amount of electricity?

b) The amounts for June/July 2000 and June/July 2001 are shown. Discuss how the hydro consumption compares and offer a possible explanation for the difference.

12. Toyah scored 72% on a mathematics test. On the same test, the other students in her class had the following marks:

58, 51, 49, 32, 80, 46, 74, 35, 78, 60, 71, 61, 49, 46, 62, 70, 68, 72, 83, 81, 70

Calculate Toyah's percentile rank.

13. a) What is the difference between primary and secondary sources of data? Give an example of each.

b) Describe the type of data and organizational structure of E-STAT.

14. Describe what is wrong with the following survey questions.

a) How would you rate your health: good, fair, or poor?

b) Since the Maple Leafs have not won the Stanley Cup in such a long time, how would you rank their ability: excellent, good, improving, or very poor?

c) At a soccer game, people were asked "Why do you eat snacks: because you are hungry, just because they are there, or as comfort food?"

15. Scrappy Fries Project Have you ever noticed how sometimes your order of French fries has lots of little scrappy fries in the bottom of the package that do not taste as good as the long ones? This project provides you with the opportunity to investigate the quality of French fries and identify which fast-food outlet is selling the most scrappy fries.

Work in a small group. Select a fast-food outlet that is convenient to you or select the school cafeteria. Decide how many packages of French fries to use in your survey and whether the time that the sample is taken might be a factor in the analysis. Measure each French fry and record the length in a table. Using statistical tools, analyse your data. As an extension, you can evaluate other qualities of the French fries, such as the number or mass per package.

Your project should use appropriate statistical concepts from Chapters 4, 5, and 6. Use the following main components as a guide to develop your project:
• Project Overview
• Data Collection
• Data Analysis
• Presentation (written and oral)

EXPLORING QUADRATIC MODELS

Specific Expectations	Sections
• Interpret a given linear, quadratic, or exponential graph to answer questions, using language and units appropriate to the context from which the graph was drawn.	7.1, 7.3, 7.4
• Communicate the results of an analysis orally, in a written report, and graphically.	7.1, 7.3, 7.4
• Factor expressions of the form $ax^2 + bx + c$.	7.2
• Solve quadratic equations by factoring.	7.2
• Evaluate any variable in a given formula drawn from an application by substituting into the formula and using the appropriate order of operations on a scientific calculator.	7.3
• Judge the reasonableness of answers to problems.	7.3, 7.4
• Enter data or a formula into a graphing calculator and retrieve other forms of the model.	7.1, 7.3, 7.4

𝒢et Ready

1. **Factors of numbers** List all possible pairs of factors for each number. (For example, for 36, one pair of factors is 2 and 18.)

   a) 36   **b)** 105   **c)** 2   **d)** 81   **e)** 74   **f)** 17

2. **Factors of numbers** Find two numbers that give these results.

   a) Their sum is 7, and their product is 10.

   b) Their sum is 3, and their product is −28.

   c) Their sum is −5, and their product is −36.

   d) Their sum is −13, and their product is 42.

3. **Degree of a polynomial** State the degree of each expression.

   a) $x^3 + 2x^2 - 9$ **b)** $9 + y - y^2$ **c)** $ax^2 + bx + c$

   d) $6k^2 + 3k - 1$ **e)** $n + n^2 - n^4$ **f)** $-5a^3 + 2a^2 - a - 7$

4. **Number skills** What number lies exactly halfway between the two given values on the real number line?

   a)

   −7 −6 −5 −4 −3 −2 −1 0 1 2 3 4 5 6 7 8 9

   b)

   −6 −5 −4 −3 −2 −1 0 1 2

5. **Number skills** Calculate. Where necessary, round answers to two decimal places.

   a) $-3 + \sqrt{16} - 4\sqrt{9}$ **b)** $-6 + \sqrt{4^2 - 4(1)(3)}$ **c)** $\dfrac{8 - (2 - 9)^2}{5(3)^2}$

   d) $\dfrac{3 - \sqrt{25}}{2}$ and $\dfrac{3 + \sqrt{25}}{2}$ **e)** $\dfrac{-4 + \sqrt{2 + 7}}{4}$ and $\dfrac{-4 - \sqrt{2 + 7}}{4}$

   f) $\dfrac{5 \pm \sqrt{81}}{-2}$ **g)** $\dfrac{-2 \pm \sqrt{7 - (-9)}}{2(1.5)}$

6. **Perimeter and area** Find each area.

   a)

   4 cm

   b)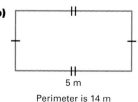

   5 m

   Perimeter is 14 m

   c)

   Perimeter is 60 m

   d)

   14 cm

7. **Simplifying expressions** Expand and simplify.

   a) $(3x - 4)(2x + 5)$ **b)** $-3(x + 5)(4x - 7)$

   c) $3(5x + 1)^2$ **d)** $-2(x - 3)^2 + 5$

8. **Evaluating expressions** Evaluate when $a = 3$, $b = -2$, $c = 5$, and $d = 0.5$.

a) $2a + 4b + (b - c)^2$ **b)** $b^2 - 4ac$ **c)** $ab - 3d + b^2$ **d)** $c^a + d$

9. **Factoring polynomials** Factor each expression by finding the greatest common factor.

a) $10y - 15$ **b)** $4x^2 + 10x - 12$ **c)** $6at + 3aw + 9az$

d) $6k^2 - 48k$ **e)** $16m + 8m^2 + 12m^3$ **f)** $35n - 15n^2$

10. **Evaluating expressions** Substitute the values given, and then, solve for the remaining variable.

a) $2l + 2w = 18$; $l = 5$ **b)** $y - 4 = 3x(4x - 2)$; $x = 5$

c) $y = 2x^2 + 6x + 7$; $x = -3$ **d)** $y = a(x - h)^2 + k$; $h = 3$, $k = 7$, $x = -1$, $y = 4$

11. **First differences** Find the first differences for each relation.

a)

x	y
2	-6
5	-1
8	4
11	9
14	14
17	19

b)

x	y
0	1
1	2
2	4
3	8
4	16
5	32

c)

x	y
-5	75
-3	27
-1	3
1	3
3	27
5	75

d)

x	5	10	15	20	25
y	49	144	289	484	729

12. **Graphing skills** Sketch the graphs of the relations in questions 11a) and 11b) on two separate sets of coordinate axes. Be sure to include all appropriate labels. Is either of these relations linear? What is special about the first differences for a linear relation?

CHAPTER PROBLEM

After completing the Financial Planning Management stream at George Brown College in Toronto, you set up a small financial service business. You have been contracted by a local sports store, SportStuff, to improve its revenue. SportStuff management wishes to determine whether they should continue to carry the Polar Bear Parka line. Using data from previous sales, and information collected from customer surveys, you are to determine a mathematical model to represent the costs and revenues associated with Polar Bear Parka.

Recognizing and Understanding Quadratic Relationships

Mathematical modelling is the process of describing a real situation in a mathematical way. This description can take on many forms, such as a graph, a table of values, an equation, or a formula.

One type of mathematical model you have already explored is the linear model. You have learned to recognize that a linear relationship exists for a set of data if the following conditions are met.

- When the first differences are calculated, they are constant.
- When the data are graphed, a straight line is formed.
- When a linear regression is done, the data fits the equation $y = mx + b$, where m is the slope and b is the y-intercept.

Not all relationships are linear. Consider the following data used by pyrotechnists in the preparation of a community fireworks display.

DISCOVER

Quadratics in Table and Graph Form

The data show the relationship between the horizontal and vertical distances of a firework launched from a barge.

Horizontal Distance (m)	0	3	6	9	12	15	18	21
Vertical Distance (m)	0	27.54	45.36	53.46	51.84	40.50	19.44	−11.34

1. Explain how you know that this is not a linear relationship.
2. Calculate the first differences. Recall that when creating a difference table, the change in the independent variable must be constant. Here the horizontal distance increases by 3 m. Describe the pattern in the first differences.
3. Second differences are calculated by subtracting the first difference values. Calculate the second differences for the data in the table. Describe the pattern.
4. a) Graph the data, plotting horizontal distance on the x-axis.
   b) Sketch a curve of best fit.
   c) Describe the shape of this curve.

5. Mathematicians call a symmetrical curve like the one shown here a parabola.

   a) Identify the coordinates of the vertex of the curve that you have drawn. Is this a maximum or minimum point?

   b) Explain what this point means in terms of the firework's projectile.

6. The point on a parabola at which the y-coordinate is zero is known as a **zero**. A parabola can have two, one, or no zeros.

   a) What are the zeros for the firework's parabola?

   b) What information does the zeros provide to the pyrotechnist?

   c) Where is the vertex in relation to the zeros?

7. The vertex of a parabola can be used to find the equation of the curve. Equations like this are important because they allow us to make predictions. Provide at least two reasons why it is important for pyrotechnists to be able to make accurate predictions.

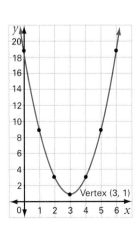

Vertex

Axis of Symmetry

EXAMPLE 1

Finding the Quadratic Equation From a Graph

Find the quadratic equation that fits the data shown.

x	y	First Differences	Second Differences
0	19		
1	9	10	
2	3	6	4
3	1	2	4
4	3	–2	4
5	9	–6	4
6	19	–10	4

Solution

Step 1: Plot the points and identify the vertex on the graph. The vertex is at (3, 1).

Step 2: The equation for a quadratic relationship in vertex form is
$$y = a(x - h)^2 + k$$
where the vertex is at (h, k).
Substitute the coordinates of the vertex, (3, 1), into the general form.
$$y = a(x - h)^2 + k$$
$$y = a(x - 3)^2 + 1$$

Vertex (3, 1)

Step 3: Substitute another point (x, y) from the curve into the equation to find the value of a.

From the table, the point $(5, 9)$ is known to be on the curve.

Substitute $x = 5$ and $y = 9$.

$$y = a(x - 3)^2 + 1$$
$$9 = a(5 - 3)^2 + 1$$
$$9 = a(2)^2 + 1$$
$$9 = 4a + 1$$
$$9 - 1 = 4a$$
$$8 = 4a$$
$$a = 2$$

The quadratic equation that fits the data is $y = 2(x - 3)^2 + 1$.

Quadratic functions do not always appear in the vertex form, $y = a(x - h)^2 + k$.

Some may appear in **standard form**:

$$y = ax^2 + bx + c$$

where a and b are called coefficients, and c is a constant.

For example, $y = -2x^2 + 3x + 10$ is a quadratic function, with $a = -2$, $b = 5$, and $c = -6$.

Its graph is as shown.

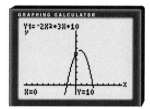

EXAMPLE 2

Changing a Quadratic From Vertex Form to Standard Form

Express $y = 2(x - 3)^2 + 1$ in standard form.

Solution

Expand the binomial part of the equation and simplify.

$$y = 2(x - 3)^2 + 1$$
$$y = 2(x - 3)(x - 3) + 1$$
$$y = 2(x^2 - 6x + 9) + 1$$
$$y = 2x^2 - 12x + 18 + 1$$
$$y = 2x^2 - 12x + 19$$

The equation $y = 2(x - 3)^2 + 1$ in standard form is $y = 2x^2 - 12x + 19$.

Language Link

The prefix *quad* usually means four. A quadrilateral is a four-sided shape. A square is the simplest four-sided figure. A quadratic equation always has a squared term.

Vertex form and standard form are just alternative ways of representing the quadratic equation for the same data. Note that regardless of the form of the quadratic equation, it always has a degree of 2. That is, the highest power in the equation is 2.

Real-life data often do not fit perfectly into a mathematical model. With linear relationships, the equation of a *line* of best fit is found because it best represents the pattern of the data. Using technology, a linear regression can be performed on the data to find quickly and accurately the equation of the line of best fit. Similarly, with quadratic models, a **quadratic regression** can be done using a graphing calculator to establish quickly and accurately a *parabola* of best fit.

EXAMPLE 3

Using Quadratic Regression to Find the Equation

Use quadratic regression to find the equation of the data in Example 1.

Solution

See Appendix B, page 427, for help with entering data.

Use a graphing calculator.

Press $\boxed{\text{STAT}}$.

Select **Edit** and press $\boxed{\text{ENTER}}$.

Input the data into **L1** and **L2**.

Press $\boxed{\text{STAT}}$.

Move the cursor to the right to select **CALC**.

Move the cursor down to select **5: QuadReg**.

Press $\boxed{\text{ENTER}}$.

The equation in standard form, with values for *a*, *b*, and *c*, is given.

Substitute $a = 2$, $b = -12$, and $c = 19$ into the standard form of a quadratic, $y = ax^2 + bx + c$.

The equation that fits the data is $y = 2x^2 - 12x + 19$.

Note that, if your calculator has **Diagnostics On**, the value of R^2 is also given. The R^2 value is an indication of how closely your data fit a quadratic model. The closer R^2 is to 1, the better the fit of the data. In this case, the data are a perfect fit. Real-life data that fit a quadratic model often result in an R^2 value that is not exactly 1.

KEY CONCEPTS

- A mathematical model may appear in a variety of forms, such as a table of values, an equation, or a graph.

- The model may be quadratic if the following conditions are met:
  - **i)** The second differences from the data are constant.
  - **ii)** The expression, in standard form, has degree 2.
  - **iii)** The data, when plotted, fit a parabolic curve.

- If a model is believed to be quadratic, the equation can be found by
  - **i)** using the vertex (h, k) of the curve and any other point (x, y) on the curve to find the value of a in the vertex form of the equation, $y = a(x - h)^2 + k$.
  - **ii)** performing a quadratic regression using a graphing calculator. Quadratic regression gives the values of a, b, and c for the quadratic equation in standard form, $y = ax^2 + bx + c$.

DISCUSS THE CONCEPTS

1. Explain two ways to determine whether data are quadratic if the equation is unknown.

2. Work with a partner. One of you will graph the fireworks data in the Discover at the beginning of this section and find the equation of the curve using the graph. The other person will use technology to complete a quadratic regression on the fireworks data and find the equation. Compare your results. Which method do you prefer and why?

PRACTISE   **A**   **1.**   Identify each set of data as quadratic, linear, or neither.

a)

x	y
0	–5
2	1
4	7
6	13

b)

x	y
–10	143
–8	99
–6	63
–4	35
–2	15

c)

Number	Cost ($)
100	5 650
125	6 900
150	8 150
175	9 400
200	10 650

d)

Magazine Price ($)	5.00	5.25	5.50	5.75	6.00
Number Sold (in 1000s)	35	80	95	80	35

e)

Number of CDs Produced	250	500	750	1000	1250	1500	1750
Cost per CD ($)	25	20	20	18	18	15	15

f)	Time (s)	0	1	2	3	4	5	6
	Height (m)	2.5	42.6	72.9	93.4	104.1	105	96.1

2. Identify the relation shown by each graph as linear, quadratic, or neither.

a)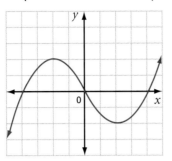

b)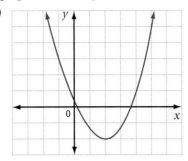

c)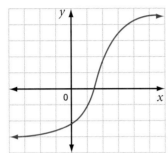

d)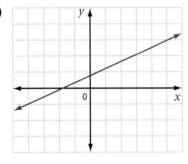

e)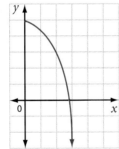

f)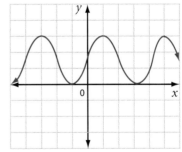

3. **Communication** Identify each relation, from its equation, as linear, quadratic, or neither. Give reasons for your answers.

a) $y = x^2 - 8x - 20$ b) $y = -5x + 7$ c) $y = 2k^2 + 6k - 30$

d) $y = -0.5x^2 + 3x - 5$ e) $y = 4x^2$ f) $y = 2x$

g) $4x - y + 2 = 0$ h) $y = 3x^3 + 2x - 5$ i) $y = 6(x - 3)^2 + 1$

4. Express each equation in standard form.

a) $y = 2(x - 1)^2 + 3$ b) $y = 4(x + 1)^2 - 7$ c) $y = (x - 2)^2 - 3$

d) $y = -(x - 5)^2 + 8$ e) $y = (x + 2)^2 - 3$ f) $y = (x - 1)^2$

g) $y = -(x + 5)^2 + 8$ h) $y = (x + 2)^2 + 3$

5. Find the equation in vertex form, $y = a(x - h)^2 + k$, of a parabola with the characteristics given.

   a) vertex at $(1, 4)$, and $(3, 8)$ is on the curve

   b) vertex at $(-2, 5)$, and $(2, 37)$ is on the curve

   c) vertex at $(4, -3)$, and a zero of 3

   d) a zero of -4, and vertex at $(1, 5)$

APPLY THE CONCEPTS

See Appendix B, page 431, for help with quadratic regression using a graphing calculator.

6. **Application** Use algebraic skills or a graphing calculator to find the quadratic equation that represents each graph or set of data.

a)

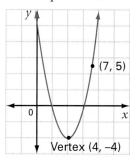

(7, 5)

Vertex (4, −4)

b)

x	y
−2	−24
−1	−25
0	−24
1	−21
2	−16
3	−9
4	0
5	11

c)

x	y
0	−240
10	630
20	1700
30	2970
40	4440
50	6110

d)

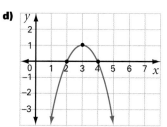

e)

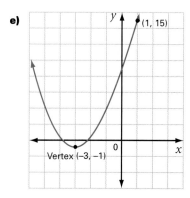

(1, 15)

Vertex (−3, −1)

f)

Number of Computers Sold	5	10	15	20	25	30
Revenue (in $100s)	36	176	416	756	1196	1736

B 7. Application, Communication In its first year of operation, the owner of an ice-cream parlour tracked sales. The data collected are shown in the table.

a) Use a pencil and paper or a spreadsheet to represent these data graphically.
(Hint: Represent each month with a number: January = 1, February = 2, and so on.)

b) What type of model do these data represent? Justify your answer.

c) Unseasonably warm temperatures are predicted for most of next year. Sketch a curve of projected sales for next year.

d) If monthly expenses are $6000, what recommendations would you make to the owner of the ice-cream parlour?

Month	Sales ($)
January	1 000
February	5 500
March	9 190
April	12 060
May	14 110
June	15 340
July	15 750
August	15 340
September	14 110
October	12 060
November	9 190
December	5 500

CHAPTER PROBLEM

8. SportStuff currently sells the Polar Bear parka for $200. An average of 90 parkas are sold each month. You have compiled this table based on previous sales and customer survey results.

Parka Price ($)	Number of Sales per Month
200	90
190	95
180	100
170	105
160	110
150	115

a) Use your own words to explain the trend in the data.

b) Could these data be a quadratic model? Explain.

c) Write an expression to represent the price of the parka, where x is the number of $10 decreases from the original price.

d) Write an expression to represent the number of parkas sold, where x is the number of $10 decreases from the original price.

e) Use the expressions found in parts c) and d) to write an equation for the expected revenue.

C 9. Thinking/Inquiry/Problem Solving Bridges often have parabolic curves. If the anchors at either end of a bridge are 42 m apart, and the maximum height of the bridge is 26 m, find an equation that models the shape of the bridge. (Hint: Draw the shape of the bridge on a Cartesian plane, labelling the zeros and the vertex.)

When data fit a quadratic model, it is often useful in problem solving to identify the *x*-intercepts—the points where the function has a value of zero. The values of *x* that result in a *y*-value of zero are called the **zeros** of the function. Quadratic functions can have one, two, or no zeros.

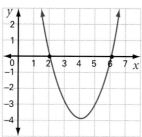

Two zeros occur, one at $x = 2$ and one at $x = 6$.

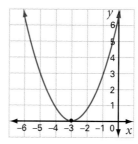

One zero occurs at $x = -3$.

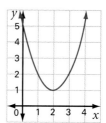

This function has no zeros.

DISCOVER

Finding the Zeros of a Quadratic

Work with a partner. Discuss your reasoning for each answer.

1. For each of the following quadratic models, determine the zeros.

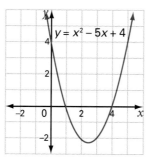

Parabola A

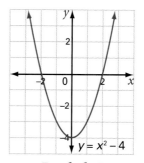

Parabola B

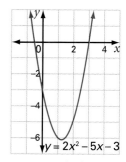

Parabola C

2. For which parabola was it most difficult to find the zeros? Why?

3. For which parabola was it the least difficult to find the zeros? Why?

4. Change each quadratic from factored form to standard form by expanding and simplifying the right side.

a) $y = (x - 1)(x - 4)$

b) $y = (x - 2)(x + 2)$

c) $y = (2x + 1)(x - 3)$

5. Based on the answers for question 4, what is another way to express the equations for parabolas A, B, and C?

6. Compare the zeros of each function with its equation in factored form. The zeros of a quadratic can be found from the equation in its factored form. Explain how.

EXAMPLE 1

Factoring $ax^2 + bx + c$, when $a = 1$

Express $y = x^2 - 3x - 18$ in factored form.

Solution

$y = x^2 - 3x - 18$

$y = (x \quad)(x \quad)$

Find two numbers whose product is -18 and whose sum is -3.

First, list numbers that have a product of 18.

Factors of 18

   1 and 18

   2 and 9

   3 and 6

Since the product is *negative* 18, one of the factors must be negative and one must be positive.

Determine which pair of factors results in a sum of -3.

$1 + (-18) \neq -3$	$-2 + 9 \neq -3$	$3 + (-6) = -3$ ✔
$-1 + 18 \neq -3$	$2 + (-9) \neq -3$	$-3 + 6 \neq -3$

Check by expanding.

$(x + 3)(x - 6) = x^2 - 6x + 3x - 18$

$\qquad\qquad\qquad = x^2 - 3x - 18$

So, $y = x^2 - 3x - 18$ expressed in factored form is $y = (x + 3)(x - 6)$.

EXAMPLE 2

Factoring $ax^2 + bx + c$, when $a \neq 1$

Express $y = 2x^2 - 5x - 3$ in factored form.

Solution

Step 1: Find two numbers that have a sum of b and a product of ac.

In $y = 2x^2 - 5x - 3$, $a = 2$, $b = -5$, and $c = -3$.

Find two numbers that have a sum of -5 and a product of $(2)(-3)$, or -6.

First, list numbers that have a product of 6.

Factors of 6

   2 and 3

   1 and 6

Since the product is -6, one of the factors must be positive and one must be negative.

$$(-2) + 3 \neq -5 \qquad\qquad (-1) + 6 \neq -5$$
$$2 + (-3) \neq -5 \qquad\qquad 1 + (-6) = -5 \checkmark$$

The only possibility is 1 and -6.

Step 2: Rewrite the equation, expanding b, using the two numbers just found as coefficients of x.

$y = 2x^2 - 5x - 3$
$y = 2x^2 - 6x + 1x - 3$
$y = (2x^2 - 6x) + (1x - 3)$ Group the first two terms and the last two terms.
$y = 2x(x - 3) + (x - 3)$ Find common factors.
$y = (x - 3)(2x + 1)$ Factor out the common binomial factor.

Check by expanding the right side.
$(x - 3)(2x + 1) = 2x^2 + x - 6x - 3$
$\qquad\qquad\qquad = 2x^2 - 5x - 3$

So, $y = 2x^2 - 5x - 3$ expressed in factored form is $y = (x - 3)(2x + 1)$.

The zeros of a function can be found readily from the equation in factored form. The solution process relies on the logic that if $A \times B = 0$, then either $A = 0$ or $B = 0$.

EXAMPLE 3

Using the Factored Form to Find the Zeros

Use the factored form (from Examples 1 and 2) to find the zeros of each quadratic function.

a) $y = x^2 - 3x - 18$ **b)** $y = 2x^2 - 5x - 3$

Solution

a) $y = x^2 - 3x - 18$
    $y = (x + 3)(x - 6)$ From Example 1
    For the zeros, $y = 0$.
    $0 = (x + 3)(x - 6)$
    For the product to equal 0, either $(x + 3) = 0$ or $(x - 6) = 0$.
    Either $x + 3 = 0$ or $x - 6 = 0$
           $x = -3$ $x = 6$
    So, the zeros of $y = x^2 - 3x - 18$ occur at $x = -3$ and $x = 6$.
    A graphing calculator may be used to confirm the zeros.

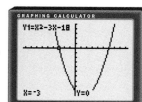

b) $y = 2x^2 - 5x - 3$
    $y = (x - 3)(2x + 1)$ From Example 2
    For the zeros, $y = 0$.
    $0 = (x - 3)(2x + 1)$

For the product to equal zero, either $(x - 3) = 0$ or $(2x + 1) = 0$.

Either $x - 3 = 0$ or $2x + 1 = 0$

$\qquad\qquad x = 3 \qquad\qquad 2x = -1$

$\qquad\qquad\qquad\qquad\qquad\qquad x = -\dfrac{1}{2}$

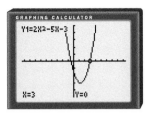

So, the zeros of $y = 2x^2 - 5x - 3$ occur at $x = 3$ and $x = -0.5$.
A graph can be used to verify these zeros.

EXAMPLE 4

Using Zeros in Problem Solving

By analysing a video, the height of a bungee jumper during a portion of the "trip" can be determined by the equation $h = -5t^2 + 40t - 45$, where h is the height above the lowest point reached, in metres, and t is the time, in seconds. At what times is the jumper even with the video camera mounted 15 m above the lowest point? Interpret the results in the context.

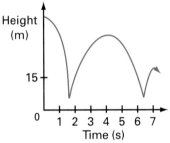

Solution

To find the time when the height of the jumper is 15 m, substitute $h = 15$.

$h = -5t^2 + 40t - 45$

$15 = -5t^2 + 40t - 45$

$0 = -5t^2 + 40t - 60$

Solve for t, by factoring the equation.

$-5t^2 + 40t - 60 = 0$ \qquad Notice that -5 is a common factor.

$-5(t^2 - 8t + 12) = 0$

$-5(t - 6)(t - 2) = 0$

Since $-5 \neq 0$, either $(t - 6) = 0$ or $(t - 2) = 0$.

So, $t - 6 = 0$ or $t - 2 = 0$

$\qquad t = 6 \qquad\qquad t = 2$

The values $t = 2$ and $t = 6$ are the zeros of the equation $0 = -5t^2 + 40t - 60$, as well as the solutions to the original equation, $15 = -5t^2 + 40t - 45$.

The jumper is even with the camera 2 s after bouncing back up from the lowest point and 4 s later, on the way back down.

KEY CONCEPTS

- The zeros of a function are the values of x for which $y = 0$. The zeros are directly related to the x-intercepts.

- For a quadratic function, there can be two, one, or no zeros.

- Factoring can be used to determine zeros of a quadratic function.
  To factor $ax^2 + bx + c$,

  i) factor out any constant that is common to all three terms.

  ii) if $a = 1$, find two numbers that have a sum of b and a product of c.

  For example, $x^2 + 5x + 6$ $b = 5$ and $c = 6$
             $= (x + 3)(x + 2)$ $3 + 2 = 5$ and $3 \times 2 = 6$

  iii) if $a \neq 1$, find two numbers that have a sum of b and a product of ac. Replace the b-term with these two numbers as coefficients of x. Group the first and last pair of terms using brackets and find the common factors.

  For example, $2x^2 + 3x - 2$ $b = 3$ and $ac = -4$
             $= 2x^2 + 4x - 1x - 2$ $4 + (-1) = 3$ and $4(-1) = -4$
             $= 2x(x + 2) - 1(x + 2)$
             $= (2x - 1)(x + 2)$

DISCUSS THE CONCEPTS

1. Can every quadratic expression be factored? Support your answer with an example. How is this related to a graph of the related quadratic function?

2. The graph of $y = x^2 - 10x + 16$ is shown. What are its x-intercepts? Explain how these values are related to the zeros of the function. What is the equation in factored form?

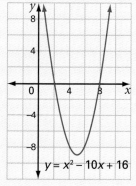

3. Show that it does not matter in which order the terms are placed when you are factoring an expression like $2x^2 - 5x - 3$ by expanding the b-term using $-6x$ and $1x$.

A **1.** Identify the zeros of each quadratic.

a)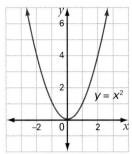

b)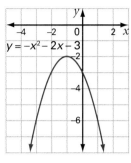

c)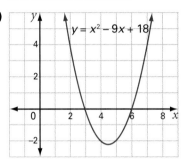

d)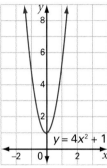

e)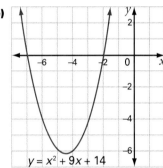

f)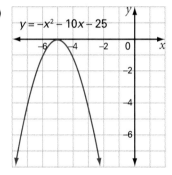

2. Find two integers that satisfy the conditions given.

	Sum	Product
a)	6	−16
b)	−7	10
c)	−11	−26
d)	17	−60
e)	21	104
f)	−19	60

3. Expand to express each equation in standard quadratic form, $y = ax^2 + bx + c$.

a) $y = (x + 4)(x - 1)$ **b)** $y = 3(x + 1)(x + 5)$ **c)** $y = (3x + 4)(2x - 1)$

d) $y = -(x - 4)(4x - 1)$ **e)** $y = 2(2x + 1)(3x + 4)$ **f)** $y = -3(x - 5)(x - 1)$

4. Factor fully. Remember to check for a common factor first.

a) $3a^2 - 9a$ **b)** $4x^2 + 16x - 18$ **c)** $-3z^2 + 24z - 21$

d) $4x(2x - 1) - 3(2x - 1)$ **e)** $3(5t + 7) - 2t(5t + 7)$ **f)** $8d(1 - 3d) + 4(1 - 3d)$

5. Group terms to factor fully.

a) $2x^2 - 2x + 3x - 6$ **b)** $4t^2 - 10t - 6t + 15$ **c)** $10x^2 + 6x - 5x - 3$

d) $2a^2 + 4a + a + 2$ **e)** $5n^2 - 15n + 2n - 6$ **f)** $56x^2 + 16x - 7x - 2$

6. Solve.

a) $(x + 2)(x - 5) = 0$ **b)** $(3x - 1)(x + 7) = 0$ **c)** $(2a + 1)(4a - 5) = 0$

d) $2x(8 - x) = 0$ **e)** $3(2x - 3)(x + 1) = 0$ **f)** $-(k + 6)(5k - 8) = 0$

7. Factor. Check your answer by expanding.

a) $x^2 + 13x + 12$ **b)** $x^2 - 7x - 30$ **c)** $x^2 - 8x + 15$

d) $x^2 - 15x + 50$ **e)** $4x^2 - 9$ **f)** $x^2 - 12x + 36$

8. Factor. Check your answer by expanding.

a) $2x^2 + 7x + 3$ **b)** $2x^2 + 9x + 10$ **c)** $3x^2 + 7x - 20$

d) $4t^2 - 13t + 10$ **e)** $8x^2 + 6x - 9$ **f)** $6x^2 + x - 1$

9. Factor completely. Remember to check for a greatest common factor first.

a) $4x^2 - 10x + 6$ **b)** $9x^2 + 33x + 30$ **c)** $-12x^2 + 10x - 2$

B **10.** Solve each equation by factoring.

a) $x^2 - 5x + 4 = 0$ **b)** $x^2 - 3x - 10 = 0$ **c)** $3x^2 + 13x - 10 = 0$

d) $4x^2 - 11x - 3 = 0$ **e)** $2x^2 + 12x - 54 = 0$ **f)** $3x^2 - 4x - 15 = 0$

11. Use factoring to find the zeros of each quadratic function.

a) $y = x^2 + 2x - 35$ **b)** $y = 16x^2 - 8x + 1$ **c)** $y = 6x^2 + 12x + 6$

d) $y = x^2 + x - 20$ **e)** $y = 30x^2 + 28x - 16$ **f)** $y = 5b^2 - 14b + 8$

APPLY THE CONCEPTS

12. Solve for x, given the value of y.

a) $y = x^2 + 13x + 26$, when $y = -14$ **b)** $y = x^2 + 7x + 15$, when $y = 3$

c) $y = 4x^2 - 12x + 16$, when $y = 7$ **d)** $y = -18x^2 + 3x - 3$, when $y = -4$

13. Application

a) Use factoring to determine the zeros of $y = 4x^2 - 4x - 15$. Express the zeros as fractions, where appropriate.

b) What value(s) of x will give a y-value of 20?

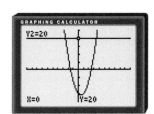

14. Graphic design A photograph measuring 10 cm by 5 cm is to be framed with a mat of uniform width surrounding the photograph. The area of the mat is to be twice the area of the photograph.

a) Calculate the area of the mat.

b) Write an expression for the length and for the width of the mat.

c) Write an expression that models the area of the mat.

d) Create an equation using your answers from parts a) and b). Solve the equation to determine the width of the mat.

15. Small business The monthly profit for a chocolate bar company can be modelled with the equation $P = -2x^2 + 240x$, where x is the number of chocolate bars produced, in hundreds, and P is the profit, in dollars.

a) Find the number of chocolate bars that should be produced for the company to break even.

b) Find the number of chocolate bars that should be produced to yield a maximum profit. Find the maximum monthly profit.

c) Last month, the company exceeded its goal of producing 6000 chocolate bars. If the result was a profit of $7000, determine the company's production levels for that month. Make a recommendation to the company regarding production levels, based on this information.

16. Landscape design Brad has a rectangular garden, 20 m by 10 m, that is to be surrounded by a paved walkway of uniform width.

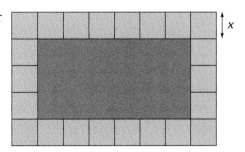

a) Write an expression for the length and for the width of the walkway.

b) Write an expression that models the area of the walkway.

c) Brad has a budget of $6000 for this project, and he estimates that it will cost $15/m^2 for paving stones. What is the maximum area of walkway he can afford?

d) Use your answers from parts b) and c) to determine the desired width of the walkway.

17. a) Last month, SportStuff adjusted the Polar Bear parka price. This resulted in a revenue of $15 600. How many parkas did SportStuff sell last month and at what price?

b) In order to cover costs associated with sales of these parkas, SportStuff needs to have a revenue of at least $140 000. What is the lowest price that the parka can be priced at so that some profit is generated? How many parkas would they expect to sell?

18. Automotive An auto parts manufacturer models their monthly profit with the equation $P = -6x^2 + 84x - 198$, where P is the profit, in thousands of dollars, and x is the number of parts produced, in thousands.

a) What is the manufacturer's profit when 4500 parts are produced?

b) What number of parts must be produced to break even?

c) How many parts should the company produce to make the maximum profit?

Career Link

An automotive marketing program is available in the engineering department of Georgian College, Barrie.

ACHIEVEMENT CHECK Knowledge/Understanding Thinking/Inquiry/Problem Solving Communication Application

19. Solve each equation twice, the first time by factoring and the second time by using another method.

a) $x^2 + 9x + 20 = 0$ **b)** $2x^2 + 12x - 54 = 0$ **c)** $x^2 - 8x + 12 = 0$

d) $2x^2 - 32 = 0$ **e)** $3x^2 - 4x - 15 = 0$

C 20. As part of your community service hours, you are planning a round-robin charity baseball tournament. Each team must play each other team exactly once. This means that the total number of games played, g, can be modelled with the equation $g = \dfrac{n(n-1)}{2}$, where n is the number of teams in the tournament.

The schedule you have designed has 7 time slots for games throughout the day, and park facilities allow 13 games to take place at the same time.

a) If you intend to use these facilities to their capacity while keeping every time slot filled, how many teams should you invite to the tournament?

b) Explain why $n(n-1)$ is divided by 2 in the equation to find the number of games.

21. Thinking/Inquiry/Problem Solving

a) Find the zeros of the function $y = x^2 - 8x + 12$.

b) Where is the vertex in relation to the zeros? Use this fact to find the coordinates of the vertex of the function. Use a graph to confirm your results.

c) Establish a general formula for finding the x-value of the vertex without a graph for any quadratic function with zeros x_1 and x_2.

Solving Quadratic Equations Without Factoring

When possible, factoring is a very effective way to find the zeros of a parabola or to solve a quadratic equation. However, many quadratic equations do not factor easily. Quadratic models for real data often are not neat enough to be solved by factoring. There are several other ways to find the zeros.

DISCOVER

Using a Graphing Calculator to Find Zeros

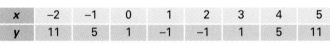

x	−2	−1	0	1	2	3	4	5
y	11	5	1	−1	−1	1	5	11

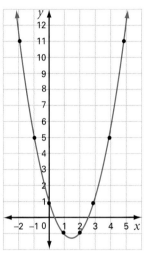

1. Estimate the zeros from the graph.

See Appendix B, page 425, for help with graphing calculator techniques.

2. Enter the data into a graphing calculator, and determine the equation using quadratic regression.

3. Explain why you cannot use factoring to find the zeros of this quadratic.

4. To find the zeros using a graphing calculator, proceed as follows:

Press (Y=) and enter the quadratic regression equation.

Press (GRAPH).

Press (2nd)(TRACE) to access the **CALC** menu.

Select **2:zero**.

Move the cursor to the left of one of the x-intercepts, and press (ENTER).

Move the cursor to the right of that same intercept, and press (ENTER).

Press (ENTER) again to see the value of the zero.

Repeat this process to find the other zero.

DISCOVER

Using a Formula to Find Zeros

1. Write the equation you determined from the data above in standard form, $y = ax^2 + bx + c$. Record the values of a, b, and c.

2. A formula has been developed to allow a quadratic equation to be solved, even if it cannot be factored. The **quadratic formula** states that the zeros of $y = ax^2 + bx + c$ are given by

$$x = \frac{-b \pm \sqrt{b^2 - 4ac}}{2a}$$

Substitute your values of a, b, and c into this formula and evaluate. Note that there will be two possibilities for x. Round your answers to two decimal places.

3. Compare these calculated answers to the answers obtained previously using a graphing calculator.

The solutions to a quadratic equation in the form $ax^2 + bx + c = 0$ are the values of x that satisfy the equation. These values of x represent the locations of the x-intercepts, or zeros, of the quadratic function $y = ax^2 + bx + c$.

The quadratic formula, $x = \dfrac{-b \pm \sqrt{b^2 - 4ac}}{2a}$, can be used any time. If you cannot find two integers that add to b and multiply to give ac in your head within a few seconds, then use the formula.

EXAMPLE 1

Solving an Equation Using the Quadratic Formula

Solve the quadratic equation $x^2 - 4x = 1$. Round answers to three decimal places.

Solution

$x^2 - 4x = 1$ Rearrange the terms so that the

$x^2 - 4x - 1 = 0$ equation is in the form $ax^2 + bx + c = 0$.

To factor, you need two numbers that have a sum of -4 and a product of -1. This is not possible.

Since this expression cannot be factored, use the quadratic formula.

Substitute $a = 1$, $b = -4$, and $c = -1$.

$$x = \frac{-b \pm \sqrt{b^2 - 4ac}}{2a}$$

$$x = \frac{-(-4) \pm \sqrt{(-4)^2 - 4(1)(-1)}}{2(1)}$$

$$x = \frac{4 \pm \sqrt{16 + 4}}{2}$$

$$x = \frac{4 \pm \sqrt{20}}{2}$$

Then, $x = \dfrac{4 + \sqrt{20}}{2}$ or $x = \dfrac{4 - \sqrt{20}}{2}$

         $x \doteq 4.236$ $x \doteq -0.236$

EXAMPLE 2

Using the Quadratic Formula to Find Zeros

Find the zeros of each quadratic function. Round answers to three decimal places.

a) $y = 3x^2 + 8x + 2$ **b)** $y = x^2 - 6x + 13$

Solution

a) For the zeros, $y = 0$.

Solve $3x^2 + 8x + 2 = 0$.

Try factoring first. Try to find two numbers that have a sum of 8 and a product of 3×2, or 6. There are no such integers.

Use the quadratic formula.

Substitute $a = 3$, $b = 8$, and $c = 2$.

$$x = \frac{-b \pm \sqrt{b^2 - 4ac}}{2a}$$

$$x = \frac{-8 \pm \sqrt{8^2 - 4(3)(2)}}{2(3)}$$

$$x = \frac{-8 \pm \sqrt{64 - 24}}{6}$$

$$x = \frac{-8 \pm \sqrt{40}}{6}$$

$x \doteq -0.279$ or $x \doteq -2.387$

The function $y = 3x^2 + 8x + 2$ has zeros at $x = -0.279$ and $x = -2.387$, to three decimal places.

A graphing calculator can be used to check that the answers are reasonable.

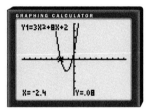

b) Solve $x^2 - 6x + 13 = 0$.

Try factoring first. You need to find two numbers that have a sum of -6 and a product of 13. No such integers exist.

Use the quadratic formula.

Substitute $a = 1$, $b = -6$, and $c = 13$.

$$x = \frac{-b \pm \sqrt{b^2 - 4ac}}{2a}$$

$$x = \frac{-(-6) \pm \sqrt{(-6)^2 - 4(1)(13)}}{2(1)}$$

$$x = \frac{6 \pm \sqrt{36 - 52}}{2}$$

$$x = \frac{6 \pm \sqrt{-16}}{2}$$

Because the square root of -16 is not a real number, there is no real solution.

The function $y = x^2 - 6x + 13$ has no zeros.

A graphing calculator confirms this; the graph has no x-intercepts.

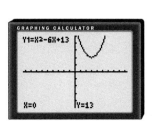

EXAMPLE 3

Business Entrepreneurship

Each month Seascape, Inc. tracks the number of pleasure boats built, *n*, and profits, in dollars, *P*. When fewer boats are built, profit is reduced. However, certain production levels lead to more overtime and more machine breakdowns, resulting in lower profits.

n	P
7	17 100
10	29 000
30	–53 000
20	21 800
17	31 000
26	–15 000
12	32 600
5	6 500
13	33 700
1	–23 500
19	25 500
22	12 200

a) Enter the data into a graphing calculator, and use quadratic regression to determine its best fit equation, to two decimal places. Identify what the variables, *x* and *y*, represent.

b) Enter this expression as (Y=), and use a graph to determine the break-even production level.

c) Verify the answer in part b) using the quadratic formula.

d) Determine the number of boats that should be produced to maximize profit. Find the maximum projected profit.

Solution

a) Press (STAT) **EDIT.**
Enter the data into columns **L1** and **L2.**

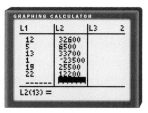

Press (STAT) **CALC.**
Select **5:QuadReg.**
Press (ENTER).

The quadratic equation for this data, to two decimal places, is $y = -340.37x^2 + 9531.49x - 32\ 681.70$, where *x* represents the number of boats built and *y* represents the profit, in dollars.

See Appendix B,
page 425, for help
with graphing.

b) At the break-even production point, the profit is 0.
Graph $y = -340.37x^2 + 9531.49x - 32\ 681.7$, and find the zeros.

 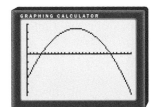

Use (2nd) (TRACE), then select **2:zero** to find each zero.
The zeros occur at approximately $x = 4.000$ and
$x = 24.003$.
The company breaks even at production levels of 4
and 24 boats.

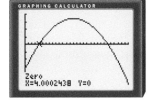

c) In the quadratic formula, substitute
$a = -340.37$, $b = 9531.49$, and $c = -32\ 681.7$.

$$x = \frac{-b \pm \sqrt{b^2 - 4ac}}{2a}$$

$$x = \frac{-9531.49 \pm \sqrt{9531.49^2 - 4(-340.37)(-32\ 681.7)}}{2(-340.37)}$$

$$x = \frac{-9531.49 \pm \sqrt{46\ 353\ 820.7}}{-680.74}$$

$x \doteq 24.003$ or $x \doteq 4.000$
The quadratic formula confirms that the break-even production levels are
4 and 24 boats.

d) Maximum profit is indicated by this curve's maximum point, the vertex.
The vertex lies midway between the zeros. Therefore, the maximum
production level is midway between 4 and 24 boats.

$$\frac{4 + 24}{2} = 14$$

To find the profit for a production of 14 boats, substitute $x = 14$ into the
profit equation.
$y = -340.37x^2 + 9531.49x - 32\ 681.7$
$y = -340.37(14)^2 + 9531.49(14) - 32\ 681.7$
$y = 34\ 046.64$
To maximize profit, 14 boats should be built each month.
The projected profit is \$34 046.64.

The maximum could also have been found using
a graphing calculator as shown.

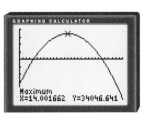

KEY CONCEPTS

- Not all quadratic equations can be solved by factoring.

- Quadratic equations can also be solved by
  - **i)** graphing the equation using a graphing calculator and determining the zeros
  - **ii)** expressing the equation in standard form, $y = ax^2 + bx + c$, and using the quadratic formula
    $$x = \frac{-b \pm \sqrt{b^2 - 4ac}}{2a}$$

- Quadratic equations may have two solutions (the parabola crosses the x-axis twice), one solution (the parabola touches the x-axis once), or no solutions (the parabola does not touch the x-axis).

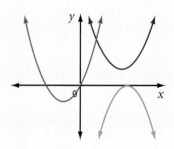

DISCUSS THE CONCEPTS

1. Match the equation with the appropriate graph. Explain your reasoning.
- **a)** $y = x^2 - 6x + 9$
- **b)** $y = x^2 - 8x + 19$
- **c)** $y = x^2 + 4x + 3$

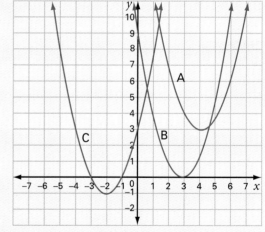

2. If a quadratic model is found using quadratic regression, which algebraic method of solving would most likely be used: factoring or quadratic formula? Explain.

3. Will the equations $3x^2 - 4x + 7 = 0$ and $-3x^2 - 4x + 7 = 0$ have the same solutions? Explain.

4. An injection moulding company produces interior car door panels. If production levels are increased from 1000 to 1500 car doors per day, a decrease in profit is projected. Give at least two possible reasons why this decrease in profit might occur.

A **1.** Express each equation in the form $ax^2 + bx + c = 0$, and identify the values of a, b, and c.

a) $5 = 3x^2 - 4x$ **b)** $-2x = 7 - 8x^2$

c) $-3 - x^2 = 17x$ **d)** $9x^2 = 7x$

e) $-2 + x = 3x^2 - 5x + 6$ **f)** $2x^2 - 5x + 8 = 3x^2 - 11$

2. Use a graphing calculator to determine the zeros of each. Round to two decimal places, where necessary.

a) $y = 3x^2 + 6x + 1$ **b)** $y = 2x^2 + 6x + 3$

c) $y = 2x^2 + 9x - 5$ **d)** $y = 5x^2 + 2x - 1$

3. **Communication** Which equations from question 2 could have been solved by factoring? Explain how you can tell.

B **4.** **Application** If possible, solve by factoring. If factoring is not possible, use the quadratic formula or a graphing calculator to solve. Round to two decimal places, where necessary.

a) $6x^2 - x - 12 = 0$ **b)** $2x^2 - 8x = 0$

c) $4x^2 - 11x - 3 = 0$ **d)** $2x^2 - 4x - 9 = 0$

e) $12x^2 + 17x - 7 = 0$ **f)** $x + 2 = 7x^2$

5. **Application** The rectangular area that can be enclosed using 120 m of fencing is given by $A = 60x - x^2$, where A is the area, in square metres, and x is the length, in metres, of one of the sides.

a) Determine the zeros of the equation, using factoring, if possible.

b) What length will yield the maximum area? Find this maximum area.

c) If the fenced area is to be 600 m^2, find the dimensions of the rectangle.

6. **Thinking/Inquiry/Problem Solving** The SuperSoaker Sprinkler sprays water out to a radius of 7 m. A competitor claims that their sprinkler, the LawnBuddy, will water an area twice as large.

a) What area is the LawnBuddy sprinkler reported to cover?

b) Let x represent the increase in radius of the LawnBuddy sprinkler. Write an expression for the area that the LawnBuddy covers.

c) Determine the radius that the LawnBuddy would have to reach for their claim to be valid.

d) What would the radius of the LawnBuddy sprinkler have to be to make a claim of covering an area three times as large as the area sprayed by the SuperSoaker?

7. The height of a football, h, is described by the function $h = -0.025d^2 + d$, where d is the horizontal distance from the point where the ball was kicked. Both variables are measured in metres.

   a) For what values of d was the football on the ground?

   b) What is the maximum height of the football?

   c) Assume the kick is for a game winning field goal. If the place kicker is 37 m away, and a height of 3 m is required to go over the cross bar of the goal post, is the kicker successful?

   d) For what values of d is the height 3 m?

   e) What is the maximum distance the kicker can be from the goal post and still have a successful kick?

8. The relation $d = 0.0056v^2 + 0.14v$ models a vehicle's stopping distance, d, in metres, and its speed, v, in kilometres per hour.

   a) What is the maximum speed that can be achieved while stopping within 50 m?

   b) A person driving at 100 km/h notices a stop sign approximately 70 m ahead and begins to break immediately. Can the vehicle stop in time?

C 9. **Thinking/Inquiry/Problem Solving** Greenthumb Growers, a local greenhouse operator, grows house plants year-round for distribution across Canada. As a result of tracking the weekly costs and revenue data over the last five years, the owner has established the following revenue and cost models. In both equations, x represents the number of plants, in hundreds.

$R(x) = x^2 - 10x + 250$

where $R(x)$ is the revenue, in hundreds of dollars, generated from selling x plants.

$C(x) = 2x^2 - 29x + 100$

where $C(x)$ is the cost, in hundreds of dollars, to produce x plants.

   a) Determine the number of plants that must be sold to break even.

   b) Create a single equation for the profit.

   c) Determine the zeros of this equation.

   d) What recommendations would you make to Greenthumb Growers in terms of production levels? Explain your reasoning.

   e) How many plants should be produced for revenue to be $5000 more than the cost?

There are many types of mathematical models. The graph showing the growth of the number of bacteria over time is an exponential model. The graph showing the depth of sea level in a harbour is a trigonometric model. However, the focus in this section is on quadratic models.

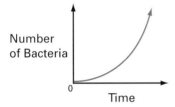

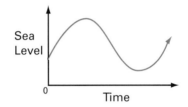

When using mathematical modelling, you need to ensure that you are using the appropriate type of model. To do this, you can look at the differences table, the shape of the graph, or the nature of the expression.

DISCOVER

Identifying the Mathematical Model

1. Copy and complete the following tables of values.

$y = 2x$		$y = 2^x$		$y = x^2 - 2x - 3$		$y = x^3$	
x	**y**	**x**	**y**	**x**	**y**	**x**	**y**
−2		−2		−2		−2	
−1		−1		−1		−1	
0		0		0		0	
1		1		1		1	
2		2		2		2	

2. Add two columns to each of your tables, and calculate the first and second differences for each.

3. Sketch a graph for each relation.

4. Which of the equations models a linear relation? quadratic? exponential? none of these? Explain.

A projectile is a general term for any object that moves through the air while its path is affected by gravity. Under ideal conditions, its height, y, in metres, can be described by the following quadratic function.

$$y = -\frac{1}{2}gt^2 + v_0t + h$$

g is acceleration due to gravity, which is 9.8 m/s^2.
v_0 is the starting vertical velocity of the projectile, in metres per second.

h is the starting height of the projectile, in metres.

t is the time, in seconds.

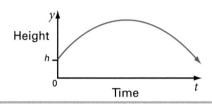

EXAMPLE

Analysing a Quadratic Model

Jared throws a baseball at 30 m/s, and releases it at a height of 1.5 m above the ground. Winnie releases the ball from the same height, but at 40 m/s.
a) Find the equation that models the height of the ball thrown by each person.
b) Find the zeros of each equation. Interpret what the zeros mean in the context.
c) What is the difference in the maximum height of the two throws?
d) Do these models give exact or approximate information? Give reasons for your answer.

Solution

a) The path of a thrown ball can be modelled by

$y = -\frac{1}{2}gt^2 + v_0t + h$, where $g = 9.8$.

So, $y = -4.9t^2 + v_0t + h$.
For Jared's throw, substitute $v_0 = 30$ and $h = 1.5$.
$y = -4.9t^2 + 30t + 1.5$
For Winnie's throw, substitute $v_0 = 40$ and $h = 1.5$.
$y = -4.9t^2 + 40t + 1.5$

For help with graphing, see Appendix B, page 425.

b) For Jared's throw, graph $y = -4.9x^2 + 30x + 1.5$, and find the zeros. Since this is a real time and distance situation, negative values have no meaning. So, use $x \geq 0$.

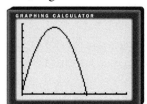

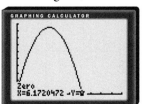

For Jared's throw, the zero is at approximately 6.2.

For Winnie's throw, graph $y = -4.9x^2 + 40x + 1.5$, and find the zeros.

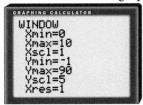 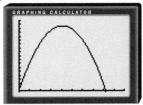 

For Winnie's throw, the zero is at approximately 8.2.

In this context, the zeros tell how long each ball is in the air. It takes 6.2 s for the ball thrown by Jared to reach the ground, and 8.2 s for the ball thrown by Winnie to reach the ground.

c) Find the maximum for each parabola.

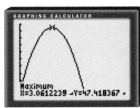 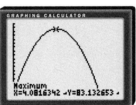

The approximate maximum heights are 47.4 m for the ball thrown by Jared, and 83.1 m for the ball thrown by Winnie. Winnie's ball goes 83.1 − 47.4, or 35.7 m higher.

d) The information obtained is approximate because all the answers are rounded. Also, the model may not reflect the true path of the balls. Air resistance and wind will cause the actual paths to differ from the ideal projectile model.

KEY CONCEPTS

- To model a real-life situation mathematically, first try to determine what type of model fits the data. It may be linear, quadratic, or some other model.

- If the model is quadratic, draw an accurate graphical model, or determine the quadratic equation that represents the data. Be sure to describe what the variables x and y represent. Then, the equation can be used to make predictions or answer questions related to the data.

DISCUSS THE CONCEPTS

1. Is it more accurate to get specific information from a graphical or algebraic model? Explain.

2. Which model, graphic or algebraic, provides a better overall picture of a situation? Explain.

APPLY THE CONCEPTS

B **1.** The Frugal Pizza Kitchen has asked you to determine the greatest number of pieces you can get from a pizza with x cuts. The pieces do not need to be the same size.

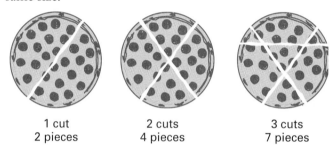

| 1 cut | 2 cuts | 3 cuts |
| 2 pieces | 4 pieces | 7 pieces |

a) Based on this scenario, copy and complete a table as started below.

Number of Cuts	0	1	2	3	4	5	6
Number of Pieces	1	2	4	7			

b) What type of model might fit this data? Justify your answer.

c) Determine an equation to represent the data.

d) Use the equation to determine the maximum number of pieces for 15 cuts.

e) If the Frugal Pizza Kitchen attempted to make a huge pizza that would give 1000 people exactly one piece (no leftovers, of course), how many cuts should be made?

See Appendix B, page 431, for help with quadratic regression using a graphing calculator.

2. A motion sensor is attached to a model rocket, and the rocket is launched. The following data are recorded.

Times (s)	1	2	3	4	5	6	7	8
Height (m)	230	310	350	360	350	300	220	

a) By hand, draw a scatter plot of the data and sketch a curve of best fit.

b) Predict the maximum height reached by the rocket.

c) Extend the curve to predict the height of the rocket at 8 s.

d) Use a graphing calculator to determine the quadratic regression equation. In this expression, what do the variables x and y represent?

e) Use the equation to extend the table to the point where the rocket falls to the ground.

f) Use technology to determine the rocket's maximum height.

g) Compare your answers from parts b) and f).

3. **Paramedics** You are managing a project to develop a new air ambulance. The set-up cost for the project is $50 000. The development cost to design and construct the air ambulance increases with time. From previous contracts, the cost is estimated to be $t^2 + 10t$, where t is the time in working days.

a) Create an equation for the total cost of the air ambulance, including set-up and development costs.

b) There is currently $650 000 allocated for this project. After how long will the project have spent all of these funds?

4. In an attempt to cross the Snake River Canyon on a motorcycle, a daredevil leaves the ramp with an initial vertical velocity of 50 m/s.

a) Assuming the ramps at both sides of the canyon are the same height (use this as $h = 0$), determine an equation for the height of the rider relative to the ground. Recall that an object moving through the air can be modelled by the equation $y = -4.9t^2 + v_0 t + h$.

b) What maximum height would the rider have achieved in ideal conditions, if his motion followed this model and the other side was reached?

c) Unfortunately, the conditions for the stunt were not ideal. The rider underestimated the wind that day. At a point 3.5 s into the jump, a maximum height of only 87 m was reached. Sketch a graph showing this data, and state the times that the rider's height above the ramp was zero.

d) Five seconds into the flight, the daredevil knew the jump was in serious trouble. The rider released the bike, and a safe descent was made into the canyon below with the help of a parachute. If the top of the ramp is 389 m from the canyon floor, how many metres did the daredevil fall from the point the bike was released?

5. **Landscaping** Babu is using square patio stones that have a side length of 3 ft to make a walkway around the outside of his square garden. The total area of the walkway and garden is 225 ft^2.

a) Find the dimensions of the garden.

b) How many patio stones will Babu require? Describe any assumptions that you need to make.

6. **Construction** The Charko family has two horses on their farm. They want to fence in a rectangular area that borders a river. The fence costs \$30/m to install, and they have a budget of \$4500.

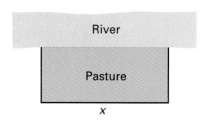

a) What length of fence can they afford to install?

b) Create an equation for the area of the pasture in terms of the length, x.

c) Graph the equation.

d) Determine the length and width that the Charkos should make the pasture to maximize the area for the horses.

7. A formula that models projectiles, using imperial measurements, is $y = v_0 t - 16t^2$, where v_0 is the initial velocity, in feet per second, t is the time, in seconds, and y is the height, in feet. Wile E. Coyote is standing on a springboard that extends over the edge of a high cliff. Roadrunner drops a boulder on the other end of the springboard, sending Wile E. up with an initial velocity of 120 ft/s.

a) What is Wile E.'s height above the cliff after 4 s?

b) Is Wile E. going up or down at this point?

c) At what other time will Wile E. Coyote be the same height as he is at 4 s?

d) When does Wile E. reach the level of the springboard again?

e) On the way down, Wile E. misses the cliff. When will he land in the river that is 350 ft below the top of the cliff?

8. **Product design** An open box is to be constructed from a square piece of cardboard by removing a square with a side length of 8 cm from each corner, and folding up the edges. The resulting box is to have a volume of 512 cm³. Find the dimensions of the original piece of cardboard.

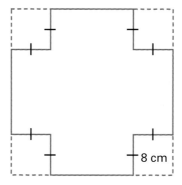

ACHIEVEMENT CHECK Knowledge/Understanding Thinking/Inquiry/Problem Solving Communication Application

9. For each of the following equations and tables, determine whether it represents a linear, quadratic, or exponential relationship. Justify your answers.

a) $5x + 2y = 23$ **b)** $5x^2 + y = 23$ **c)** $y = -2x^2 - 3x + 5$

d)

Time (s)	1	2	3	4	5
Cost ($)	13	20	27	34	41

e)

Time (s)	1	2	3	4	5
Distance (m)	1	5	25	125	625

f)

Time (s)	1	2	3	4	5
Height (m)	3	3	5	9	15

C **10.** **Pharmacology** An Olympic gymnast was stripped of her gold medal after testing positive for a banned stimulant. The team's doctor claimed that the athlete's petiteness—147 cm tall, 37.2 kg—contributed to the positive test. The doctor did not say when the athlete took the medication. Suppose the following readings describe the amount of cold medication in the gymnast's bloodstream.

Hours Since Peak Concentration of Cold Medicine	Milligrams of Cold Medicine per Millilitre of Blood
1	14.0
2	11.5
3	9.4
4	7.7
5	6.3
6	5.2
7	4.3
8	3.5
9	2.9

a) Draw a scatter plot of this data, and sketch a curve of best fit.

b) Interpolate to predict the amount of cold medicine in the gymnast's blood at 2.5 h.

c) Is this a quadratic model? Explain.

d) When would her blood be drug-free?

*C*AREER PROFILE

Computer Programmer Analyst

Computer programmers write the logical steps that computers must follow to complete their tasks. The steps must be logical and efficient for the computer to process the vast amounts of information that they are required to handle. Programmers often interact with employees to ensure the computers are programmed to meet the company's needs. Once a system has been adopted by a company, the computer systems analyst (CPA) works closely with programmers to ensure the smooth running of the system by ridding it of any errors.

Programming graduates who are creative, logical, and clear communicators are in high demand. To meet this demand, many colleges including Seneca and Centennial in Toronto have CPA programs. These programs integrate concepts related to program design, system analysis, networking, and data management, which are used in software development.

Humans rely on multiplication to combine amounts quickly. However, for a computer, multiplication is more time-consuming, so programmers try to use addition instead. Often polynomials can be rewritten in **nested form** to reduce the multiplication involved in evaluating them.

For example, $-3x^3 + 8x^2 - 7x + 1$ means $-3 \times x \times x \times x + 8 \times x \times x - 7 \times x + 1$. There are six multiplications and three additions. The number of multiplications can be reduced by using common factoring in a nested way, as shown.

$$-3x^3 + 8x^2 - 7x + 1 = x(-3x^2 + 8x - 7) + 1$$
$$= x(x[-3x + 8] - 7) + 1$$

This nested form of the polynomial has three multiplications and three additions. If this polynomial were to occur in an application in business, where x is defined for different input values, much computer time, and hence money, would be saved by re-writing the expression in nested form.

1. Express each polynomial in nested form.

   a) $7x^3 + x^2 + 4x - 2$

   b) $4x^3 - 2x^2 + 8x - 5$

2. Expand to write the following in the more familiar polynomial form.

   a) $x[x(x + 2) + 3] - 6$

   b) $x[x(4x - 1) - 5] + 3$

CHAPTER PROBLEM WRAP-UP

After completing the Financial Planning Management stream at George Brown College in Toronto, you set up a small financial service business. You have been contracted by a local sports store, SportStuff, to help it improve its revenue. SportStuff management wishes to determine whether they should continue to carry the Polar Bear Parka line. Using data from previous sales, and information collected from customer surveys, you are to determine mathematical models to represent the costs and revenues associated with Polar Bear Parka.

Refer back to any work you have done so far on this problem. Use your answers from Section 7.1, question 8 and Section 7.2, question 17. Continue the analysis by answering the following.

1. **a)** What should the selling price be for the parka at SportStuff in order to maximize revenue? What is this maximum expected revenue?
   b) The sales manager has challenged staff to sell at least $17 000 worth of Polar Bear parkas in the next month. Determine the selling price and expected number of sales associated with this level of revenue. Give recommendations to the sales staff based on your findings.

2. With further research, you conclude that for each $10 increase in the price of the parka, SportStuff could expect to lose five parka sales per month. Create an expression to model the expected revenue with this scenario.

3. Write a report for SportStuff management advising what changes, if any, should be made to the selling price of the Polar Bear parka. Provide mathematical reasoning wherever possible, including graphs, diagrams, and charts.

Review

7.1 Recognizing and Understanding Quadratic Relationships, pages 308–315

1. Determine whether each model is linear, quadratic, or neither. Include reasons.

a)

x	−5	−1	3	7	11
y	−16	−4	8	20	32

b)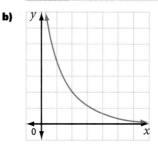

c)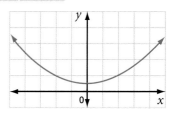

d) $(x - y)^2 = 2y$

e) $y = -3(x + 5)(x - 2)$

f)

x	−7	−2	3	8	13
y	25	0	25	100	225

g)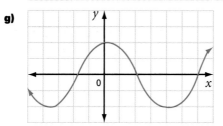

h)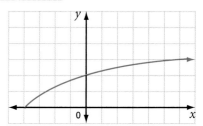

2. Expand and simplify.

a) $y = -2(x + 1)^2 - 3$ **b)** $y = 0.4(x - 1.5)^2 + 12.7$ **c)** $y = -(x - 8)^2$

3. Find the quadratic equation for each graph. Express each one in standard form and in vertex form.

a)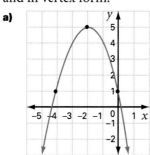

b)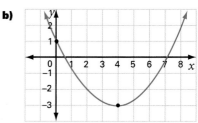

See Appendix B, page 431, for help with quadratic regression using a graphing calculator.

4. a) Use a graphing calculator to find the quadratic equation that models each set of data. Round to two decimal places.

i)

x	2	4	6	8	10
y	7	23	55	103	167

ii)

x	10	20	30	40	50
y	840	3500	8000	14 400	22 100

b) For which set of data is the equation a better fit? Explain.

7.2 Factoring to Find the Zeros of a Quadratic, pages 316–324

5. Factor fully.

a) $x^2 - 3x - 4$ **b)** $6x^2 + 13x - 5$ **c)** $-5x^2 - 17x + 12$

d) $x^2 - 2x - 63$ **e)** $x^2 + 16x + 64$ **f)** $-4x^2 - 20x + 144$

6. Solve by factoring.

a) $x^2 + 7x + 12 = 0$ **b)** $2x^2 + 15x - 50 = 0$ **c)** $x^2 - 2x - 11 = 4$

d) $2x^2 + 11x + 3 = -2$ **e)** $x^2 - 2x = 15$ **f)** $2x^2 + 3 = 5x + 1$

7. **Graphic design** A photograph measuring 4 in. by 6 in. is to be surrounded by a mat before framing. The width of the mat is to be the same on all sides of the photograph. The area of the mat around the photograph is to equal the area of the photograph. Find the width of the mat.

7.3 Solving Quadratic Equations Without Factoring, pages 325–332

8. Express in the form $y = ax^2 + bx + c$, and list the values of a, b, and c.

a) $y - 3 = (4x + 7)^2$

b) $y + 7 - 6x = -\dfrac{1}{2}x^2 - 3x + 2$

c) $y = -2(x + 3)^2 - 4$

d) $y = \dfrac{3}{4}(x + 12)^2 - \dfrac{3}{2}$

9. Solve each equation using technology or the quadratic formula. Round answers to two decimal places.

a) $2x^2 + 3x - 7 = 0$ **b)** $x^2 - 3x = -1$

c) $x^2 - 2x - 5 = 0$ **d)** $(n - 4)(n - 2) = 12$

e) $-0.2x^2 + 2.5x + 1 = -7$ **f)** $1.2x^2 = 1.4x + 1$

10. The profit for a chair manufacturer can be modelled by the equation $P = -0.124x^2 + 105x + 3990$, where x is the total number of chairs made and P is the profit, in dollars.

a) How many chairs must be made to start to make a profit?

b) What is the maximum profit and how many chairs need to be made to yield this profit?

c) What range of production of chairs should this company avoid to ensure that they do not experience a loss?

7.4 Modelling With Quadratics, pages 333–339

11. Psychologists have determined that there is an optimum level of anxiety needed to perform well on a mathematics test. Too much or too little anxiety is said to result in more careless errors. Prior to a math test, the 28 students in a class were asked to rate their own level of anxiety using a self-assessment scale—a rating of 1 being low anxiety and a rating of 9 being high anxiety. The anxiety ratings for these students and the corresponding number of errors made are shown.

Student	Anxiety Rating	Number of Errors Made
Hakim	3	1
Shanita	5	2
Jaffer	2	4
Sydney	7	10
Fran	6	5
Tyler	3	1
Alysha	1	9
Jayden	4	1
Dixie	8	16
Luther	6	5
Tina	5	1
Albany	2	5
Jordan	2	5
Asia	4	1
Sasha	2	6
Xavier	8	18
Jamed	5	2
Alex	1	10
Tamara	7	10
Tovah	5	2
Francesca	3	3
Danielle	5	3
Meredith	5	1
Frank	5	1
Ernest	9	26
Jake	6	6
Audra	1	10
Nick	3	3

a) Find the average number of errors, to the nearest whole number, made for each of the nine anxiety ratings.

b) Organize your results from part a) in a table, and graph them. Does this appear to be a quadratic relationship? Explain.

c) Based on these results, what is the optimum level of anxiety for writing a math test?

d) Find the equation that relates anxiety score, x, to the number of errors made, y.

12. The Ambassador Bridge in Windsor is the longest international suspension bridge. The suspension cables form a shape that resembles a parabola, as does the roadway supported by these cables.

a) The roadway spans a length of 2800 m and reaches a maximum height of 46 m above each end. Sketch the graph of the corresponding parabola, using the y-axis as the axis of symmetry and placing the highest point of the roadway at the origin.

b) Determine the quadratic equation that represents the roadway.

c) If the roadway is to cross each of the support pillars at a height of 30 m, how far from each end should these pillars be placed?

d) Assume that the suspension cables, at their lowest point, are 2 m above the roadway. At their highest point, where they attach to the top of the pillars, they are 53 m above the roadway. Sketch the parabola that approximates the cables on the same axes as in part a). Determine the quadratic equation that models the cables.

13. A golf pro uses technology to analyse your golf swing. She determines that the golf ball leaves your 4-cm tee at an initial vertical velocity of 20 m/s.

a) Assume that the golf ball follows the path of a projectile, modelled by the equation $y = -4.9t^2 + v_0 t + h$. At its maximum height, will the ball clear a tree that is 12.5 m tall?

b) How long is the ball in the air, to the nearest second?

Practice Test

1. Determine whether each relationship is linear, quadratic, or neither.

a)

x	−3	−1	1	3	5	7
y	10	6	2	−2	−6	−10

b) $y = (x - 2)(x + 4)$

c)

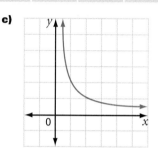

d)

x	−5	−2	1	4	7	10
y	20	3	0	14	48	90

2. Determine the quadratic expression that fits the data given.

a)

x	−4	−2	0	2	4	6
y	57	17	1	9	41	97

b)

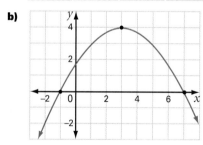

3. Factor fully.

a) $x^2 + 2x - 15$ **b)** $6x^2 + 13x - 5$ **c)** $-12x^2 + 90x - 42$

4. Solve for x algebraically or using technology. Round answers to two decimal places, where necessary.

a) $x^2 - 5x + 4 = 0$ **b)** $2x^2 - 5x - 3 = 0$ **c)** $0 = 4x^2 + 16x + 15$

d) $y = x^2 + 3x - 4, y = 14$ **e)** $4(x + 1)^2 - 1 = 0$ **f)** $y = 3x^2 + 12x + 11, y = 10$

5. Kayla wants to fence a rectangular plot of land beside a lake. Only three sides must be fenced, with the lakeshore being the fourth side. Kayla will use 100 m of fencing, and wants the plot of land to have an area of 1050 m^2. Find the possible dimensions of the plot of land. Draw a diagram to illustrate your solution.

6. As part of a summer festival, a water balloon is launched from a catapult. For precise targeting purposes, stop-motion photography is used and the data are recorded as shown.

Horizontal Distance (m)	Height (m)
0	0.0
2	4.2
4	8.1
6	11.6
8	14.9
10	17.8
12	20.4
14	22.7
16	24.7
18	26.4
20	27.7
22	28.7
24	29.5
26	29.9
28	29.9
30	29.7

a) Construct a scatter plot of the data.

b) Sketch a curve of best fit.

c) Extend the curve to show both zeros. At what horizontal distance will the water balloon hit the ground?

d) What evidence suggests that this is a quadratic relationship?

e) Using technology, determine the quadratic equation that models this data.

f) Verify the zeros determined from the curve of best fit.

g) What is the maximum height reached by the water balloon, and how far from the launch site does this take place?

h) How far from the launch site must a target be if the water balloon is to hit it, on the way down, at a height of 2 m?

ACHIEVEMENT CHECK Knowledge/Understanding Thinking/Inquiry/Problem Solving Communication Application

7. a) Describe a situation that could be modelled with a quadratic equation. Provide suggested input data to support your model, determine the representative quadratic equation, and find the solution.

b) Which method did you use to solve your equation: factoring, the quadratic formula, or a graphing calculator? Why?

c) What changes, if any, would be needed in your situation so that you could use one of the other methods? Describe why these changes are necessary.

INTERPRETING GRAPHICAL AND ALGEBRAIC MODELS

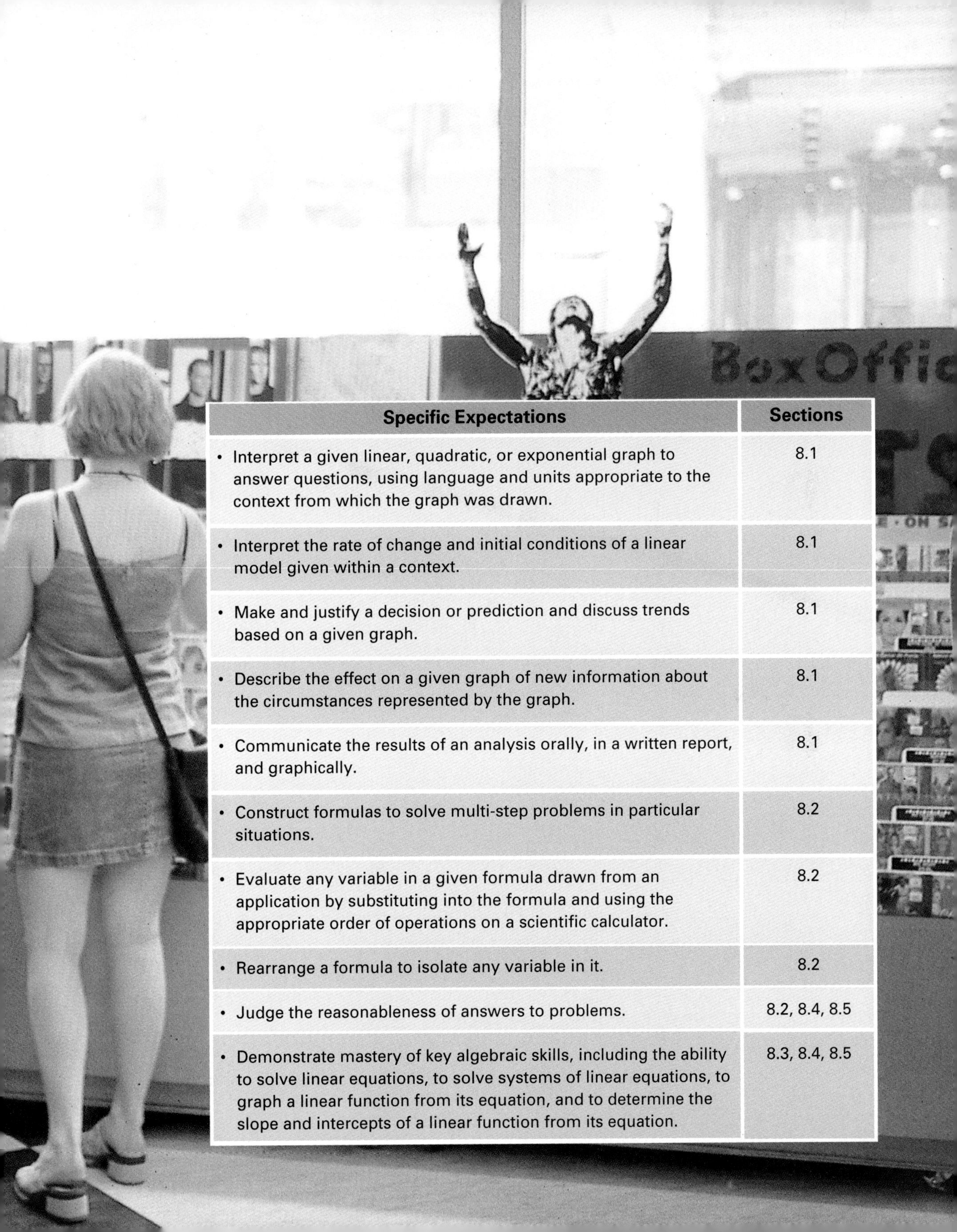

Specific Expectations	Sections
• Interpret a given linear, quadratic, or exponential graph to answer questions, using language and units appropriate to the context from which the graph was drawn.	8.1
• Interpret the rate of change and initial conditions of a linear model given within a context.	8.1
• Make and justify a decision or prediction and discuss trends based on a given graph.	8.1
• Describe the effect on a given graph of new information about the circumstances represented by the graph.	8.1
• Communicate the results of an analysis orally, in a written report, and graphically.	8.1
• Construct formulas to solve multi-step problems in particular situations.	8.2
• Evaluate any variable in a given formula drawn from an application by substituting into the formula and using the appropriate order of operations on a scientific calculator.	8.2
• Rearrange a formula to isolate any variable in it.	8.2
• Judge the reasonableness of answers to problems.	8.2, 8.4, 8.5
• Demonstrate mastery of key algebraic skills, including the ability to solve linear equations, to solve systems of linear equations, to graph a linear function from its equation, and to determine the slope and intercepts of a linear function from its equation.	8.3, 8.4, 8.5

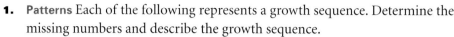

*G*et Ready

1. **Patterns** Each of the following represents a growth sequence. Determine the missing numbers and describe the growth sequence.

   a) 7, 10, 13, …, …, …

   b) 3, 6, 12, …, …, …

   c) 1, 4, 9, 16, …, …, …

   d) …, …, 100, 10, 1, …, …

   e) 5, …, …, …, 25, 30, …, …

   f) …, 0.49, …, 0.81, …

2. **Number skills** Use a calculator and the standard order of operations to evaluate each of the following expressions.

   a) $2[3 + 4(7 - 5)]$

   b) $(3 - 7)^2$

   c) $3^2 - 7^2$

   d) $5 - 3(8 + 6^2)$

   e) $\dfrac{(16 - 12)(14 - 10)}{2}$

   f) $(-5)(-6) - (-4)(10)$

   g) $-3[(5 + 1)(6 - 1)]^2$

   h) $\dfrac{-2(4)^2 - 8}{-5}$

3. **Trigonometry** Use a calculator to evaluate each trigonometric expression. Round to four decimal places, where necessary.

   a) $\sin 30°$

   b) $\cos 45°$

   c) $\tan 15°$

   d) $\dfrac{2 \sin 67°}{5}$

   e) $-0.8 \cos 100°$

   f) $(\tan 30°)^2$

   g) $-3 \cos^{-1} 0.1298$

   h) $\dfrac{1}{2} \tan^{-1} 1.2734$

   i) $\dfrac{2}{3 \cos^{-1} 0.9436}$

4. **Evaluating expressions** Evaluate each expression by substituting $a = 2$, $b = -3$, and $c = -5$.

   a) $2a - 3b + c$

   b) ab^2

   c) $(ab)^2$

   d) abc

   e) $bc - a^2$

   f) b^a

   g) $(a + b)(a - c)$

   h) $(c - b)^3$

5. **Solving equations** Find the value of the variable in each equation.

   a) $2x + 3 = 9$

   b) $7 - 5y = -18$

   c) $3k + 7k - 8 = 22$

   d) $2(6m + 4) = 4$

   e) $6c + 7 = 4c - 9$

   f) $\dfrac{x}{9} = -6$

   g) $\dfrac{h}{6} = \dfrac{7}{3}$

   h) $\dfrac{3}{x} = \dfrac{6}{7}$

   i) $\dfrac{5}{9} = \dfrac{-8}{y}$

   j) $\dfrac{3x - 7}{2} = 4$

   k) $7 = -\dfrac{5e + 3}{6}$

   l) $\dfrac{y + 3}{5} = \dfrac{y - 4}{6}$

6. **Adding or subtracting polynomials** Combine the following expressions by adding or subtracting as indicated. Describe the results.

a) Add
$$3x + 2y$$
$$+ 3x + 7y$$

b) Add
$$2k - 8m$$
$$+ 6k + 8m$$

c) Subtract
$$5x + 6y$$
$$- 4x + 6y$$

d) Subtract
$$8e - 3f$$
$$- 6e - 9f$$

7. **Linear graphs** For each equation,

   i) graph, labelling at least two points

   ii) give the x- and y-intercepts of the line

a) $y = 2x - 5$

b) $y = -\dfrac{1}{2}x + 3$

c) $2x + 3y = 6$

d) $5x - 6y = 30$

8. **Quadratic graphs** Graph each of the following quadratic equations. Label the vertex and at least two other points.

a) $y = x^2$

b) $y = 2x^2$

c) $y = x^2 - 5$

d) $y = -3x^2 + 2$

e) $y = (x - 1)^2$

f) $y = 5(x + 3)^2 - 1$

9. **Exponential graphs** Graph each of the following exponential equations. Label the y-intercept and at least two other points.

a) $y = 2^x$

b) $y = 3^x$

c) $y = 2(10)^x$

d) $y = \left(\dfrac{1}{2}\right)^x$

*C*HAPTER PROBLEM

Remote Recordings are launching the marketing campaign for a new CD. Total revenue, in dollars, can be modelled using this equation:

$R = 100x^2 + 5000x$

The total cost of the campaign, in dollars, can be modelled using this equation:

$C = 5000x + 10\ 000$

In both equations, x is the number of days of advertising in a 45-day campaign. In this

chapter, you will use graphing technology and algebraic techniques to analyse the profit or loss. Then, you will write a report to the advertising executives.

8.1 Interpreting Graphical Models

Graphs can play a major role in modelling a situation. You can interpret graphs by reading information from them, by extending them, and by making predictions based on the model.

Change occurs in daily life. In most mathematical models, minor changes have little or no effect on the model. This is because, when developing an equation to represent the model, the curve-fitting methods take those changes into consideration. However, more significant changes can occur that were unforeseen when developing the model. In models of human populations, major climate changes, war, and leaps in technology all have major impacts on the modelling process. In science, changes in temperature, pressure, or cleanliness may affect the results of an experiment and, by extension, the graph of the results.

DISCOVER

Interpreting Graphs Using a Motion Detector

1. Describe the motion of the person who walked in front of a motion detector to produce this graph. Refer to the person's starting position, time, distance, and speed.

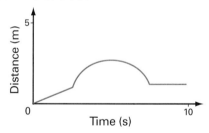

2. Working in groups of two or three, discuss the motion required to reproduce these distance-time graphs.

a)

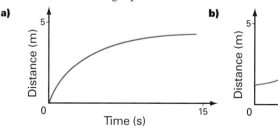

b)

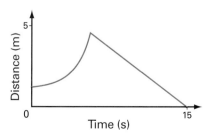

3. Use a motion detector or a computer-based ranger (CBR) attached to a graphing calculator to reproduce the graphs. Start the timer and walk away from or toward the sensor, based on the motion you discussed in question 2. Make at least two attempts to reproduce the graphs accurately.

4. Describe your motion, and your changes in motion, for each graph.

5. How accurate were your graphs on the first attempt? The second attempt? What did you change in your methods between attempts?

DISCOVER

Height of a Bouncing Ball

An experiment measures how high three different balls can bounce when dropped from various heights. The results of the experiment are illustrated in these Fathom™ graphs. A line of best fit has been drawn for each ball.

Average Bounce Heights in cm

	Drop...	Red	Blue	Green	<new>
1	10	2.02	4.95	7.4	
2	20	3.81	10.2	15.07	
3	30	6.12	15.01	23.43	
4	40	7.98	20.31	30.22	
5	50	9.48	24.62	37.13	
6	60	12.1	30.39	44.76	

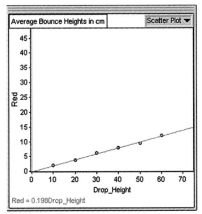

Red = 0.198Drop_Height

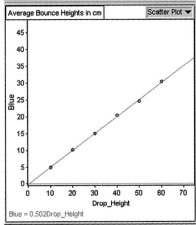

Blue = 0.502Drop_Height

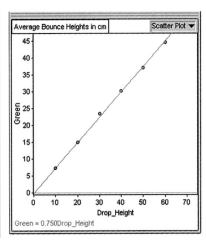

Green = 0.750Drop_Height

1. What is the *y*-intercept for each graph? What does it represent?
2. Which graph represents the bounciest ball? Give reasons.
3. Use interpolation to confirm your answer. Choose a drop height and compare the bounce heights of each ball.
4. Use extrapolation to predict how high each ball should bounce when dropped from 100 cm. Copy the scatter plots, extend each line, and read how high each ball should bounce.
5. If a number were to represent the "bounciness rating" of a ball, what quantity would you use? Give reasons.
6. What is the "bounciness rating" of each ball?

EXAMPLE 1

Height Clearance Under a Bridge

The profile of a bridge archway over a road is modelled in the graph. All measurements are in metres.

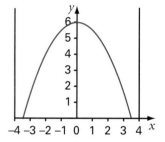

a) Describe the shape of the graph.
b) What is the maximum height of the bridge?
c) How wide is the bridge at its base?
d) Could a truck that is 2.4 m wide and 4.6 m tall fit under the bridge?
e) What is the maximum height of a 2.4 m-wide truck that could fit under the bridge?
f) Suppose the road under the bridge is two lanes wide. What is the maximum height of a 2.4-m wide truck that could fit under it, while remaining in its own lane?

Solution

a) The graph is parabolic in shape.
b) The maximum height of the bridge is 6 m.
c) The bridge is about 7 m wide at its base.
d) One half of the truck is on each side of the road. Interpolate at $x = 1.2$. At this value of x, $y \doteq 5.3$, which is greater than 4.6.

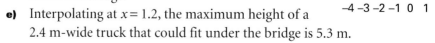

So, a truck that is 2.4 m wide and 4.6 m tall could fit under the bridge.

e) Interpolating at $x = 1.2$, the maximum height of a 2.4 m-wide truck that could fit under the bridge is 5.3 m.
f) Interpolate at $x = 2.4$. At this value of x, $y \doteq 3.1$.
The maximum height of a 2.4 m-wide truck remaining in its own lane is 3.1 m.

EXAMPLE 2

Predicting Population Growth

The graph shows the growth of the world's population, in millions, between the years 900 and 2000.

a) Describe the type of growth.

b) Use interpolation to estimate the world's population in 1492.

c) Estimate the world's population in 1975.

d) Use extrapolation to predict the world's population in 2025.

e) What implications do these predictions have?

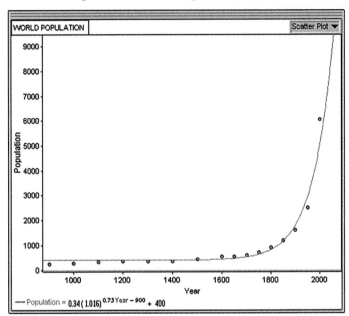

Solution

a) The population is growing exponentially.

b) The world's population in 1492 was approximately 500 million.

c) The world's population in 1975 was approximately 4 billion.

d) The world's population in 2025 will be approximately 6.7 billion.

e) The world's population may become too large to sustain itself.

EXAMPLE 3

Effects of Changes in Weather

The population of an invasive weed increases over an eight-year period. The graph models the growth, assuming no pesticides are used. The growth factor is 1.5, so the equation is

$$y = 100(1.5)^x$$

where x is the number of years and y is the number of weed plants in a particular area.

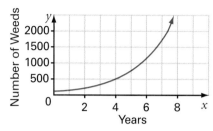

a) Describe the type of growth that is modelled by this graph.

b) Suppose that, in the fifth year, there is major frost damage to the native plant population. This allows the weed population to grow by a factor of 2 for one year. The weed population then returns to its regular growth pattern. How would the graph of this model change?

Solution

a) This graph is modelling exponential growth.

b) During the sixth year, the graph would increase by a greater amount than before. In the following years, the graph would have the same shape as before, but it would be stretched vertically.

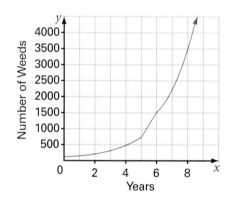

KEY CONCEPTS

- Graphs can be used to model situations, if they are based on statistical trends or on accurate calculations.

- Predictions may be made from graphs if the model can be accurately extended. To interpolate a graph, estimate values between accurate readings. To extrapolate a graph, extend the graph and make a prediction.

- New information or major changes in data can change a graph, usually by shifting it or stretching it.

DISCUSS THE CONCEPTS

1. Describe the differences between linear, quadratic, and exponential graphs. Include diagrams.

2. Describe how a graph would change if the quantity being modelled increased at a greater rate than expected. Explain your reasoning.

PRACTISE

A 1. A cyclist must travel up a steep hill at the beginning of a race. Which graph represents the cyclist's distance from the starting point? Why?

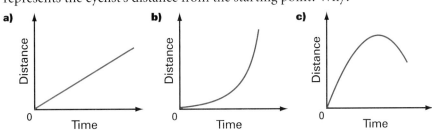

2. Chloë is riding a Ferris wheel. Which graph best represents Chloë's distance above the ground? Why?

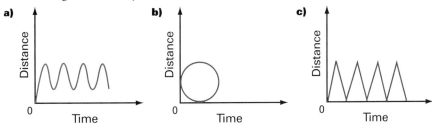

APPLY THE CONCEPTS

3. **Target heart rate** A graph in a fitness club shows target heart rates, in beats per minute, for fit people doing aerobic exercise.

   a) Why does the graph show only target heart rates for people between the ages of 20 and 70 years?

   b) What is the target heart rate for a person 35 years of age?

   c) Extrapolate the graph to find the target heart rate for a person 18 years of age.

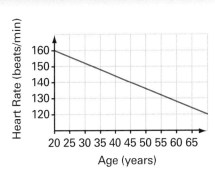

4. Use a motion detector or a computer-based ranger (CBR) to reproduce the following velocity-time graphs. Begin the timer and walk away from or toward the sensor. Describe your movements in each case.

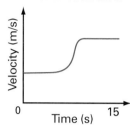

5. Converting temperatures This graph is used in a radio station as a quick reference to convert temperatures between Fahrenheit and Celsius.

a) Why would a graph be used instead of a formula?

b) What is the equivalent Fahrenheit temperature of 0°C?

c) What is the equivalent Celsius temperature of 90°F?

d) Use extrapolation to determine the temperature at which both scales give the same value.

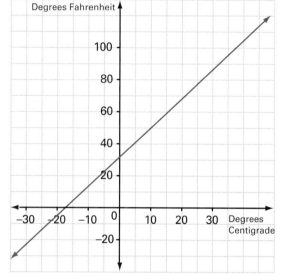

6. Predicting fuel economy The graph shows a scatter plot and a curve that models the average fuel economy of a particular car, in litres per 100 km, at various speeds, measured at a test track.

a) What is the fuel economy at 30 km/h?

b) At what speed does this car have its best fuel economy? What is that fuel economy?

c) What would happen to this curve for a driver who accelerates hard?

d) This car does not have very good fuel economy. What would a graph look like for a car with better fuel economy? Explain.

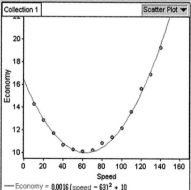

7. The period of a pendulum (the time for one full swing, back and forth) depends on the length of the pendulum. This graph shows the period, in seconds, for pendulums of different lengths, in centimetres.

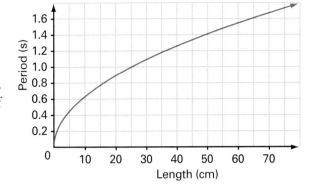

a) What is the period of a pendulum 6 cm long?

b) What is the length of a pendulum with a period of 1 s?

c) Extrapolate to find the length of a pendulum with a period of 2 s.

d) Many clocks have an adjustable pendulum to correct the speed of the clock. What would you need to do to the length of a pendulum in order to speed up a clock? Why?

8. Communication Advertisers use mathematical models to estimate the effectiveness of an advertising campaign. The graph provides an "Effectiveness Index" of a television advertisement after viewing it a number of times.

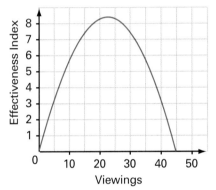

a) What is the Effectiveness Index for an advertisement viewed 15 times?

b) Why does the graph have the shape shown?

c) Is it possible to have a negative Effectiveness Index? Explain.

9. Problem solving The graph models the trajectory of a thrown football, where h is the height in metres, and d is the distance in metres from the quarterback.

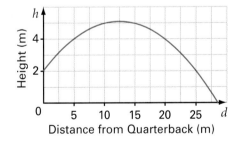

a) Describe the trajectory of the football.

b) What is the vertical intercept? What does it represent?

c) What is the height of the football when it is 20 m from the quarterback?

d) The ball was caught by a diving receiver, 0.5 m above the ground. How far from the quarterback was the ball caught?

10. Thinking/Inquiry/Problem Solving During the 1990s, the world's population grew at a rate of 1.4% per year. In 1999, the world's population reached 6 billion people. During the 1990s, the world's urban population grew at a rate of 2.5% per year, reaching about 3 billion people in 1999. The graph illustrates these data.

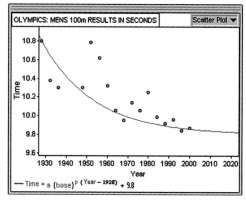

a) Use extrapolation to predict when the world's population will be entirely urban.

b) Use a graphing calculator to confirm your prediction in part a). Use the equations $y = 6(1.014)^x$ and $y = 3(1.025)^x$, where x is the number of years after 1999 and y is the population in billions.

c) Do you think this is a valid prediction? Explain.

11. Communication The scatter plot shows the winning times for the mens' 100-m race at the Olympics since 1928. A curve modelling the times has been drawn.

	Year	Time	<ne\
1	1928	10.8	
2	1932	10.38	
3	1936	10.3	
4	1948	10.3	
5	1952	10.79	
6	1956	10.62	
7	1960	10.32	
8	1964	10.06	
9	1968	9.95	
10	1972	10.14	
11	1976	10.06	
12	1980	10.25	
13	1984	9.99	
14	1988	9.92	
15	1992	9.96	
16	1996	9.84	
17	2000	9.87	

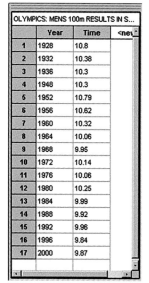

OLYMPICS: MENS 100m RESULTS IN S...

OLYMPICS: MENS 100m RESULTS IN SECONDS — Scatter Plot

Time = a (base)$^{P \, (Year - 1928)}$ + 9.8

a) Describe the shape of the curve.

b) Use the curve to predict the winning time for the mens' 100 m at the 2004 Olympics.

c) Use the curve to predict the winning time for the mens' 100 m at the 2020 Olympics.

d) Discuss the accuracy of the curve, and thus, the accuracy of your predictions.

12. Stopping distances The graph on the left models the average stopping distance for cars travelling at various speeds on dry asphalt. The graph on the right models the average stopping distance for cars travelling at various speeds on wet asphalt. The average stopping distance is measured in metres, and speed is measured in kilometres per hour.

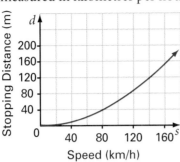
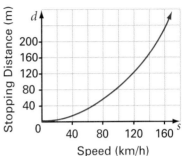

a) Compare the stopping distances on dry and wet asphalt for cars travelling at
   i) 40 km/h
   ii) 100 km/h
   iii) 120 km/h

b) How fast would a car have been travelling on wet asphalt if it had the same stopping distance as a car travelling at 90 km on dry asphalt?

c) Can there be a speed at which the stopping distances on both surfaces are the same? Explain.

***C*HAPTER PROBLEM**

13. Remote Recordings are launching the marketing campaign for a new CD. Total revenue, in dollars, can be modelled using this equation:
$R = 100x^2 + 5000x$
The total cost of the campaign, in dollars, can be modelled by this equation:
$C = 5000x + 10\ 000$
In these equations, x is the number of days completed out of a 45-day advertising campaign.

a) Use appropriate technology to graph the two equations for 45 days of advertising.

b) Use the graph to estimate the profit or loss after 7 days of advertising.

Throughout this course, you have evaluated formulas by substituting known values and calculating the results. You have also rearranged formulas after first substituting values for some of the variables.

Often, it is important to rearrange the entire formula in order to solve for a different variable.

In the last section, you used a graph to convert temperatures from Fahrenheit to Celsius. You can also describe the relation between these two scales using this formula:

$$C = \frac{5}{9}(F - 32)$$

In this form, the formula has the variable C isolated, so you can calculate Celsius temperatures from Fahrenheit temperatures. To convert in the opposite direction, you need to rearrange the formula to isolate the variable F.

DISCOVER

Isolating the Variable

1. For each equation, state the first operation that needs to be used in order to isolate the x-term. Explain why you chose that operation.

    a) $2x + 1 = 7$ **b)** $7x - 8 = 2$

    c) $9 + 3x = 12$ **d)** $6 - 4x = 10$

    e) $-5x = 10$ **f)** $x^2 - 5 = 11$

    g) $3(-4x + 1) = 6$ **h)** $\dfrac{2x - 6}{8} = 7$

    i) $2x + y = 8$ **j)** $y = mx + b$

2. In these formulas, state the first operation needed to isolate the variable named, and perform this operation.

a) $\dfrac{3-7y}{8} = 7 + 8x$; isolate y

b) $5p + 7q = 0$; isolate q

c) $V = \dfrac{1}{3}\pi r^2 h$; isolate r

d) $12 - \dfrac{V}{Q} = \dfrac{R}{t}$; isolate t

3. In each rearrangement, describe the operation used in each step.

a)
$$v^2 = u^2 + 2as$$
$$v^2 - u^2 = 2as$$
$$\dfrac{v^2 - u^2}{2a} = s$$
$$s = \dfrac{v^2 - u^2}{2a}$$

b)
$$v^2 = u^2 + 2as$$
$$v^2 - 2as = u^2$$
$$\sqrt{v^2 - 2as} = u$$
$$u = \sqrt{v^2 - 2as}$$

EXAMPLE 1

One-Step Rearranging

A farmer's field is to be subdivided into rectangular housing plots of various sizes and dimensions. The area of a rectangle is given by the formula $A = lw$.

a) What is the length of a housing plot with an area of 700 m^2 and a width of 10 m?

b) Rearrange the formula, solving for l.

c) What is the length of a housing plot with an area of 600 m^2 and a width of 15 m?

d) What is the length of a housing plot with an area of 850 m^2 and a width of 18 m?

Solution

a) Substitute $A = 700$ and $w = 10$ into the formula. Then, solve for l.

$$A = lw$$
$$700 = l(10)$$

$10l = 700$ Exchange sides to get l on the left.

$l = \dfrac{700}{10}$ Divide both sides by 10.0.

$l = 70$

A plot with an area of 700 m^2 and a width of 10 m will have a length of 70 m.

b)

$A = lw$ You need to isolate l.

$\dfrac{A}{w} = l$ l is multiplied by w, so divide both sides by w (inverse of multiplication).

$l = \dfrac{A}{w}$ Exchange left and right sides.

c) Substitute for A and w in the formula developed in part b).

$$l = \frac{A}{w}$$

$$l = \frac{600}{15}$$

$$l = 40$$

A plot with an area of 600 m² and a width of 15 m will have a length of 40 m.

d)
$$l = \frac{850}{18}$$

$$l \doteq 47.2$$

A plot with an area of 850 m² and a width of 18 m will have a length of 47.2 m, to the nearest tenth of a metre.

When evaluating an expression, you use this order of operations, often referred to as "BEDMAS":

Brackets

Exponents

Division
Multiplication } in the order written

Addition
Subtraction } in the order written

When rearranging a formula, you use the same basic steps as when solving a simple equation. However, you must reverse the steps and the operations, "undoing" the order of operations:

Subtraction
Addition } in the order written

Multiplication
Division } in the order written

Exponents

Brackets

EXAMPLE 2

Two-Step Rearranging

The velocity of an object thrown upward into the air can be modelled with the equation $v = -9.8t + u$

where v is the upward velocity, in metres per second, at time t, in seconds, and u is the initial upward velocity, in metres per second.

a) Rearrange the formula and solve for t.

b) When would a rock thrown upward at 12 m/s reach its maximum height?

c) How long would it take for a rock to reach a velocity of −10 m/s (or 10 m/s *downward*), if it were thrown upward at 15 m/s?

Solution

a)

$$v = -9.8t + u \quad \text{Isolate } 9.8t \text{ first, then } t.$$

$$9.8t + v = u$$

$$9.8t = u - v$$

$$t = \frac{(u - v)}{9.8} \qquad t \text{ is multiplied by 9.8, so divide by 9.8.}$$

b) At maximum height, the velocity will be 0 m/s. Substitute $v = 0$ and the initial velocity, $u = 12$, into the equation for t.

$$t = \frac{(u - v)}{9.8}$$

$$t = \frac{(12 - 0)}{9.8}$$

$$t \doteq 1.2$$

The rock would reach its maximum height after 1.2 s, to the nearest tenth.

c)

$$t = \frac{[15 - (-10)]}{9.8}$$

$$t = \frac{25}{9.8}$$

$$t \doteq 2.6$$

The rock would reach a velocity of −10 m/s after 2.6 s, to the nearest tenth.

EXAMPLE 3

Multi-Step Rearranging

To determine a person's keyboarding speed on a word processor, the formula

$$s = \frac{w - 10e}{t}$$

is used, where s is the keyboarding speed in words per minute, w is the number of words keyed, e is the number of errors, and t is the time in minutes.

a) Rearrange the formula, solving for the number of errors, e.

b) A person types 600 words in 6 min, and has a keyboarding speed of 75 words per minute. How many errors were made?

Solution

a) First, you need to isolate $w - 10e$, then the term $10e$, and finally the variable e.

$$s = \frac{w - 10e}{t}$$

$$st = w - 10e \qquad \text{Dividing by } t \text{ is the final operation, so multiply by the variable } t.$$

$$st + 10e = w \qquad \text{Add } 10e \text{ to make the } e\text{-term positive and on the left side.}$$

$$10e = w - st \qquad st \text{ is now added to the } e\text{-term, so subtract } st.$$

$$e = \frac{w - st}{10} \qquad e \text{ is multiplied by 10, so divide by 10.}$$

b)

$$e = \frac{w - st}{10}$$

$$e = \frac{600 - (75)(6)}{10}$$

$$e = 15$$

The person made 15 keyboarding errors.

EXAMPLE 4

Formulas Involving Exponents

The area of a circle can be found using the formula $A = \pi r^2$, where A is the area and r is the radius.

a) Rearrange the formula and solve for r.

b) What is the radius of a circle with an area of 25 cm^2?

c) Use a spreadsheet to determine the radii of circles with areas between 10 cm^2 and 50 cm^2, in increment steps of 5 cm^2.

Solution

a) $A = \pi r^2$ First isolate r^2, then r.

$\dfrac{A}{\pi} = r^2$ r^2 is multiplied by π, so divide by π.

$r^2 = \dfrac{A}{\pi}$

$r = \sqrt{\dfrac{A}{\pi}}$ The opposite of raising to an exponent 2, or squaring, is taking the square root.

b) $r = \sqrt{\dfrac{A}{\pi}}$

$r = \sqrt{\dfrac{25}{\pi}}$ Use the exact value of π on your calculator, not 3.14.

$r \doteq 2.8$

The radius of a circle with an area of 25 cm^2 is 2.8 cm, to the nearest tenth.

c) In the formula bar, enter the appropriate formula:

Quattro Pro: @SQRT(*Cell reference*/(@Pi))

Excel: =SQRT(*Cell reference*/Pi())

Then, fill the appropriate cells with the formula and appropriate cell references:

Quattro Pro: Highlight the cells and select Edit…Fill…Quick Fill

Excel: Highlight the cells and select Edit…Fill…Down

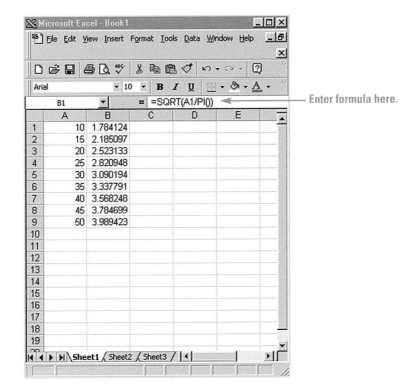

Enter formula here.

=SQRT(A1/PI())

	A	B
1	10	1.784124
2	15	2.185097
3	20	2.523133
4	25	2.820948
5	30	3.090194
6	35	3.337791
7	40	3.568248
8	45	3.784699
9	50	3.989423

EXAMPLE 5

Solving for the Denominator

The formula to find the average mark of a class of n students is

$$A = \frac{T}{n}$$

where A is the average mark and T is the sum of all the marks.

a) Rearrange the formula and solve for n.

b) How many students are in a class with an average mark of 68 and a class total of 1428 marks?

Solution

a) $A = \dfrac{T}{n}$

$nA = T$ Since n is in the denominator, multiply both sides by n.

$n = \dfrac{T}{A}$ n is multiplied by A, so divide by A.

b) $n = \dfrac{1428}{68}$

$n = 21$

There are 21 students in the class.

EXAMPLE 6

Rearranging Trigonometric Formulas

The width of the image on the film in a camera can be calculated using the formula $W = 2F \tan\left(\dfrac{A}{2}\right)$, where W is the width in millimetres, A is the field of view in degrees, and F is the focal length of the lens in millimetres.

a) Rearrange the formula, solving for the field of view, A.

b) To the nearest degree, what field of view is necessary for a lens with a focal length of 50 mm to provide an image 35 mm wide?

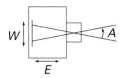

Solution

a)
$$W = 2F \tan\left(\frac{A}{2}\right)$$ You need to isolate $\tan\left(\dfrac{A}{2}\right)$ first, then $\dfrac{A}{2}$, and then A.

$$\frac{W}{2F} = \tan\left(\frac{A}{2}\right)$$ $\tan\left(\dfrac{A}{2}\right)$ is multiplied by $2F$, so divide by $2F$.

$$\tan^{-1}\left(\frac{W}{2F}\right) = \frac{A}{2}$$ To isolate $\dfrac{A}{2}$, use the inverse of tan.

$$2\tan^{-1}\left(\frac{W}{2F}\right) = A$$ Multiply by 2 to isolate A.

$$A = 2\tan^{-1}\left(\frac{W}{2F}\right)$$

b)
$$A = 2\tan^{-1}\left(\frac{W}{2F}\right)$$

$$A = 2\tan^{-1}\left(\frac{35}{2 \times 50}\right)$$

$$A \doteq 39°$$

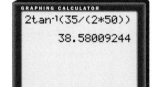

The field of view for a 50-mm lens and a 35-mm image would be 39°.

KEY CONCEPTS

- When rearranging a formula, "undo" the order of operations:

  Subtraction
  Addition $\Big\}$ in the order written

  Multiplication
  Division $\Big\}$ in the order written

  Exponents

  Brackets

DISCUSS THE CONCEPTS

1. Describe the sequence of steps to rearrange the formula $y = mx + b$ and solve for x.

2. When rearranging the formula $I = \dfrac{k}{d^2}$ to solve for d, a student performed the following steps. Find the error(s) and make the appropriate correction(s). Explain your reasoning.
$$d^2 = kI$$
$$d = kI^2$$

PRACTISE

Ⓐ 1. Substitute the values given and then, solve for the indicated variable.
  a) $y = mx + b$; $y = 10$, $m = 3$, $b = 4$; solve for x
  b) $I = Prt$; $I = \$30$, $P = \$1000$, $t = 0.5$ years; solve for r
  c) $S = \dfrac{d}{t}$; $S = 120$ km/h, $t = 4$ h; solve for d
  d) $P = 2(l + w)$; $P = 100$ m, $l = 30$ m; solve for w
  e) $x^2 + y^2 = r^2$; $y = 5$ mm, $r = 13$ mm; solve for x
  f) $a = b \sin C$; $a = 12$ cm, $b = 20$ cm; solve for $\angle C$

2. Rearrange each of the following formulas and solve for the indicated variable.
  a) $y = mx + b$; solve for x **b)** $I = Prt$; solve for r
  c) $S = \dfrac{d}{t}$; solve for d **d)** $P = 2(l + w)$; solve for w
  e) $x^2 + y^2 = r^2$; solve for x **f)** $a = b \sin C$; solve for $\angle C$

APPLY THE CONCEPTS

3. Photographic developing costs The cost, C, in dollars, to develop and print n photographs is given by the equation $C = 0.28n + 4$.
  a) Rearrange the formula and solve for n.
  b) What size roll of film could be developed and printed for \$10.72?
  c) What size roll of film could be developed and printed for under \$15?

Ⓑ 4. Exercise safety It is not safe for an adult to surpass her or his maximum heart rate. This maximum heart rate, M, in beats per minute, is modelled by the equation $M = 230 - 1.2A$
  where A is the age of the adult, in years.
  a) Rearrange the formula, solving for A.
  b) At what age should a person's maximum exercising heart rate be
    i) 194 beats per minute? **ii)** 134 beats per minute?

5. The cost, C, in dollars, of producing a school yearbook is given by the formula
$$C = S + 4n$$
where S is the set-up cost and n is the number of yearbooks printed.
   a) Solve the formula for n.
   b) If the set-up cost is $925, how many yearbooks can be printed for $5000?
   c) If the set-up cost is $1500, how many yearbooks can be printed for $10 000?

6. **Application** To determine the amount of an investment with simple interest, you use the formula $A = P(1 + rt)$
where A is the amount in dollars, P is the principal in dollars, r is the interest rate, and t is the time in years.
   a) Rearrange the formula and solve for P.
   b) What principal must be invested at an interest rate of 5% for 8 years to end up with $2000?
   c) Using a spreadsheet or a calculator, make a chart to calculate the principals invested at different rates, r, to end up with $10 000 after 10 years. Vary r between 3% and 10%, in steps of 0.5%.

7. **Currency conversion** The formula to convert from a foreign currency to Canadian dollars is $C = Fr$, where C is the value in Canadian dollars, F is the foreign currency, and r is the exchange rate.
   a) What is the value, in Canadian dollars, of $500 U.S., if the exchange rate is $1.524 per U.S. dollar?
   b) Rewrite the formula, solving for r.
   c) Adam bought 7200 Jamaican dollars for $250 Canadian. What is the exchange rate, in Canadian dollars per Jamaican dollar?
   d) Maria bought 860 000 Italian lire for $600 Canadian. What is the exchange rate, in Canadian dollars per lira?
   e) How could you express the exchange rate in part d) in lire per Canadian dollar, instead of Canadian dollars per lira? Explain. What is the advantage of expressing the exchange rate in this way?

8. The period of a pendulum (time for one full swing back and forth) can be calculated using the formula $p = 0.2\sqrt{l}$
where l is the length of the pendulum in centimetres and p is the period in seconds.
   a) What is the period of a pendulum that is 30 cm long?
   b) Rewrite the formula, solving for l.
   c) How long does a pendulum need to be for the period to be 1 s?
   d) How long does a pendulum need to be for the period to be 2 s?
   e) Explain why the answer in part d) is not double the answer in part c).

9. **Application** The height of a stone dropped from the top of a cliff can be found using the formula $h = -9.8t^2 + k$

where h is the height, in metres, after t seconds, and k is the initial height, in metres.

a) A stone is dropped from a 20-m cliff. After how many seconds will the height of the stone be 10 m? Substitute for h and k in the formula and calculate.

b) Rearrange the formula to solve for t.

c) Confirm your answer to part a), using the rearranged formula from part b).

d) When will a stone dropped from a 50-m cliff land?

e) Using a spreadsheet or a calculator, make a chart showing the times taken for a stone, dropped from a 50-m cliff, to reach heights from 50 m down to 0 m, in steps of 5 m.

10. When an object is pulled by a force, F, measured in newtons, the force applied in a horizontal direction (the horizontal component) is given by the formula

$F_H = F \cos A$

where A is the angle the force makes with the horizontal.

a) Calculate the horizontal force for a force of 50 N applied at each of these angles.

   i) $\angle A = 25°$ **ii)** $\angle A = 10°$

b) Rewrite the formula to solve for A.

c) Calculate the angle at which a force of 100 N is applied, given these horizontal components in newtons.

   i) $F_H = 80$ **ii)** $F_H = 25$

11. To calculate the average speed of a moving object, use the formula

$$S = \frac{d}{t}$$

where S is the speed in kilometres per hour, d is the distance in kilometres, and t is the time in hours.

a) Rewrite the formula and solve for t.

b) How long will it take to drive 400 km at an average speed of 80 km/h?

c) How long will it take to walk 500 m at a speed of 6 km/h?

12. Machining dimensions Machinists often need to taper steel rods that are to be used in tools. The formula

$$T = \frac{D - d}{L}$$

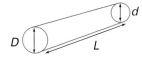

calculates the average taper of a rod of length L, where D is the diameter of the larger end and d is the diameter of the smaller end.

a) Calculate the average taper of a rod 50 cm long, with end diameters of 2.8 cm and 2.2 cm.

b) Solve the formula for D. Then, calculate the diameter of the larger end of a 40-cm rod with a smaller diameter of 3.7 cm and an average taper of 0.01.

c) Solve the formula for d. Then, calculate the diameter of the smaller end of a 32-cm rod with a larger diameter of 4 cm and an average taper of 0.04.

d) Solve the formula for L. Then, calculate the length of a rod that has an average taper of 0.005 and end diameters of 2.1 cm and 1.8 cm.

13. If a baseball were hit on different planets, it would travel different distances, because the gravitational pull varies from planet to planet. Ignoring the effects of the atmosphere, the following formula estimates the distance, d, in metres, that a ball will travel with an initial velocity of 100 km/h.

$$d = 771.6 \sin\left(\frac{2A}{g}\right)$$

In this formula, A is the angle of elevation at which the ball is struck, and g is the acceleration due to gravity, in metres per second squared.

a) Rewrite the formula, solving for A.

b) If a ball is to be hit a distance of 150 m, estimate the angle at which the ball would need to be hit on each planet, given the gravitational pull in metres per second squared.

    i) Earth, where $g = 9.8$ **ii)** Mercury, where $g = 3.7$

    iii) Jupiter, where $g = 23.1$ **iv)** Pluto, where $g = 0.6$

ACHIEVEMENT CHECK   Knowledge/Understanding   Thinking/Inquiry/Problem Solving   Communication   Application

14. Each of these graphs shows how the height of water in a vase rises as it is filled. Each vase holds the same amount of water and the rate of pouring is always the same. Sketch a possible shape for each vase and explain how you developed your sketch.

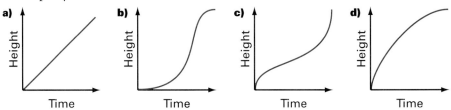

C **15. Problem Solving** The surface area of a cylindrical tank is given by the formula

$$S = 2\pi(rh + r^2)$$

where r is the radius, h is the height, and S is the surface area.

a) Rearrange the formula to solve for h.

b) What is the height of a cylindrical tank with a surface area of 113 m^2 and a radius of 2 m?

c) What is the height of a cylindrical tank with a surface area of 200 m^2 and a radius of 5 m?

d) A can of paint will cover a surface area of 10 m^2. Write a formula for n, the number of cans of paint required to paint a given area, S.

e) Combine the formula in part d) with the formula for the surface area of a cylinder. What is the tallest cylinder with a radius of 2 m that could be painted with just eight cans of paint?

Career Link

Jewellers need a good grasp of optics to create eye-catching work. For example, the higher the refractive index of a gemstone, the more rays of light tend to bounce back and forth inside the gem, creating its sparkle. Diamond's high refractive index is thus part of its beauty. Find out about the training and education needed to work in the jewellery industry.

16. Refraction When a ray of light is shone through different substances, the light refracts by different amounts. Most transparent substances have a constant refractive index. The following formula can be used to find the refractive index, angle of incidence, or angle of refraction.

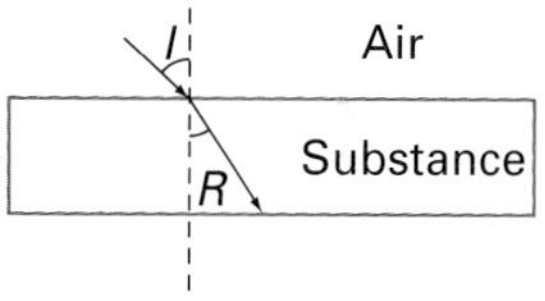

$$n = \frac{\sin I}{\sin R}$$

In this formula, n is the refractive index, I is the angle of incidence, and R is the angle of refraction, as shown in the diagram.

a) When the angle of incidence in air is 25.0°, the angle of refraction in water is 18.5°. Determine the refractive index of water.

b) Solve the formula for I.

c) The refractive index of quartz in air is 1.54. Determine the angle of incidence for a ray of light whose angle of refraction is 32.3°.

d) Solve the formula for R.

e) The refractive index of diamond in air is 2.42. Determine the angle of refraction for a ray of light with an angle of incidence of 60.0°.

f) Go to *www.mcgrawhill.ca/links/MPCA12* and follow the links for this question. Use the information you find to write a brief report on the use of mathematics in gemmology.

Graphing and Interpreting Linear Models

Linear models are some of the most common mathematical models studied. They give a mathematically simple description of a situation. Their uses vary from predicting life expectancy to determining the best buy from a range of pricing options.

In Chapter 5, you generated linear models based on sample data. In Section 8.1, you began to study and interpret linear graphs. You can enhance your interpretive skills by exploring the slope and intercepts of linear graphs, based on their equations.

Manipulating linear equations allows you to identify these key features.

- The **slope** of a linear graph is calculated as "rise divided by run." It has many practical applications; for example, if the horizontal axis represents time, the slope represents the rate of change of the quantity plotted on the vertical axis.
- The **x-intercept** is the distance from the origin of the point where the graph crosses the x-axis. It can be important in predicting when a quantity will be zero.
- The **y-intercept** is the distance from the origin of the point where the graph crosses the y-axis. It often represents an initial value in a process; for example, a business start-up cost.

DISCOVER

Slope and Intercepts

Two linear models are compared in this graph.

1. The general formula for a linear model is $y = mx + b$. Compare the values of m and b for each of the models in the graph.
2. Explain what m and b represent in the general formula.
3. Which of the lines in the graph has the greater value of m, and how does this affect its shape?

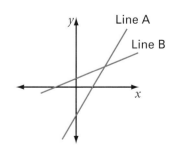

4. **a)** State the value of m for the line $y = 2 - 3x$.
   b) Is the slope of this line positive or negative?
   c) How could you determine the sign of the slope, from looking at the graph?

5. Describe a method to determine the slope and y-intercept of a linear equation given in any form. Then, apply your method to the following equations.
   a) $y - 3x = -7$ **b)** $5x = 1 + 2y$ **c)** $2x + 0.5y + 1 = 0$

6. To find the x-intercept of a line, you can use the fact that when the y-coordinate is zero, the x-coordinate is equal to the x-intercept.
   a) Describe a method of finding the x-intercept of the line $2y = -9x + 18$.
   b) Apply your method to find the x-intercept.
   c) Check your answer by plotting the line and finding the x-intercept graphically.

7. Make a sketch of each of these linear models, by plotting the x- and y-intercepts and drawing a line through them.
   a) $y = 7x - 2$ **b)** $40x - 5y = 15$ **c)** $4x = 10 - 8y$

8. Compare the slopes of the three graphs in question 7, discussing their direction and steepness.

EXAMPLE 1

Using Intercepts to Draw a Graph

A tray of ice cubes was taken out of the freezer and accidentally left to melt. The melting process follows a linear model. Initially, the ice cubes were 25 mm thick. They were completely melted after 125 min.

a) Make a graph of the linear model for the melting process. Mark both intercepts on your graph.
b) Use interpolation to determine the time at which the ice was 15 mm thick.
c) What is the rate of change in the thickness of the ice, in millimetres per minute?
d) Write the equation of the graph in the form $y = mx + b$.

Solution

a) From the information given, the y-intercept is $(0, 25)$ and the x-intercept is $(125, 0)$. Plotting these two points and drawing a line through them gives the graph shown.

b) By interpolation, the ice was 15 mm thick after 50 min of melting.

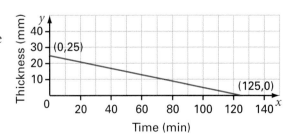

c) Between the *x*- and *y*-intercepts, the rise is −25 and the run is 125. Thus,

$$\text{slope} = \frac{\text{rise}}{\text{run}}$$

$$\text{slope} = \frac{-25}{125}$$

slope = −0.2

The slope gives a rate of change in the thickness of the ice of −0.2 mm/min. That is, the ice melted at a rate of 0.2 mm each minute.

d) From part a), *b* = 25. From part c), *m* = −0.2. So, the equation of the line is $y = -0.2x + 25$.

EXAMPLE 2

Comparing Pricing Options by Graphing

You want to install a beverage vending machine in a college dorm hallway. There are two pricing options to consider.

Option A: A monthly rental fee of $30.00 is charged for the unit, and you are responsible for buying drinks to fill the machine. You make $0.75 on each can sold. So, for *x* cans sold, your profit, *y*, is $y = 0.75x - 30.00$.

Option B: No rental fee is charged because the supplier maintains and refills the machine. You receive $0.25 per can sold as part of the agreement, so your profit is $y = 0.25x$.

Graph the equations representing these options. Use your graph to determine the minimum number of beverage sales per month that makes option A the better deal.

Solution

Method 1: Pencil and Paper

Begin with the equation for option A. Plot the *y*-intercept, −30, on the *y*-axis of your graph. To get a second point on the line, you can find the *x*-intercept:

$$0 = 0.75x - 30 \qquad \text{Substitute } y = 0.$$

$$0.75x = 30$$

$$x = \frac{30}{0.75}$$

$$x = 40$$

Or, you can use the slope value of $0.75 = \frac{75}{100}$. The graph rises 75 units for every 100 units along the *x*-axis. This gives a further point at $(100, -30 + 75) = (100, 45)$.

Now, plot a line representing option B. The y-intercept is zero; that is, the line passes through the point $(0, 0)$. The slope value, $0.25 = \dfrac{25}{100}$, tells you that the graph rises 25 units for every 100 units along the x-axis, giving a point at $(100, 25)$.

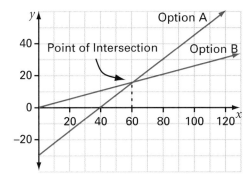

The point of intersection of these two lines is at $(60, 15)$. So, options A and B give the same profit when 60 cans are sold per month. For values of x greater than 60, the line representing option A lies above the line for option B. So, option A is the better deal when beverage sales are more than 60 cans per month.

Method 2: Graphing Calculator
Press MODE and check that the fourth line has **Func**.
Press Y= . Enter the equation for option A as Y_1.
Enter the equation for option B as Y_2.

Use the **Window** settings shown.

Refer to page 427 in the Technology Appendix for more information on the **Intersect** function.

Press GRAPH . Use the **Intersect** function to find the point of intersection, $(60, 15)$.

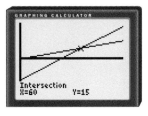

KEY CONCEPTS

- The general form of a linear equation is
  $y = mx + b$
  where m represents the slope of the line and b is the y-intercept. This form of the equation can be found from any other form by solving for the variable y.

- The slope and intercepts of a linear graph can be determined from the graph, or used to help draw the graph.

- A linear graph with a positive slope points up and to the right. A linear graph with a negative slope points down and to the right.

- The value of the slope determines the steepness of the line. The graph of a line with a slope of 3 (or -3) is much steeper than the graph of a line with a slope of 0.3 (or -0.3).

DISCUSS THE CONCEPTS

1. When an equation is not in the form $y = mx + b$, how can you determine the slope and y-intercept without graphing?

2. Give examples of linear models for real-life situations in which
   a) the slope represents a rate of change
   b) the y-intercept represents an initial value
   c) the x-intercept represents the endpoint of a process
   Justify your answers.

PRACTISE

(A) 1. For each equation, state the slope and y-intercept. Rearrange the equation first, if necessary.

   a) $y = \dfrac{1}{2}x - 4$ **b)** $2y = 4x - 2$ **c)** $7x + 14y = 21$

2. For each equation in question 1, find the x-intercept. Using pencil and paper, graph the equations on the same set of axes.

3. Graph each of the following pairs of equations. Find the point of intersection.

a) $y = 2x - 5$
   $y = 3x - 9$

b) $y = 4x - 3$
   $y = 2x + 1$

c) $y = -4x + 1$
   $y = 4x - 7$

d) $4x + y = -3$
   $7x + 3y = 1$

APPLY THE CONCEPTS

B **4.** **Swimming plans** The Pleasantvale Aquatic Facility has two options for swimming:
- Plan A costs $2.50 per swim: $y = 2.50x$
- Plan B costs $60 for a six-month membership, plus $1.25 per swim:
  $y = 1.25x + 60$

a) Graph the two plans.

b) Determine which plan is the better deal. State your answer as a function of how often you plan to go swimming in a six-month period. Support your conclusions.

5. **Self-sufficient heating** The total cost to heat a house using solar panels or an oil-burning generator is given by the equations

Solar: $y = 200x + 9000$
Oil: $y = 900x + 2700$

where y is the total cost, in dollars, and x is the number of years.

a) Graph both equations.

b) After how many years will the two heating methods cost the same? What is that cost?

6. **Problem Solving, Application** A new set of tires costs $90 per tire. The set comes with a 200 000-km warranty for any defective features. Sometime after buying new tires, a customer noticed considerable vibrations through the steering column. Under the terms of the warranty, the defective tire is not replaced free. Instead, the tire is pro-rated, so the customer is credited only for the unused portion of tread. Thus, if half the tread remains, a refund of $45 is credited.

a) Write an equation, modelling the amount of the credit compared to the portion of unused tread on the tire. (Hint: The portion should vary between 1 (good as new) and 0 (completely bald tire).)

b) Graph your linear model.

c) In this case, two thirds of the tread is found to be unused. What amount of credit will the customer receive? What amount will the customer have to pay toward a new tire?

Solving Systems of Equations: Substitution Method

In Section 8.3, you studied linear models and their graphs, and saw how drawing two linear graphs accurately on the same set of axes could determine their point of intersection. What does this point of intersection represent? Consider the point (*x, y*) at which the lines cross. If you substitute this pair of *x*- and *y*-values into either of the equations, you should find that they satisfy both the equations. The ordered pair (*x, y*) is, therefore, the solution of this system of equations.

Systems of equations have many applications. In business, for example, the solution of a system of equations modelling cost and revenue gives a break-even point. This section covers the first of two algebraic methods for solving systems of equations.

DISCOVER

Break-Even Point

In a small bird feeder business, the start-up cost for a new model is $37, and the cost of materials for each feeder is $4. The cost, in dollars, to make *n* bird feeders is given by the equation

$$C = 4n + 37$$

The first feeder sold to a retailer is priced at $10, and each subsequent feeder in the order is sold for $5. The revenue, in dollars, from the sale of the bird feeders is given by the equation

$$R = 5n + 5$$

Determine the number of bird feeders required in order to break even. Use both of the following methods.

Method 1: Graphing

1. Graph both equations. If you use a graphing calculator, enter *x* in place of *n* and *y* in place of *C* and *R*.
2. Find the point of intersection. On a graphing calculator, use the **Intersect** feature under the **Calc** menu.
3. How many bird feeders are required to break even? What is the break-even cost?

Method 2: Algebraically

1. Because you are looking to break even, set the cost and revenue variables to be equal.
2. Now that you have a single equation in terms of n, solve for n.
3. Substitute the value of n into both original equations, and solve for C and R.
4. How many bird feeders are required to break even? What is the break-even cost?

EXAMPLE 1

Simple Substitution

Solve the following system of equations.

$$y = 4x - 1 \qquad ①$$
$$y = 3x + 1 \qquad ②$$

Solution

Both equations are already in the "$y =$" form. To substitute equation ① into equation ②, simply set the two right-hand sides of the equations to be equal. Then, solve for x:

$$4x - 1 = 3x + 1$$
$$4x - 3x = 1 + 1$$
$$x = 2$$

Substitute the value for x into ① and solve for y:

$$y = 4(2) - 1$$
$$y = 7$$

Because you substituted into ①, check in ②:

L.S. $= y$ **R.S.** $= 3x + 1$
L.S. $= 7$ **R.S.** $= 3(2) + 1$
                 R.S. $= 7$
      L.S. $=$ **R.S.**

The solution of the system of equations is $x = 2$, $y = 7$.

EXAMPLE 2

Substitution With Rewriting

Find the point of intersection of the lines $y - 2x = 1$ and $2y + 6x = 12$.

Solution

Start by labelling the equations:

$$y - 2x = 1 \qquad ①$$
$$2y + 6x = 12 \qquad ②$$

- *Rewrite* equation ❶ and relabel it as ❸. Equation ❶ is easier to rewrite because the coefficient of y is 1.

$$y - 2x = 1 \qquad \text{Add } 2x \text{ to isolate } y.$$
$$y = 1 + 2x \qquad ❸$$

- *Substitute* equation ❸ as the "value" for y into equation ❷.

$$2(1 + 2x) + 6x = 12$$

- *Solve* for x.

$$2 + 4x + 6x = 12$$
$$2 + 10x = 12$$
$$10x = 12 - 2$$
$$10x = 10$$
$$x = 1$$

- *Solve* for y by substituting $x = 1$ into ❸.

$$y = 1 + 2(1)$$
$$y = 3$$

- *Check* $x = 1$ and $y = 3$ in ❷ to verify the solution for both equations. Use ❷ to check, since you used ❶, rearranged as ❸, to solve.

L.S. $= 2y + 6x$ **R.S.** $= 12$

L.S. $= 2(3) + 6(1)$

L.S. $= 6 + 6$

L.S. $= 12$

                L.S. $=$ **R.S.**

So, the point of intersection of the two lines is $(1, 3)$.

EXAMPLE 3

Earnings With Commission

The following equations represent the monthly earnings, salary plus commission, for Josh and Sonya.

Josh: $y = 0.07x + 500$ ❶

Sonya: $y = 0.05x + 800$ ❷

In these equations, x is monthly sales, in dollars, and y is monthly salary, in dollars.

a) Express the commission rates Josh and Sonya earn, as percents.

b) Use the substitution method to solve this system of equations.

c) For what value of monthly sales will Josh's earnings equal Sonya's? What is this monthly income?

Solution

a) Josh earns a commission of 7% on the value of all sales he makes. Sonya earns a commission of 5%.

b) There is no need to rewrite the equations, as both have the y-term isolated.

Substitute ❶ into ❷:

$$0.07x + 500 = 0.05x + 800$$

$$0.07x - 0.05x = 800 - 500 \quad \text{Collect all } x\text{-terms on the left and all numerical terms on the right.}$$

$$0.02x = 300$$

$$x = \frac{300}{0.02} \quad \text{Divide by 0.02 to isolate } x.$$

$$x = 15\,000$$

Substitute $x = 15\,000$ into Josh's equation:

$$y = 0.07(15\,000) + 500$$

$$y = 1550$$

Check. Substitute $x = 15\,000$ and $y = 1550$ into Sonya's equation:

L.S. $= y$ **R.S.** $= 0.05x + 800$

L.S. $= 1550$ **R.S.** $= 0.05(15\,000) + 800$

                           R.S. $= 1550$

         L.S. $=$ **R.S.**

The solution is $x = 15\,000$ and $y = 1550$.

c) Josh's and Sonya's earnings will be the same, $1550 per month, with sales of $15 000.

> **Career Link**
>
> A career in sales requires strong communication and organizational skills, a good memory for names and faces, a thorough understanding of your product, and the ability to think on your feet.

KEY CONCEPTS

- The steps of the substitution method are as follows:
    - *Rewrite* one equation (the easier equation) to isolate y and relabel it as ❸.
    - *Substitute* ❸ for y into the other equation.
    - *Solve* for x.
    - *Solve* for the other variable, y, by substituting the solved variable x into ❸.
    - *Check* by substituting the solution into the unchanged original equation.
      In some situations, it is easier to isolate x first, then substitute for x, and solve for y, finally solving for x.
- Solutions can also be checked using a graphing calculator.

DISCUSS THE CONCEPTS

1. Explain why the method used in this section is called the substitution method.

2. $x + 3y = 40$ ❶
$2x + 5y = 10$ ❷
To solve this system of equations, which variable would you isolate first? Why?

Ⓐ 1. Solve each system of equations, using the substitution method.

a) $y = 2x - 3$
$y = 5x + 9$

b) $y = 5x + 4$
$y = 7x - 9$

c) $y = -4x + 1$
$9x + 2y = 2$

2. Solve each system of equations, using the substitution method.

a) $3x + y = 12$
$2x + 5y = 21$

b) $5x + y = -7$
$2x - 4y = 6$

c) $3x + 6y = 4$
$x - 2y = 1$

d) $2x - 3y = 5$
$2x - y = 3$

e) $2x - 5y = 7$
$x + 2y = 8$

f) $3a + 5b = 1$
$a - 2b = 4$

g) $2m - 3n = 4$
$3m + n = 17$

h) $12c + 7d = 16$
$c - 9d = 78$

i) $4x + 5y = 6$
$3x - 2y = -7$

Ⓑ 3. **Retail break-even** The owner of a second-hand computer store has a sideline business selling refilled printer cartridges. The start-up cost for this operation is $2500 and it costs $8 to refill a cartridge. Therefore, the equation modelling the cost of this business is
$C = 8x + 2500$
The cartridges sell for $56 each, so the equation modelling the revenue is
$R = 56x$
In each equation, x represents the number of cartridges. How many printer cartridges does the owner need to sell in order to break even?

4. **Catering costs** The weekly costs, C, in dollars, of operating two school cafeterias are modelled by the equations below, where n is the number of students.
$C = 2.5n + 1500$ ❶
$C = 1.8n + 2060$ ❷

a) Explain how these equations model the weekly costs.

b) For how many students will the costs be the same? What is that cost?

5. Salary scales Some companies pay their employees using a salary scale that depends on the number of years employed. The equations representing the annual salaries for two such companies are

Cantech: $S = 25\ 000 + 2000n$

Wordcom: $S = 30\ 000 + 1500n$

where S is the annual salary, in dollars, and n is the number of years employed.

a) Explain how these equations model the salaries.

b) After how many years would an employee's salary be the same at either company? What is that salary?

6. Problem Solving, Communication A system of equations is defined by

$y = 1.5x - 6$

$3x - 2y = 4$

a) Apply the substitution method to this system.

b) What happened in part a)? Use a graph to help explain the result.

c) Make a general statement about this finding.

7. Communication Based on a statistical study, these equations were developed to model the percents of the female and male populations that smoke, in the years from 1981 to 2001.

Women: $y = -0.71x + 37$

Men: $y = -1.14x + 47$

In these equations, x is the number of years after 1981 and y is the percent of the gender's population.

a) After how many years will the percents of men and women who smoke be equal? What is this percent?

b) Graph the equations to show the point of intersection.

c) How well do you think these equations model smoking behaviour? Explain your answer.

You might use a drawing package such as *AutoCAD LT*® or *Corel DRAW*® for design work such as in question 8. See Appendix B, page 446, for more information on these drawing packages.

8. Design modelling A graphic designer is modelling the cross section of the roof of a house by assigning an equation to each sloped surface as shown.

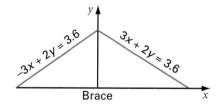

a) Graph the cross section of the roof, placing the horizontal brace along the x-axis. Label the three vertices and the equations.

b) Calculate the point of intersection of the two sides of the roof, using the substitution method.

c) How high is the peak of the roof above the brace?

d) How wide is the roof?

9. Two pizza companies sell similar pizzas using different price structures. Piece-a-Pizza sells a small pizza for $7.95 plus $0.75 a topping. Flying Pizza sells a small pizza for $8.40, charging an additional $0.60 per topping.
   a) Write a system of equations, modelling the price of a small pizza at each company.
   b) Use the substitution method to find the number of toppings that makes the prices the same. What is that price?

10. Car operating costs The average monthly cost to operate Anne's sedan is $350 plus $0.35 per kilometre. The average monthly cost to run Tien's hatchback is $285 plus $0.41 per kilometre.
   a) Write a system of equations, modelling the average monthly cost for each car.
   b) Use the substitution method to determine how many kilometres per month make Anne's and Tien's costs the same. What is that cost?

CHAPTER PROBLEM

11. When launching the marketing campaign for a new CD, Remote Recordings models its total revenue, R, and total costs, C, using the equations
$R = 100x^2 + 5000x$
$C = 5000x + 10\ 000$
where x is the number of days into the 45-day campaign.
   a) Write an equation, representing the profit after x days of advertising.
   b) After how many days will the company break even?
   c) After how many days will the profit reach $100 000?
   d) What is the total expected profit from this campaign?

Ⓒ12. Problem Solving, Communication Write a system of two equations that gives the solution $x = 2$, $y = 5$.
   a) Use the substitution method to verify that $x = 2$, $y = 5$ is the solution of your system of equations.
   b) Describe a real-life situation that could be modelled with these equations.

EXTEND THE CONCEPTS

13. a) Problem Solving, Communication, Application Describe the shapes of the graphs of the following equations.
$y = x^2 + 5$
$3x - 2y + 15 = 0$
   b) How many intersection points would they have?
   c) Solve the system of equations algebraically.
   d) Graph the results to illustrate the solutions.

Solving Systems of Equations: Elimination Method

For some systems of equations, the substitution method may involve some awkward algebra. In some instances, another technique, the elimination method, is much easier to use.

DISCOVER

Methods for Finding a Point of Intersection

1. Graph these lines:

$2x + 3y = 9$ ❶

$5x - 3y = 12$ ❷

Estimate the point of intersection.

2. Solve the system of equations, using the substitution method.

3. Add ❶ and ❷. Using the resulting equation ❸, solve for x. Then, use either ❶ or ❷ to solve for y.

4. Compare your methods in questions 1, 2, and 3.

In question 3 of the Discover, when you added the equations, the y-term was eliminated. This is because the coefficients of the y-terms are $+3$ and -3, respectively. With the elimination method, if you can find a common coefficient for the same variable in the two equations, you can eliminate that variable first. It does not matter which variable is eliminated. The method works equally well with either one. You can also multiply to create a common coefficient, as the next example shows.

EXAMPLE 1

The Elimination Method

Use the elimination method to solve the system of equations $2x + 3y = 6$ and $5x + 2y = -7$.

Solution

Label the equations:

$2x + 3y = 6$ ❶

$5x + 2y = -7$ ❷

- *Multiply* equation ❶ by 5 and equation ❷ by 2, to get a common coefficient of 10 for the x-terms. Then, relabel the new equations. (Note: You could get a common coefficient for the y-terms in order to eliminate them instead.)

    $5 \times$ ❶ $10x + 15y = 30$ ❸

    $2 \times$ ❷ $10x + 4y = -14$ ❹

- *Subtract* the equations, because the x-coefficients are both $+10$. Then, solve for y.

$$\begin{array}{r} ③ \quad 10x + 15y = 30 \\ -④ \quad 10x + 4y = -14 \\ \hline 0x + 11y = 44 \\ 11y = 44 \\ y = 4 \end{array}$$ You have eliminated the x-term by subtracting.

- *Substitute* $y = 4$ into ① and solve for x.

$$2x + 3y = 6$$
$$2x + 3(4) = 6$$
$$2x + 12 = 6$$
$$2x = 6 - 12$$
$$2x = -6$$
$$x = -3$$

- *Check.* Substitute $x = -3$ and $y = 4$ into ②.

L.S. $= 5x + 2y$ **R.S.** $= -7$
L.S. $= 5(-3) + 2(4)$
L.S. $= -15 + 8$
L.S. $= -7$
       L.S. = R.S.

The solution of this system of equations is $x = -3$, $y = 4$.

EXAMPLE 2

Point of Intersection

Use the elimination method to find the point of intersection of the lines $6x - 4y = -8$ and $5x + 3y = 25$.

Solution

$6x - 4y = -8$ ①
$5x + 3y = 25$ ②

Use multiplication to obtain a common coefficient of 12 for the y-terms.

$3 \times ①$ $18x - 12y = -24$ ③
$4 \times ②$ $20x + 12y = 100$ ④

Because the coefficients are of different signs, add the equations to eliminate the y-terms.

$$\begin{array}{r} ③ \quad 18x - 12y = -24 \\ +④ \quad 20x + 12y = 100 \\ \hline 38x + 0y = 76 \\ 38x = 76 \\ x = 2 \end{array}$$ You have eliminated the y-term by adding.

Substitute $x = 2$ into ❶:
$$6x - 4y = -8$$
$$6(2) - 4y = -8$$
$$12 - 4y = -8$$
$$-4y = -8 - 12$$
$$-4y = -20$$
$$y = 5$$

Check. Substitute $x = 2$ and $y = 5$ into ❷:

L.S. $= 5(2) + 3(5)$ 　　　**R.S.** $= 25$

L.S. $= 10 + 15$

L.S. $= 25$

　　　　　L.S. = R.S.

The point of intersection is $(2, 5)$.

EXAMPLE 3

Driving Distances

The distance for one route from Cambridge to Ottawa is 480 km. Ted is able to drive at an average speed of 100 km/h on Highway 401, but can only average 60 km/h on secondary highways. The distance data can be modelled with this equation:

$x + y = 480$

Suppose the trip took 6 h. The driving speed and time data can be modelled with this equation:

$$\frac{x}{100} + \frac{y}{60} = 6$$

In these equations, x is the distance travelled on Highway 401 and y is the distance travelled on secondary highways. How far did Ted travel on Highway 401, and how far on secondary highways?

Solution

Solve the system of equations to determine the values of x and y.

$$x + y = 480 \qquad \textbf{①}$$

$$\frac{x}{100} + \frac{y}{60} = 6 \qquad \textbf{②}$$

Multiply to eliminate the fractions and get a common coefficient of 6 for the x-terms:

$$6 \times \textbf{①} \qquad 6x + 6y = 2880 \qquad \textbf{③}$$

$$600 \times \textbf{②} \qquad \frac{600x}{100} + \frac{600y}{60} = 600 \times 6$$

$$6x + 10y = 3600 \qquad \textbf{④}$$

$$\textbf{③} - \textbf{④} \qquad \overline{}$$

$$-4y = -720$$

$$y = 180$$

Substitute $y = 180$ into $\textbf{①}$:

$$x + y = 480$$

$$x + 180 = 480$$

$$x = 480 - 180$$

$$x = 300$$

Check. Substitute $x = 300$ and $y = 180$ into $\textbf{②}$:

$$\textbf{L.S.} = \frac{300}{100} + \frac{180}{60} \qquad \textbf{R.S.} = 6$$

$$\textbf{L.S.} = 3 + 3$$

$$\textbf{L.S.} = 6$$

$$\textbf{L.S.} = \textbf{R.S.}$$

Ted drove 300 km along Highway 401 and 180 km along secondary highways.

KEY CONCEPTS

- The steps of the elimination method are as follows:
  - *Multiply* each equation, if necessary, to get a common coefficient for one term.
  - *Add* or *subtract* the equations to eliminate this term. Solve for the remaining variable.
  - *Substitute* this value into one of the original equations, and solve for the eliminated variable.
  - *Check* by substituting your solution into the other original equation.

- Solutions can also be checked using a graphing calculator.

DISCUSS THE CONCEPTS

1. Explain why this method is called the elimination method.

2. Discuss the advantages and disadvantages of the elimination method, compared to the substitution method.

PRACTISE

Ⓐ 1. Solve each system of equations, using the elimination method.

a) $x + 6y = 8$
$x + 4y = 6$

b) $3x + y = 13$
$5x - y = 3$

c) $2x - 3y = -4$
$2x + 2y = 16$

d) $5x - 3y = 3$
$2x - 3y = -6$

2. By what number(s) would you multiply, if you were to use the elimination method

i) to eliminate the x-terms?

ii) to eliminate the y-terms?

(Do not solve the equations.)

a) $3x - 5y = 8$
$2x + y = 10$

b) $7x + 3y = 9$
$x - 8y = 12$

c) $5x + 7y = 6$
$3x + 11y = 14$

d) $9x + 3y = 4$
$2x - 4y = 5$

3. Solve each system of equations, using the elimination method.

a) $x + 9y = 14$
$2x + 7y = 6$

b) $5x - y = 16$
$4x + 3y = 9$

c) $8x + 5y = 4$
$x - 3y = -14$

d) $12x - 5y = 3$
$8x - y = -5$

Ⓑ 4. Solve, using the elimination method.

a) $5x - 3y = 1$
$4x + 2y = 14$

b) $-2x + 8y = 3$
$3x - 5y = 15$

c) $3x + 2y = 1$
$2x - 5y = -12$

d) $4x + 3y = 3$
$3x - 2y = -19$

e) $2m - 5n = 29$
$7m - 3n = 0$

f) $3p - 4q = 10$
$2p + 6q = 11$

5. Solve, using the elimination method.

a) $x + y = 50$

$0.25x + 0.5y = 21$

b) $x + y = 45$

$\dfrac{x}{20} - \dfrac{y}{5} = 1$

c) $\dfrac{x}{3} + \dfrac{y}{7} = -\dfrac{2}{21}$

$x + y = 2$

d) $0.01x + 0.05y = 170$

$0.03x + 0.02y = 120$

APPLY THE CONCEPTS

6. **Sports club fees** A sports club charges a one-time start-up fee and a monthly fee. At the end of five months, a member had paid a total of $170, and at the end of ten months, she had paid a total of $295. This can be modelled with the equations

$x + 5y = 170$

$x + 10y = 295$

where x is the start-up fee, in dollars, and y is the monthly fee, in dollars. What are the start-up and monthly fees?

7. In some hardware stores, nails can be purchased in bulk, by mass. A box containing 50 nails has a total mass (including the box) of 300 g. A box of 40 nails has a total mass of 250 g. The two boxes weigh the same empty. This can be modelled with the equations

$50x + y = 300$

$40x + y = 250$

where x is the mass of a nail and y is the mass of the box. Find the mass of the empty box and the mass of a single nail.

8. **Construction** At a construction site, five large trucks and three small trucks can carry 112 m^3 of earth, whereas four large trucks and five small trucks can carry 113 m^3 of earth. This can be modelled with the equations

$5x + 3y = 112$

$4x + 5y = 113$

where x is the capacity of a large truck and y is the capacity of a small truck. What is the load capacity of each truck?

9. **Problem Solving, Communication** A system of equations is defined by

$2x + 3y = 8$

$4x + 6y = 16$

a) Apply the elimination method to this system.

b) What happened in part a)? Use a graph to help explain the result.

c) Make a general statement about this finding.

10. Transport costs To take the ferry across a river, there is a cost for the vehicle and a cost for each occupant (including the driver). A trip for a vehicle and two occupants costs $7.50. A trip for a vehicle and five occupants costs $12.00.

   a) Write a system of equations, modelling the cost of each trip.

   b) Use the elimination method to determine the cost for a vehicle and the cost per person.

11. Yen drove 220 km, averaging 90 km/h on the main highways and 70 km/h on the secondary highways. The total driving time was 3 h.

   a) Write a system of equations, modelling the distance, speed, and time data.

   b) Use the elimination method to determine the distance driven on each type of highway.

12. Break-even projection Sumiko and Glen are planning to start a business, painting the exteriors of houses during the summer months. Start-up costs are estimated to be $6500 and operating costs are $450 per house. They expect revenue to be $1800 per house. These estimates can be modelled with the equations

Cost: $C = 6500 + 450h$

Revenue: $R = 1800h$

where h is the number of houses. After how many houses can they expect to break even?

13. Thinking/Inquiry/Problem Solving

   a) Write a system of two equations giving the solution $x = 2$, $y = -5$. The coefficients of the x- and y-terms should not be 1.

   b) Use the elimination method to show that $x = 2$, $y = -5$ is the solution of your system of equations.

   c) Describe a real-life situation that would be modelled with these equations.

CAREER PROFILE

Sales Representative for Magazine Advertising

When selling advertising space to companies, magazines rely on dependable, people-oriented, and knowledgeable sales representatives. Sales managers create an advertising rate grid that establishes how much space should be sold in the magazine. This grid begins with a one-time rate, and then, rates decrease for higher volumes of advertising. Thus, the more advertising a customer purchases, the less each advertisement costs. Costs also reflect the size of the advertisement. A full page might cost $12 000. However, a half page will typically cost more than $6000, or half the full-page price; perhaps $9000, or 75% of a cost of the full page.

Sales representatives earn their income by selling advertising on behalf of the company. The sales manager sets a quota or target sales for each representative to sell; for example, a sales representative could have a quota of $1 000 000 in sales for one year. She would earn a 5% commission on that amount.

Many colleges provide programs in advertising, business, sales, and marketing to help prepare an individual for this challenging field.

1. Verna sold $1 376 000 worth of advertising for "Your Neighbourhood" magazine. If the quota set by her sales manager was 5% for $1 000 000 of advertising sold and 8% for the amount above the quota, how much did Verna earn that year (before deductions)?

2. Tools 'R Us want to advertise in "Your Neighbourhood" magazine. It has budgeted $75 000 for advertising this year. The magazine comes out bimonthly. Using the grid, recommend the size and the location of advertising the company should buy.

National	1st advert	3rd advert	6th advert	9th advert	12th advert	15th advert
Full Page	12 863	12 602	12 363	12 180	11 991	11 709
$\frac{2}{3}$ Page	10 775	10 562	10 359	10 203	10 043	9 802
$\frac{1}{2}$ Page	8 697	8 522	8 354	8 226	8 094	7 896
$\frac{1}{3}$ Page	5 927	5 801	5 681	5 590	5 496	5 354
$\frac{1}{2}$ Spread	14 931	14 643	14 367	14 057	13 940	13 615

Your Neighbourhood National Advertising Rates for 2001, in Dollars

CHAPTER PROBLEM WRAP-UP

Remote Recordings are launching the marketing campaign for a new CD. Total revenue, in dollars, can be modelled using this equation:
$R = 100x^2 + 5000x$
The total cost of the campaign, in dollars, can be modelled by this equation:
$C = 5000x + 10\ 000$
In both equations, x is the number of days of advertising in a 45-day campaign.

1. a) Use technology to graph the two equations for 45 days of advertising.

   b) Use the graph to estimate the profit or loss after 7 days of advertising.

2. Write an equation to represent the profit after x days of advertising.

3. a) After how many days does the company expect to break even?

   b) After how many days does the company expect the profit to reach $100 000?

   c) What is the total expected profit from this campaign?

4. Describe a situation in which the cost equation would change to $C = 5000x + 20\ 000$. How would this affect the profit?

5. Explain what might happen to the revenue curve if the marketing plan were more successful than expected. How would this affect the profit?

6. Write a one- to two-page report to the advertising executives, presenting your analysis of this model.

Review

8.1 Interpreting Graphical Models, pages 352–361

1. The graph shows the motion of a person who is travelling along a road. Describe the person's motion. Refer to the person's starting position, time, distance, and speed, relative to this person's home.

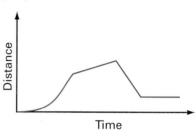

2. The graph models the height, in metres, of a figure skater, t seconds after beginning her jump, for a successful jump (a full rotation and landing on one skate without falling).

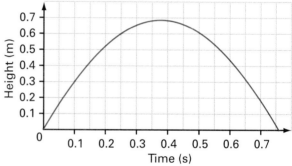

  a) What is the maximum height of the figure skater's jump?

  b) For how many seconds is the figure skater in the air?

  c) If the maximum height of the figure skater's next jump were 0.5 m, would the jump be successful? Explain.

3. **Long-term investment growth** The growth of a $1000 investment is modelled in the graph, using interest rates of 5%, 8%, and 10% per year, compounded annually.

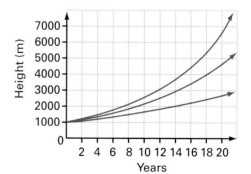

  a) Describe the type of growth modelled by these graphs.

  b) How long will it take for each investment to double in value?

  c) Extrapolate the graphs to determine the values of these investments after 30 years.

  d) How would the graph change if the money were invested at 4%?

  e) How would the graph change if $2000 were invested, instead of $1000?

4. This scatter plot shows the length of a spring when different masses are hung from it. A line of best fit has also been drawn.

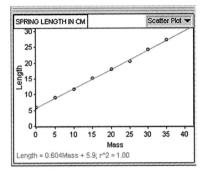

SPRING LENGTH IN CM			
	Mass	**Length**	<ne
1	0	6	
2	5	9.1	
3	10	11.7	
4	15	15.3	
5	20	18	
6	25	20.5	
7	30	24.2	
8	35	27.3	

SPRING LENGTH IN CM   Scatter Plot ▼

Length = 0.604Mass + 5.9; r^2 = 1.00

a) Determine the length of the spring when a mass of 6 kg is used.

b) What is the vertical intercept? What does it represent?

c) What mass creates a spring length of 22 cm?

d) Extrapolate to determine the length of the spring for a mass of 50 kg. Is this answer realistic? Explain why or why not.

8.2 Evaluating Formulas for Any Variable, pages 362–373

5. Substitute the given values into the formula and solve for the variable indicated.

a) $V = \frac{1}{3}\pi r^2 h$; $V = 56$ cm^3, $r = 4$ cm; solve for h, in centimetres

b) $I = \frac{V}{R}$; $I = 64$ amps, $V = 320$ volts; solve for R, in ohms

c) $a \cos C = b$; $a = 2$, $b = 1.732\ 05$; solve for $\angle C$, in degrees

d) $A = P(1 + rt)$; $A = \$1800$, $P = \$1500$, $r = 0.04$; solve for t, in years

6. **Rental costs** The cost, in dollars, to rent power tools from Ray's Rental is given by the formula
$C = I + rh$
where I is the initial cost, r is the hourly rate, and h is the number of hours that the tool is rented.

a) Rearrange the formula and solve for h.

b) Ray would like to make sure the cost of a 4-h rental is at least $30. Use a spreadsheet or calculator to make a chart, showing the minimum hourly rental rates that will ensure this return, based on initial costs of $0, $5, $10, $15, $20, and $25.

7. **Water tank dimensions** A company makes cylindrical water tanks of various sizes. The volume of a cylindrical water tank is given by the formula $V = \pi r^2 h$.
   a) Rearrange the formula and solve for r.
   b) A tank is to have a volume of 50 m³ and a height of 6 m. What radius should the company use for this tank?

8. The area of a triangle can be calculated if you know two sides and the angle between them, using the formula

   $$A = \frac{1}{2} ab \sin C$$

   where A is the area of the triangle, a and b are two sides of the triangle, and $\angle C$ is the contained angle.

   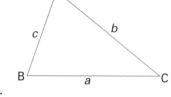

   a) Rearrange the formula and solve for $\angle C$.
   b) The area of a triangle is 100 cm², and $\angle C$ is between two sides of length 40 cm and 10 cm. Determine the measure of $\angle C$.

8.3 Graphing and Interpreting Linear Models, pages 374–379

9. The Mount Edward Recreation Centre offers two yearly fee plans for playing squash. Plan A costs $120 for a membership plus $3 for every squash game. Plan B charges a flat fee of $195, with no extra charge per squash game. How many games would you have to play in a year before plan B is the better deal?

8.4 Solving Systems of Equations: Substitution Method, pages 380–386

10. **Communication** Describe the steps you would use to solve a system of equations by the substitution method.

11. Solve each system of equations, using the substitution method.
   a) $y = 2x + 3$
      $y = 5x - 12$
   b) $3x + 4y = 8$
      $x - 5y = 9$
   c) $6m + n = -9$
      $5m - 4n = 7$
   d) $2a - 7b = -8$
      $5a + 2b = 21$
   e) $y = 0.4x - 5$
      $0.2x + 0.3y - 2 = 0$
   f) $\frac{1}{2}c + d = 6$
      $\frac{3}{2}c - 4d = 11$

12. Driving times Farouk left Windsor, driving at 80 km/h. Two hours later, Sandra left Windsor, driving at 90 km/h. Their distances travelled can be found using the equations

Farouk: $d = 80(t + 2)$

Sandra: $d = 90t$

where t is the time, in hours, after Sandra leaves, and d is the distance, in kilometres, from Windsor after Sandra leaves. How far from Windsor does Sandra catch up to Farouk?

13. Printing costs Eastview Elementary School wants to purchase stickers for a fundraising campaign. The prices quoted by two printing services are modelled with the equations

Fast Print: $C = 50 + 0.135n$

Quick Print: $C = 248 + 0.09n$

where C is the cost, in dollars, and n is the number of stickers purchased. How many stickers need to be purchased for the prices to be the same? What is that price?

8.5 Solving Systems of Equations: Elimination Method, pages 387–393

14. Communication Describe the steps you would use to solve a system of equations by the elimination method.

15. Solve each system of equations, using the elimination method.

a) $5x + 3y = 1$
$5x + 8y = 11$

b) $8x - 3y = 5$
$7x + 6y = 13$

c) $2x + 5y = 4$
$3x + 7y = 5$

d) $9a - 4b = 2$
$8a - 5b = 9$

e) $0.5x + 0.2y = 1$
$0.3x - 0.5y = 3.7$

f) $\dfrac{m}{5} + \dfrac{n}{3} = 6$
$2m - 3n = 3$

16. Ticket sales analysis For a school musical production, the cost of a student ticket was $4 and an adult ticket was $5. A total of $1950 was raised through ticket sales and a total of 450 tickets were sold. This can be modelled with the equations

$4x + 5y = 1950$

$x + y = 450$

where x is the number of student tickets sold and y is the number of adult tickets sold. Use the elimination method to determine the numbers of student and adult tickets sold.

Practice Test

1. The graph shows the length, in metres, of a person's shadow at a certain time, relative to the person's height, in metres.

a) Describe the graph.

b) What is the value of the slope and what does it mean?

c) What is the length of a shadow of a person who is 1.6 m tall?

d) What is the height of a person whose shadow is 0.55 m long?

e) Extrapolate to determine the length of a shadow for an object 3 m tall.

f) How would the graph change if the sun were higher in the sky?

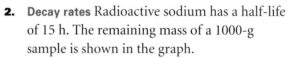

2. **Decay rates** Radioactive sodium has a half-life of 15 h. The remaining mass of a 1000-g sample is shown in the graph.

a) Describe the graph.

b) What is the vertical intercept and what does it represent?

c) Describe the change of the slope and what it means.

d) What fraction of the original sample remains after 30 h?

e) Extrapolate to determine the remaining mass after 2 days.

f) How would the graph change if the half-life were 5 h?

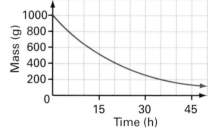

3. Substitute the value(s) given, and then, solve for the variable indicated.

a) $A = \dfrac{1}{2}bh$; $h = 40$, $A = 500$; solve for b

b) $V = A - Dn$; $V = 50$, $A = 74$, $n = 8$; solve for D

c) $L = \dfrac{6000}{w + 1}$; $L = 150$; solve for w

4. Solve each system of equations, using the substitution method.

a) $y = 5x - 8$
$y = 3x + 6$

b) $4x + 7y = 13$
$x - 5y = -17$

5. Solve each system of equations, using the elimination method.

a) $6x - y = 7$
$5x + 4y = 1$

b) $3x + 8y = 13$
$2x + 9y = 16$

6. A parking meter contains 140 coins, all quarters and loonies. The total amount of money is $97.25. This can be modelled with the equations
$x + y = 140$
$25x + 100y = 9725$
where x is the number of quarters and y is the number of loonies. How many of each coin are there?

7. **Engine dimensions** The displacement of an engine is given by the formula
$$D = \pi n s \frac{b^2}{4}$$
where D is the displacement, in cubic centimetres, n is the number of cylinders, s is the length of the stroke of each cylinder, in centimetres, and b is the cylinder bore, in centimetres.

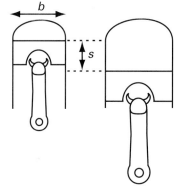

a) Find the displacement of a four-cylinder engine that has a stroke of 8 cm and a bore of 7.5 cm.

b) Rearrange the formula and solve for b.

c) Determine the bore of a six-cylinder engine, with a displacement of 2500 cm^3 and a stroke of 7.8 cm.

ACHIEVEMENT CHECK   Knowledge/Understanding Thinking/Inquiry/Problem Solving Communication Application

8. Barb raised twice as much money as Alan in the Terry Fox Run/Walk for Cancer Research. Barb decided to jog, paid an entry fee of $3, and raised $16/km in sponsors. Alan wanted to walk, so he paid an entry fee of $15 and raised $7.25/km in sponsors. They both travelled the same distance.

a) Set up equations to model this situation.

b) Find how much each person raised, using either the elimination or the substitution method.

c) Check your results using the other method.

d) Which method do you prefer for this situation? Justify your choice.

*C*umulative Review

Chapters 1–8

1. Find the measures, to one decimal place, of all missing angles and sides in each triangle.

a)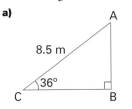

b)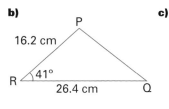

c)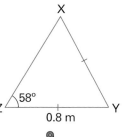

2. The movable arm of a crane can rotate vertically through an angle of 80°. How far above the ground is the end of the crane when the arm is at its highest point?

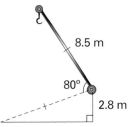

3. A rectangular room measures 6.1 m by 3.8 m. How much will it cost to cover the floor with carpeting at $35.99/yd² and underpadding at $19.00/yd²?

4. A washer should to have a diameter of 225 mm. If the manufacturer allows a tolerance of ±0.75 mm, what is the range of allowable diameters?

5. Tents can be purchased in many shapes and sizes.
   a) Make a list, including sketches, of at least three different types of tents.
   b) Choose one particular type. Draw three orthographic views of the tent. Indicate appropriate measures for the actual tent.
   c) Make a scale drawing of the pattern pieces that would be needed to make the tent.
   d) Construct a scale model of the tent using paper or fabric.

6. A teacher's report included the following data.
   a) Prepare a histogram to represent the distribution of students' scores.
   b) Describe the type of distribution.

Mark Range	Number of Students
40–49	6
50–59	15
60–69	17
70–79	21
80–89	14
90–100	8

7. Environment Canada spends millions of dollars each year collecting data about our weather. The data are stored in a central database that weather forecasters can use to predict the weather for newspapers and television reports. The data shown were taken from a national newspaper.

Toronto Almanac

24 h ending 5 p.m. yesterday

High	15.6 (normal 19.5)
Low	10.6 (normal 11.6)
Precipitation	1.2 mm

For Today

Record High (city)	30.6 in 1914
Record Low (city)	1.7 in 1889
Sunrise	5:43 a.m.
Sunset	8:48 p.m.
UV Index	6.8

(0–2 minimal, 3–4 low, 5–6 moderate, 7–8 high, 9+ very high)

Air pollution	25

(less than 32 is satisfactory)

Ontario Forecast

Bracebridge	Showers	19/11
Brantford	Showers	18/12
Cornwall	Thunderstorm	21/13
Dryden	Showers	17/7
Hamilton	Showers	20/12
Huntsville	Showers	19/11
Kenora	Thunderstorm	16/8
Kingston	Showers	20/12
Kitchener	Showers	18/12
London	Thunderstorm	19/11
Niagara Falls	Showers	20/12
North Bay	Showers	19/12
Ottawa	Thunderstorm	21/13
Owen Sound	Showers	16/11
Parry Sound	Showers	19/11
Pembroke	Thunderstorm	20/9
Peterborough	Thunderstorm	20/9
St.Catharines	Showers	20/12
Sudbury	Showers	19/10
St.Ste.Marie	Showers	20/12
Thunder Bay	Showers	16/7
Timmins	Showers	21/10
Windsor	Thunderstorm	17/12

a) The chart for Ontario does not have column headings. Give a suitable title for each column.

b) Which city had the greatest high temperature forecast? Which had the lowest?

c) What month do you think this data is for? Explain how you decided.

d) For the Ontario forecast, find the mean, median, and mode high temperature.

e) Airplane pilots Bev and Catherine do not have a night rating which means they must land no later than 30 min after sunset. A return flight from Ottawa to Peterborough takes approximately 1.5 h. What is the latest time they can depart Ottawa and legally land in Peterborough?

8. Art Doyle operates a custom farm north of Whitby. His business involves planting, fertilizing, and harvesting for local area farmers who do not have the money or the time to maintain the costly farm equipment. The area north of Whitby is growing very quickly and new subdivision developers are purchasing a lot of valuable farmland. Realizing that there may come a time when the farm operation will need to relocate to a more agricultural area, Art needs to research new areas for his business.

Σ-STAT

a) Using E-STAT, follow the path below to obtain a chart and map showing the number of farm operations in Ontario.

Agriculture
  Farmers
    Characteristics of Farm Operations
      Geographical
        Ontario
          Characteristics
            Total number of operations

b) Prepare a report describing three suitable areas, within a 200-km radius north of Whitby, where Doyle Farms might find new business.

9. The table shows the temperature of Earth's atmosphere at various altitudes on a particular day.

Altitude (km)	Temperature (°C)
0	13
1	0
2	−1
4	−13
5	−20
6	−27
8	−41
10	−55

a) Draw a scatter plot, with Altitude on the horizontal axis and Temperature on the vertical axis.
b) Describe the relationship between altitude and temperature.
c) Draw a line of best fit. Describe your steps.
d) What is the slope of your line of best fit? What does it represent?
e) Write an equation relating temperature to altitude.
f) What would the temperature be at an altitude of 12 km?

10. A technician is studying a test that measures the difference signal and the depth of cracks in aluminum. The results are shown in the table. One mil is one thousandth of one inch, or 0.0254 mm.

Crack Depth (mil)	Difference Signal (mV)
5	0.27
8	0.61
20	1.50
22	1.75
26	2.05
30	2.39
33	2.55
36	2.70
40	3.15

a) Prepare a scatter plot of the data.

b) Using technology, find the equation of the line of best fit.

c) What is the regression coefficient? Is this a strong correlation?

d) What is the depth of the crack tested, in millimetres, for a difference signal of 2.50 mV?

11. The relation between a man's age and his blood pressure is given by the equation $y = 1.5x + 78.9$, where x represents age and y represents blood pressure.

a) Create a table of values for this relation. Start at age 25 and use increments of 1, up to age 50.

b) What is the expected blood pressure of a 42-year-old man?

12. Solve by factoring.

a) $x^2 + 5x + 6 = 0$

b) $4x^2 - 9 = 0$

c) $2x^2 + x - 15 = 0$

d) $x^2 + 8 = 6x$

e) $3x^2 + x = 2$

f) $x^2 + 4x - 8 = 13$

13. Find the quadratic equation for each graph. Express each equation in standard form and in vertex form.

a)

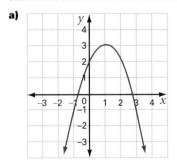

b)

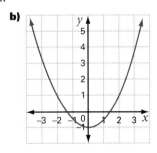

14. Rearrange each equation into $y = mx + b$ form. State the slope and the y-intercept of each line.

a) $6 = 4x + 3y$ **b)** $0 = 2 - 3x + y$ **c)** $2y = 4x - 8$

15. The graph on the left models the height of a projectile fired at an initial upward velocity of 10 m/s, from an initial height of 1.8 m. The graph on the right models the height of a projectile fired at an initial upward velocity of 50 m/s, from an initial height of 1.8 m.

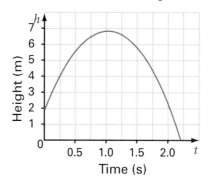 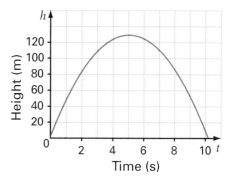

a) Compare the heights of the two projectiles after 0.5 s.

b) What is the vertical intercept and what does it represent?

c) After how many seconds does each projectile land?

d) How would the graph on the left change if the initial height were changed to 3 m?

16. Did you know that your fingernails grow twice as fast as your toenails? The equation to model the growth of your fingernails is $y = 0.32x$, where x represents the number of months of growth and y is the length of the nail, in centimetres.

a) If your fingernail is currently 0.5 cm long, graph the relation that models its growth, $y = 0.32x + 0.5$.

b) State the slope and both intercepts.

c) How long will it take to grow your fingernail to 1 cm in length?

17. Find the coordinates of the point of intersection of each pair of lines. Choose the most appropriate method, substitution or elimination, in each case.

a) $2x + 3y = 4$
$4x - 3y = -10$

b) $y = x - 1$
$y = 2x + 3$

c) $2x - 5y = 3$
$3x + 2y = 14$

18. Bank service charges vary considerably. Consider the following options.

A: A monthly fee of $8, plus $0.55 per transaction (includes cheques and withdrawals), with a minimum balance of $1000 to avoid being subject to greater charges

B: $1.25 per transaction with no minimum balance.

a) Write an equation to represent the service charges for each option. Let x represent the number of transactions and y represent the service charge, in dollars.

b) Graph the two equations. For how many transactions is the cost the same for both options?

c) Discuss which option might be better and why.

19. Olympics Project The modern Olympic Games have been held every four years since 1896. Records have been kept for the fastest time, longest jump, and so on, for men and women in all the competing sports. Times and distances have improved a great deal since the beginning of the modern Olympics. Some people try to predict winning results based on trends. Although men and women do not compete side-by-side in most sports, the differences between their results have narrowed over the years. Is there a trend towards men's and women's results being equal in the future? Will women's results ever match men's results? Your task is to analyse Olympic results for one particular event and make a prediction based on your analysis.

a) Obtain data for an Olympic sport in which both men and women have competed since at least 1948. Select a sport in which men's and women's results have narrowed over the years. Use the **Data Bank** (see page 460).

b) Construct a scatter plot of the winning results, in each of the Olympic Games, using separate sets of axes for the men's and women's competitions. Use technology to help you graph the data.

c) Develop a formula to model each set of Olympic results. Provide reasons for your choice of model type.

d) Extrapolate the model to predict the winning results in the next Olympics.

e) Extrapolate the model to predict when men and women will achieve the same results in this event.

f) Present your findings in a written report or make an oral presentation to your class. Include appropriate graphs, equations, and explanations. Include your sources.

APPENDIX A

REVIEW OF PREREQUISITE SKILLS

Adding or subtracting polynomials

The sum of two polynomials is found by adding like terms.

$$
\begin{array}{r}
2x + 3y \\
+\ 7x - 8y \\
\hline
9x - 5y
\end{array}
$$

The difference of two polynomials is found by subtracting like terms.

$$
\begin{array}{r}
5x + 7y \\
-\ 10x - 7y \\
\hline
15x + 14y
\end{array}
$$

1. Find the sum of each pair of polynomials.

a) $6x + 2y$
   $4x + 3y$

b) $9x - 3y$
   $4x + 3y$

c) $-3x - 4y$
   $3x - 7y$

2. Find the difference of each pair of polynomials.

a) $12x + 5y$
   $5x + 8y$

b) $8x + 2y$
   $6x + 2y$

c) $5x - 6y$
   $-2x - 6y$

Angles of a triangle

The sum of the angles in any triangle is $180°$. Subtract the measures of two given angles from $180°$ to find the measure of the third.

$\angle C = 180° - 45° - 80°$

$\angle C = 55°$

1. Find the measure of the third angle in each triangle.

a)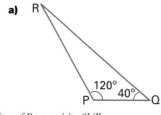

b)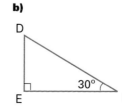

c)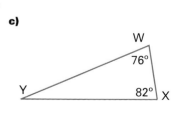

Degree of polynomials

The degree of a term is the sum of the exponents on the variables. The degree of a polynomial is the greatest degree of the terms.

The polynomial $2x^3 - 3x^4y^2 + 5y - 7$ has degree 6.

The first term, $2x^3$, has degree 3.

The second term, $3x^4y^2$, has degree 6.

The third term, $5y$, has degree 1.

The fourth term, 7, is a constant and has degree 0.

1. State the degree of each polynomial.

a) $5x^5 + 9x^4 + 6x^3$ **b)** $3x - 8$ **c)** $8y^2 + 4y - 2$

d) $5xy + 3y - 10x$ **e)** $-6x^2y^2 + 7xy^3 - 9x^3y$ **f)** 12

Drawing diagrams

When drawing diagrams, identify clues as to the shape and dimension. For example, a square-based pyramid is a three-dimensional solid with a square base and four triangular faces.

1. Draw a diagram to represent each situation. Label the diagrams.

a) A 15-m tall building casts a shadow that is 8 m long.

b) The angle of depression from the top of a cliff to a building in the valley is 23°.

c) The peaked roof of a house is symmetric and is inclined at an angle of 40°.

d) A piece of lumber 3 m long is leaning against a wall. It touches the wall 2 m above the ground.

e) A cylindrical tank has a diameter of 5 m and a height of 8 m.

f) A tent in the shape of a triangular prism has a height of 2 m and a length of 8 m.

Evaluating expressions

To evaluate the expression $2x^2 - 7xy + 5$ for $x = 2$ and $y = -3$, substitute 2 for x and -3 for y. Then, simplify using the order of operations.

$$2x^2 - 7xy + 5 = 2(2)^2 - 7(2)(-3) + 5$$
$$= 8 + 42 + 5$$
$$= 55$$

1. Evaluate for $x = -2$, $y = 5$, and $z = 4$.

a) $x^2 + 5x - y^2$ **b)** $(2x + y)(3z - 2y)$ **c)** $8xy + 3y^3 - 6z$

d) $-5y - 4x^2y^2 + 3$ **e)** $2z - y(3x^2 - 4y)$ **f)** $8 + 6yx - 7y^2$

g) $xyz - xy - xz - yz$ **h)** $5x^2 - 9z^2y + 1$ **i)** $(xz - xy)(xz + xy)$

Exponential graphs

In the exponential function $y = 0.3(2)^x$, the base is 2, the numerical coefficient is 0.3, and the exponent is x. Use the equation to make a table of values for selected values of x. Sketch the graph by plotting the points and drawing a smooth curve through the points.

x	y
−2	0.075
−1	0.15
0	0.3
1	0.6
2	1.2
3	2.4
4	4.8
5	9.6

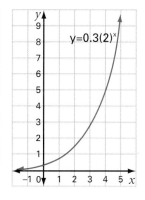

1. Graph each function.

   a) $y = 2^x$
   b) $y = 2^{-x}$
   c) $y = 1.5(2)^x$

   d) $y = 3(0.5)^x$
   e) $y = 0.1(4)^x$
   f) $y = 81\left(\dfrac{1}{3}\right)^x$

Factors of numbers

24 can be expressed as the product of its prime factors: $2 \times 2 \times 2 \times 3$.
The factors of 24 are formed by taking combinations of the prime factors.
The factors of 24 are 1, 2, 3, 4, 6, 8, 12, and 24.
Two numbers whose sum is 14 and whose product is 24 are 6 and 8.

1. List the prime factors and all possible factors of each number.

   a) 32   b) 10   c) 50
   d) 21   e) 115   f) 13

2. Find two numbers that give these results.
   a) Their sum is 12 and their product is 27.
   b) Their sum is 10 and their product is 16.
   c) Their sum is 8 and their product is 15.
   d) Their sum is −12 and their product is 27.
   e) Their sum is −10 and their product is 27.
   f) Their sum is −8 and their product is 15.
   g) Their sum is 6 and their product is −27.
   h) Their sum is −6 and their product is −27.
   i) Their sum is 6 and their product is −16.
   j) Their sum is −6 and their product is −16.
   k) Their sum is 2 and their product is −15.

Factoring polynomials

The polynomial $6x^2y^3 + 8xy^4 - 2x^2y^2$ has a common factor of $2xy^2$. Each term is divided by $2xy^2$ to find the other factor.

$6x^2y^3 + 8xy^4 - 2x^2y^2 = 2xy^2(3xy + 4y^2 - x)$

1. Factor each polynomial by finding the greatest common factor.

a) $3m - 6n + 9$

b) $18c^3 + 36c^2 - 9c$

c) $14jk + 16jm - 10jn$

d) $200x^3y^5 + 30x^4y^2 + 50x^3y^2$

e) $6a - 12a^3$

f) $5cd + 12c^2d - 7cd^2$

First differences

First differences of relations are found by subtracting successive y-values in the table for equal steps in the x-values.

x	y	First Differences
−3	7	
1	2	−5
5	−3	−5
9	−8	−5

1. Find the first differences for each relation.

a)

x	y
−6	3
−1	0
4	−3
9	−6

b)

x	y
2	9
3	15
4	21
5	27

c)

x	y
−3	2
−1	4
1	8
3	16

d)

x	y
−4	15
−2	3
0	−1
2	3
4	15

Geometric shapes and solids

A regular polygon has equal angles and sides of equal length.

Equilateral Triangle   Square   Pentagon   Hexagon   Octagon

1. Describe each of the polygons above.

2. Draw a top view and a side view of a cylinder.

3. The diagram is a net of a solid. Name the solid.

Graphing and interpreting data

The table gives the age of 80 cars sold at a used car lot.

The percents were calculated by dividing each frequency by 80 and then, multiplying by 100. Of the cars sold, the most frequent age was 6 years old, at 22.5% of the total. The least frequent age was 2 years old, at 2.5% of the total.

These data can be graphed in two ways: a bar graph or a circle graph.

Age (years)	Frequency	Percent
1	3	3.75
2	2	2.5
3	7	8.75
4	15	18.75
5	13	16.25
6	18	22.5
7	10	12.5
8	6	7.5
9	3	3.75
10	3	3.75

To find the angle for each sector of the circle graph, find the percent of 360°. For example, the sector angle for one-year-old cars is $\dfrac{3.75}{100} \times 360° = 13.5°$.

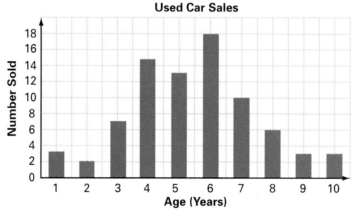

1. The table shows the number of tickets sold for a two-week theatre production of *Little Shop of Horrors*.
  a) Draw a line graph of the ticket sales.
  b) What types of graphs would not be suitable for these data?

Day	Number of Tickets Sold
1	2350
2	2350
3	2350
4	2350
5	2189
6	2012
7	1850
8	1878
9	1504
10	920
11	1267
12	1422
13	998
14	835

2. Grade 12 students were asked to provide the number of colleges to which they were considering applying. The results are recorded in the table.

Number	Frequency
0	35
1	25
2	56
3	27
4	18
5	12
6	3

a) How many students were surveyed?

b) Display the data in a bar graph.

c) Determine the percent of total responses for each number.

d) Construct a circle graph for the data.

e) What type(s) of graph would not be suitable for these data?

Graphing skills

Points can be plotted on a coordinate grid, either on an axis or in quadrant I, II, III, or IV, as shown.

Point	Location
A	Quadrant I
B	y-axis
C	Quadrant II
D	x-axis
E	Quadrant III
F	Quadrant IV

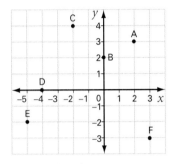

1. Plot these points on a coordinate grid. Identify the location of each point.

a) $G(3, 7)$ **b)** $H(-2, 5)$ **c)** $I(0, -6$ **d)** $J(0, 7)$

e) $K(4, 0)$ **f)** $L(-4, -4)$ **g)** $M(3, -3)$ **h)** $N(0, 0)$

2. a) Plot these points on a coordinate grid: $(2, 5)$, $(4, 7)$, $(6, 9)$, and $(-2, 1)$.

b) Describe the relation.

c) Give the coordinates of three other points that satisfy this relation.

Interpreting graphs

The graph shows the results of a survey of voters' rating of their local government. The total number of respondents was 48. The most frequent response was Fair, with 13 people giving this response. The least frequent response was both Excellent and Very Poor, with 4 people responding to each. The difference between the number responding Excellent, Very Good, or Good and the number responding Fair, Poor, or Very Poor is $25 - 23$, or 2. The results indicate that the voters are split on their rating of the government.

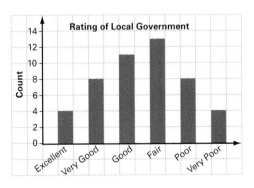

1. The graph shows the number of students enroled in a grade 12 math class during the first two weeks of September.

    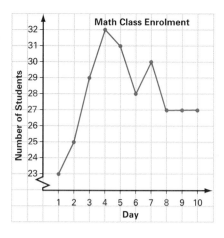

    a) Describe the enrolment trend in this class.

    b) On which day was the enrolment the greatest? the least?

    c) Between which two days did the enrolment increase the most?

    d) Between which two days did the enrolment decrease the least?

    e) Explain why a line graph is used to display this information.

2. The graph shows the spread of marks on a quiz out of 10.

    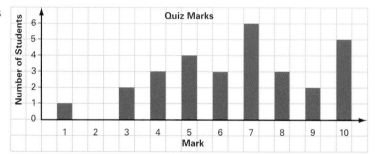

    a) How many students wrote the quiz?

    b) What is the most frequent mark?

    c) How many students passed the quiz (scored 5 or more)?

    d) What percent of students got a perfect mark on the quiz?

    e) What is difference between the number who got perfect and the number who did not pass the quiz?

    f) What is the average mark on the quiz?

    g) Explain why a bar graph was used to display this information.

Linear graphs

To graph the line $y = \frac{2}{3}x - 1$ using the slope and y-intercept form $y = mx + b$, identify the slope, $m = \frac{2}{3}$, and the y-intercept, $b = -1$. Plot the y-intercept first. Then, plot a second point by moving 2 units upward (rise) and 3 units to the right (run) to the point $(3, 1)$. Draw a line through both points.

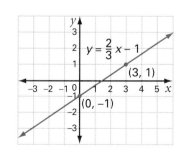

1. Graph each line using the slope and y-intercept.

a) $y = \dfrac{5}{3}x + 2$
   b) $y = 4x - 5$
   c) $y = -\dfrac{1}{2}x + 2$

d) $y = -x - 2$
   e) $2x + y = 8$
   f) $5x - 6y = 9$

Metric and imperial measures

1 in. $\doteq$ 2.5 cm   1 cm $\doteq$ 0.4 in.
So, 8 in. $\doteq$ 8 × 2.5, or 20 cm   So, 30 cm $\doteq$ 30 × 0.4, or 12 in.

1. a) 1 m $\doteq$ 1.09 yd. Convert 6 m to yards.
   b) 1 yd $\doteq$ 0.91 m. Convert 4 yd to metres.
   c) 1 mi $\doteq$ 1.61 km. Convert 45 mi to kilometres.
   d) 1 km $\doteq$ 0.62 mi. Convert 100 km to miles.
   e) 1 L $\doteq$ 0.22 gal. A gas tank holds 56 L. Convert this to gallons.
   f) 1 kg $\doteq$ 2.2 lb. A man has a mass of 68 kg. What is his mass in pounds?
   h) A woman is 140 cm tall. How tall is she in feet and inches?

Metric measurement

Metric prefixes can be placed on a line to show their relationships. As you move to the right, the measures become smaller. To convert units to the right, you multiply, and to convert units to the left, you divide.

Prefix	mega	kilo	hecto	deca	UNIT	deci	centi	milli
Symbol	M	k	h	da		d	c	m
	10^4	10^3	10^2	10^1	1	10^{-1}	10^{-2}	10^{-3}

To convert 45 kL to millilitres, you move six spaces to the right. So, multiply by 10^6.
$45 \times 10^6 = 45\ 000\ 000$
45 kL = 45 000 000 mL

To convert 6.2 cm to kilometres, you move five spaces to the left. So, divide by 10^5.
$6.2 \div 10^5 = 0.000\ 062$
6.2 cm = 0.000 062 km

1. Complete each conversion.
   a) 5250 mm to metres
   b) 5 809 000 cg to kilograms
   c) 24 ML to litres
   d) 2.4 km to centimetres
   e) 56 ms to seconds
   f) 0.45 m to millimetres

2. Measure each line segment to the nearest millimetre. Then, convert each measure to metres.
   a) ————————————————
   b) ————————

Number skills

To evaluate expressions, use the order of operations (BEDMAS).

$$(-2)(-6) - 5(-3)^2$$
$$= 12 - 5(9)$$
$$= 12 - 45$$
$$= -33$$

$$\frac{2}{3} + \frac{1}{2}\left(2\frac{3}{4} - 1\frac{1}{2}\right) = \frac{2}{3} + \frac{1}{2}\left(\frac{11}{4} - \frac{3}{2}\right)$$
$$= \frac{2}{3} + \frac{1}{2}\left(\frac{11}{4} - \frac{6}{4}\right)$$
$$= \frac{2}{3} + \frac{1}{2}\left(\frac{5}{4}\right)$$
$$= \frac{2}{3} + \frac{5}{8}$$
$$= \frac{16}{24} + \frac{15}{24}$$
$$= \frac{31}{24} \text{ or } 1\frac{7}{24}$$

1. Evaluate each expression.

a) $-3(5) + (-6)(-8)$
b) $(12)(11) - (-9)(7)$
c) $-10(3)(2) + (-12)(7)(-2)$
d) $\frac{3}{5} + 2\frac{2}{3}$
e) $8 - 2\left(\frac{7}{8} + 3\frac{3}{4}\right)$
f) $\left(\frac{3}{4}\right)^2\left(\frac{8}{3}\right)$
g) $\frac{5}{2} - \frac{4}{5}\left(\frac{3}{2}\right)^3$
h) $3.2\left(\frac{1}{2}\right) - 2.5\left(\frac{3}{5}\right) + 1.4\left(\frac{2}{7}\right)$
i) $0.3(5.5)^2 - (-6.7)(2.1)^3 + 4.2(-1.1)^5$

2. Order each set of fractions from least to greatest.

a) $\frac{5}{8}, \frac{2}{3}, \frac{3}{5}, \frac{7}{12}, \frac{1}{2}$
b) $3\frac{3}{4}, 3\frac{7}{8}, 3\frac{5}{9}, 3\frac{6}{7}$

3. Explain why the two expressions are not equal.

a) $\sqrt{9 + 16}, \sqrt{9} + \sqrt{16}$
b) $(x + y)^2, x^2 + y^2$
c) $\frac{2}{3} + \frac{5}{6}, \frac{7}{9}$

Patterns

A pattern can be described by identifying the operation(s) needed to find successive terms. 44, 33, 22, ... is a sequence of terms found by subtracting 11. The next three terms are 11, 0, and −11.

1. Describe each pattern. Find the next three terms.

a) $15, 12, 9, \ldots$
b) $4, 12, 36, \ldots$
c) $1, 4, 9, \ldots$
d) $2, 4, 6, \ldots$
e) $9, -3, 1, \ldots$
f) $80, -40, 20, \ldots$
g) $p, pq, pqq, pqqq, \ldots$
h) $p, pq, pqq, pqqqq, \ldots$
i) $p, q, pp, qq, ppp, qqqq, \ldots$

Percents

The table shows how equivalent fractions, percents, and decimals can be expressed.

Fraction	Percent	Decimal
$\dfrac{63}{100}$	63%	0.63
$\dfrac{8}{100} = \dfrac{2}{25}$	8%	0.08
$\dfrac{5}{1000} = \dfrac{1}{200}$	0.5%	0.005
$\dfrac{150}{100} = \dfrac{3}{2}$	150%	1.5

1. Complete the following table. Express all fractions in lowest terms.

	Fraction	Percent	Decimal
a)	$\dfrac{75}{100}$		
b)	$\dfrac{1}{2}$		
c)	$8\dfrac{2}{5}$		
d)		34%	
e)		0.03%	
f)		5.6%	
g)			0.45
h)			0.03
i)			2.68

Perimeter and area

Perimeter is the distance around the outside of a figure. The perimeter of a circle is called the circumference.

To find the perimeter and area of the shape shown, consider its two parts: a semicircle and a rectangle.

The circumference of the semicircle is found by dividing the circumference of a circle by 2.

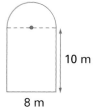

10 m

8 m

$C = 2\pi r \div 2$

$C = \pi r$

$C = \pi(4)$

$C \doteq 12.6$

The perimeter of the shape is found by adding the lengths of all the sides:

$P = 10 + 10 + 8 + 12.6$

$P = 40.6$

The perimeter of the shape is approximately 40.6 m.

The area of the semicircle is found by dividing the area of a circle by 2.

$A = \pi r^2 \div 2$

$A = \pi(4)^2 \div 2$

$A = 8\pi$

$A \doteq 25.1$

The area of the rectangle is

$A = l \times w$

$A = 10 \times 8$

$A = 80$

To find the total area of the shape, add the area of the semicircle and the area of the rectangle.

$A = 25.1 + 80$

$A = 105.1$

The area of the shape is approximately 105.1 m^2.

1. Find the area and the perimeter of each shape. Use the measurement formulas on page 54.

a)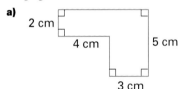

b)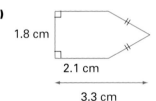

Proportion

Proportions can be solved by cross-multiplying.

$\dfrac{x}{4} = \dfrac{7}{5}$

$x = \dfrac{7}{5} \times 4$

$x = \dfrac{28}{5}$

$\dfrac{5}{9} = \dfrac{10}{y}$

$5y = 10 \times 9$

$y = \dfrac{90}{5}$

$y = 18$

1. Solve each proportion.

a) $\dfrac{k}{7} = \dfrac{6}{5}$

b) $\dfrac{8}{w} = \dfrac{12}{5}$

c) $\dfrac{3}{4} = \dfrac{c}{10}$

d) $\dfrac{9}{15} = \dfrac{6}{x}$

e) $\dfrac{f}{2.1} = \dfrac{5.6}{0.7}$

f) $\dfrac{4.8}{1.2} = \dfrac{6.3}{y}$

Pythagorean theorem

The Pythagorean theorem can be used to find the length of any side of a right triangle, if the lengths of the other two sides are known.

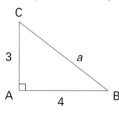

$$a^2 = b^2 + c^2$$
$$a^2 = 3^2 + 4^2$$
$$a^2 = 9 + 16$$
$$a^2 = 25$$
$$a = \sqrt{25}$$
$$a = 5$$

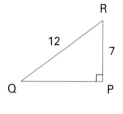

$$p^2 = q^2 + r^2$$
$$12^2 = 7^2 + r^2$$
$$r^2 = 144 - 49$$
$$r^2 = 95$$
$$r = \sqrt{95}$$
$$r \doteq 9.7$$

1. Use the Pythagorean theorem to find the length of the unknown side in each triangle. Round your answers to the nearest tenth.

a)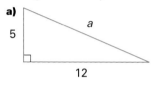

b)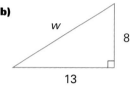

c)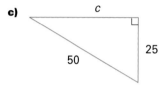

Quadratic graphs

To graph $y = 2(x - 3)^2 - 7$, plot the vertex at $(3, -7)$. Sketch the parabola opening upward, stretched vertically by a factor of 2.

To find the y-intercept, substitute $x = 0$ and evaluate. The y-intercept is 11.

From the graph, the x-intercepts are approximately 1.1 and 4.9.

You may use a graphing calculator.
One x-intercept is at approximately 4.9.

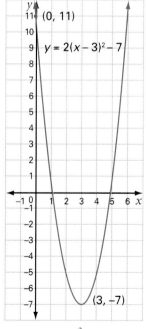

1. Graph the following functions and find the y-intercept. Estimate any x-intercepts.

a) $y = 3x^2$

b) $y = -3x^2$

c) $y = 2(x - 1)^2 + 3$

d) $y = 3(x + 2)^2 + 1$

e) $y = -(x - 1)^2 + 2$

f) $y = -2(x + 1)^2 + 3$

g) $y = \frac{1}{2}(x - 3)^2 - 4$

h) $y = (x + 5)^2 - 3$

i) $y = -\frac{1}{2}(x - 4)^2 + 6$

Reducing ratios

A ratio can be reduced to lowest terms by dividing all terms by the greatest common factor.

$80:24 = 10:3$ $\qquad$ $55:95:75 = 11:19:15$

1. Express each ratio in lowest terms.

a) $63:81$ **b)** $12:36$ **c)** $49:14:35$ **d)** $40:24:56:160$

e) $39:91$ **f)** $77:44:132$ **g)** $12:57:34:83$ **h)** $200:100:450:50$

Scientific notation

Numbers are converted to scientific notation by moving the decimal point to the left or right, so that the decimal point is immediately following the first non-zero digit. The exponent on 10 is the number of decimal places that must be moved to obtain the original number.

$265\ 000\ 000 = 2.65 \times 10^8$

$0.000\ 000\ 000\ 000\ 14 = 1.4 \times 10^{-13}$

1. Convert each number from standard form to scientific notation.

a) $65\ 310\ 000\ 000\ 000$ **b)** 25.7 **c)** $8\ 143\ 000\ 000$

d) $0.000\ 000\ 237$ **e)** $0.000\ 041$ **f)** $0.002\ 15$

g) $789\ 234$ **h)** $0.000\ 000\ 000\ 000\ 000\ 000\ 000\ 765$

2. Convert each number from scientific notation to standard form.

a) 3.4×10^5 **b)** 9.34×10^{-3} **c)** 5.066×10^2

d) 2.83×10^{-4} **e)** 7.54×10^6 **f)** 8.3333×10^3

Simplifying expressions

The expression $(x - y)(2x - 3y)^2$ can be simplified by expanding and then, collecting like terms.

$$
\begin{aligned}
(x - y)(2x - 3y)^2 &= (x - y)(4x^2 - 12xy + 9y^2) \\
&= x(4x^2 - 12xy + 9y^2) - y(4x^2 - 12xy + 9y^2) \\
&= 4x^3 - 12x^2y + 9xy^2 - 4x^2y + 12xy^2 - 9y^3 \\
&= 4x^3 - 16x^2y + 21xy^2 - 9y^3
\end{aligned}
$$

1. Expand and simplify.

a) $(x + 2y)(x - 2y)$ **b)** $(2x - 5)^2$ **c)** $(x^2 + y)^2$

d) $(5x - 2y)^2$ **e)** $3x + (2x + 1)^2$ **f)** $(5y - 2)^2 - 3y^2$

g) $(k - 3)^2 + 4$ **h)** $(5m - 2n)(5m + n)$ **i)** $2(5y + 4x)^2$

Solving equations

To solve this equation, expand the brackets, then collect like terms on one side of the equation.

$$5(x + 3) = 4(x + 7)$$
$$5x + 15 = 4x + 28$$
$$5x - 4x = 28 - 15$$
$$x = 13$$

To solve the following equation, take the root of each side.

$$x^2 = 64$$
$$x = \pm\sqrt{64}$$
$$x = \pm 8$$

1. Solve for x.

   a) $4x - 5 = 3x - 8$
   b) $4x + 3 = 2x - 9$
   c) $7x + 5 = 2x + 10$

   d) $8(x - 3) = 3(2x + 4)$
   e) $\dfrac{x}{9} = 3x + 7$
   f) $\dfrac{3x - 2}{5} = \dfrac{2x + 1}{3}$

   g) $\dfrac{5x - 2}{4} = \dfrac{3x + 1}{2}$
   h) $x^2 = 36$
   i) $x^2 = 121$

Trigonometry

Scientific calculators can be used to find trigonometric ratios of given angles. Make sure your calculator is in degree mode. Trigonometric values are generally rounded to four decimal places.

$$\sin 40° = 0.6428 \qquad \cos 52° = 0.6157 \qquad \tan 75° = 3.7321$$

If the trigonometric ratio is given, the angle can be found by first pressing the $\boxed{\text{2nd}}$ or $\boxed{\text{INV}}$ key on your scientific calculator.

Given $\cos A = 0.9659$.
$$\angle A = \cos^{-1}(0.9659)$$
$$\angle A = 15°$$

1. Determine each value, to four decimal places.

   a) $\sin 25°$
   b) $\sin 65°$
   c) $\cos 68°$

   d) $\cos 20°$
   e) $\tan 30°$
   f) $\tan 88°$

2. Determine each angle measure, to one decimal place.

   a) $\sin A = 0.5672$
   b) $\cos B = 0.2134$
   c) $\cos C = 0.6509$

   d) $\tan D = 1.2761$
   e) $\tan E = 0.5638$
   f) $\sin F = 0.0204$

3. Evaluate each trigonometric expression. Round your answers to two decimal places.

   a) $5.4\sin 30°$
   b) $0.5\tan 70°$
   c) $2\cos 45°$

   d) $\dfrac{\sin 43°}{6.5}$
   e) $\dfrac{\cos 55°}{20}$
   f) $\dfrac{12}{\tan 18°}$

TECHNOLOGY APPENDIX

CONTENTS

TI-83 and TI-83 Plus Graphing Calculators

Spreadsheets

Fathom™ Statistical Software

The Geometer's Sketchpad® Geometry Software

CorelDRAW® Interactive Drawing Software

AutoCAD LT® Interactive Drawing Software

TI-83 AND TI-83 PLUS GRAPHING CALCULATORS
THE BASICS

The Keyboard

The keys are colour-coded to make it easier to locate them.

The Home Screen or Graph Viewing Window

Each key has three functions:
- The **primary function** is on the key. For example, the primary function of x^2 is x^2. To evaluate 3^2, press 3 x^2 ENTER.
- The **secondary function** is above the key in yellow. For example, the secondary function of x^2 is $\sqrt{\ }$. To evaluate $\sqrt{9}$, press 2nd x^2 9) ENTER.
- The **alpha function** is also above the key but in green. For example, the alpha function of x^2 is the letter I. To display I on the screen, press ALPHA x^2.

To access the **Finance Applications** on the TI-83 Plus, press APPS. On the TI-83, press 2nd x^{-1}.

These five blue keys are used when graphing.

The four blue arrow keys are used to move the cursor.

The $\wedge$ key is used for exponents. For example, to evaluate 3^4, press 3 $\wedge$ 4 ENTER.

These five blue keys are the math operations. Note the following:
- ENTER is the same as = on other calculators.
- When you press $\div$, a slash, /, appears on the screen.
- When you press $\times$, an asterisk, *, appears on the screen.

The twelve light grey keys include the numbers, decimal point, and negative sign.

Don't confuse these two keys:
- − is for subtraction
- (-) is for negative values

Contrast—Increase and Decrease

To increase the contrast, press 2nd, and then, hold ▲. To decrease the contrast, press 2nd, and then, hold ▼. Alternatively, press 2nd ▲ or 2nd ▼ to increase or decrease the contrast one increment at a time.

Enter and Graph an Equation

- Press Y= . Enter the equation.
- To display the graph, press GRAPH .

 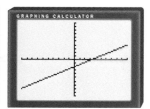

For example, enter $y = \frac{3}{5}x - 2$ by pressing

Y= (3 ÷ 5) X,T,θ,n — 2.
Press GRAPH .

Set the Viewing Window

The WINDOW key defines the appearance of the graph. You can identify the *x*- and *y*-axis minimums/maximums, scales, and pixel resolution.

The standard (default) window settings are shown.

To change the window settings:

- Press WINDOW . Enter the desired window settings.

In the example shown, you will have a minimum *x*-value of –47, a maximum *x*-value of 47, and the scale of the *x*-axis will be in increments of 10. Likewise, for the *y*-axis, the minimum is –31, the maximum is 31, and the scale is also 10. The resolution is 1. This means functions are evaluated and graphed at each pixel on the *x*-axis.

Change Mode Settings

The MODE key is used to configure the settings of the TI-83 Plus.

The **Default Settings**, shown here, have all the features on the left "turned on."

To change the plotting mode:
When plotting graphs, the TI-83 Plus will graph a line continuously, connecting all the points (**Connected**) or graph only the points (**Dot**).

- Press MODE . Cursor down four lines and right to **Dot**. Press ENTER .
- Press 2nd MODE for [**QUIT**] to return to the home screen.

Adjust the Viewing Window Using Zoom

The (ZOOM) key is used to change the area of the graph that is displayed in the graphing window.

```
GRAPHING CALCULATOR
ZOOM MEMORY
1:ZBox
2:Zoom In
3:Zoom Out
4:ZDecimal
5:ZSquare
6:ZStandard
7↓ZTrig
```

To set the size of the area you want to zoom in on:
- Press (ZOOM). Select **1:Zbox**. The graph screen will be displayed, and the cursor will be flashing. If you cannot see the cursor, use the (◀), (▶), (▲), and (▼) keys to move the cursor until you see it.
- Move the cursor to an area on the perimeter of where you would like a closer view. Press (ENTER) to mark that point as a starting point.
- Press the (◀), (▶), (▲), and (▼) keys as needed to move the sides of the box to enclose the area you want to look at. Press (ENTER) when you are finished. The area will now appear larger.

To zoom in on an area without identifying a boxed-in area:
- Press (ZOOM). Select **2:Zoom In**.

To zoom out of an area:
- Press (ZOOM). Select **3:Zoom Out**.

To display the viewing area where the origin appears in the centre and the *x*- and *y*-axes intervals are equally spaced:
- Press (ZOOM). Select **4:ZDecimal**.

To reset the axes range on your calculator:
- Press (ZOOM). Select **6:ZStandard**.

Trace a Graph

The **Trace** instruction is used to determine the coordinates of a graph.

For example, to determine *x*- and *y*-values along the graph of $y = 2^x$, press (TRACE), and then, use the (◀) and (▶) keys:

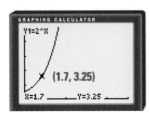
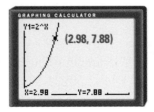

The *x*- and *y*-coordinates of the current cursor location are displayed at the bottom of the window.

Determine the Point of Intersection

Using TRACE and ZOOM:
- Once you have entered the two equations, press [GRAPH].
- Press [TRACE] to start the trace option. Your cursor will be blinking. In the top left-hand corner, you will see one of the two equations you entered; that is, the line your cursor is on. At the bottom of the screen, you will see values for *x* and *y*. These values will change as you move along the line.
- Use the [◄] and [►] keys to move toward the intersection point.
- To get a closer look, press [ZOOM]. Select **2: Zoom In**, then press [ENTER].
- Move your cursor so that you are closer to the point of intersection.
- Repeat the "**Zoom-In** and move" procedure several times until you are on the intersection point.

The values on the bottom of your screen represent the point of intersection.

Using the Intersect function:
- Enter the two equations.
- Press [2nd] [TRACE] for the [**CALC**] menu.
- Select **5:intersect**.
- You will be presented with the graph and a series of questions in the lower left-hand corner.
  - "**First curve**?" The cursor will be flashing and positioned on the first equation. Press [ENTER].
  - "**Second curve**?" The cursor will be flashing and positioned on the second equation. Press [ENTER].
  - "**Guess**?" Press [ENTER].
  - "**Intersection**." The *x*- and *y*-coordinates of the point of intersection will appear at the bottom of the screen.

STATISTICAL FUNCTIONS

Enter and Sort Data Using the STAT Editor

For statistical work, you can enter, edit, and sort data sets in easily accessed lists. Many TI-83 Plus statistical functions can be applied directly to these lists.

To enter data:
- Press [STAT]. The cursor will highlight the **EDIT** menu.
- Press **1** or [ENTER] to select **Edit...** mode. This allows you to enter new data, or edit existing data, in lists L_1 to L_6.

For example, press $\boxed{\text{STAT}}$ **1** or $\boxed{\text{STAT}}$ $\boxed{\text{ENTER}}$, and enter six test scores in $\mathbf{L_1}$.

- Use the cursor to move around the editor screen.
- Complete each data entry by pressing $\boxed{\text{ENTER}}$.
- Press $\boxed{\text{2nd}}$ $\boxed{\text{MODE}}$ for [**QUIT**] to exit the list editor when the data are entered.

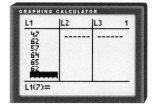

Data can be sorted in ascending or descending order. For example, to sort $\mathbf{L_1}$ in ascending order:

- Press $\boxed{\text{STAT}}$ **2** to select **SortA(**.
- Press $\boxed{\text{2nd}}$ **1** for [$\mathbf{L_1}$], then press $\boxed{\text{)}}$ $\boxed{\text{ENTER}}$.
- Press $\boxed{\text{STAT}}$ **1** to view $\mathbf{L_1}$ as sorted.
- Alternatively, press $\boxed{\text{2nd}}$ **1** for [$\mathbf{L_1}$], then press $\boxed{\text{ENTER}}$ to display $\mathbf{L_1}$ directly.

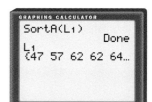

You may need to clear a list before you enter data into it. For example, to clear list $\mathbf{L_1}$:

- Press $\boxed{\text{STAT}}$ **4** to select **ClrList**.
- Press $\boxed{\text{2nd}}$ **1** for [$\mathbf{L_1}$], and press $\boxed{\text{ENTER}}$.

Generate Random Numbers

To generate a random number between 0 and 1:

- Press $\boxed{\text{MATH}}$ $\boxed{\text{ ◄}}$ to select the **PRB** menu.
- Press **1** or $\boxed{\text{ENTER}}$ to select **Rand**.
- Press $\boxed{\text{ENTER}}$ to display a random number between 0 and 1.

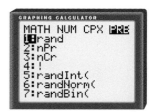

The **MATH PRB** menu also has a function, **RandInt(**, which generates an integer in a set range. For example, to obtain a random integer between 5 and 20 inclusive:

- Press $\boxed{\text{MATH}}$ $\boxed{\text{ ◄}}$ **5** to select **RandInt(**.
- Press **5** $\boxed{\text{ ,}}$ **2 0** $\boxed{\text{)}}$ $\boxed{\text{ENTER}}$.

You can also generate sequences of random numbers:

- Press $\boxed{\text{MATH}}$ $\boxed{\text{ ◄}}$ **1** to select **Rand**.
- Press $\boxed{\text{ (}}$ **1 0** $\boxed{\text{)}}$ to perform ten trials.
- Press $\boxed{\text{ENTER}}$ to display the ten random numbers on the screen.

You can store a sequence of random numbers in a list. For example, to choose ten random integers between 5 and 20:

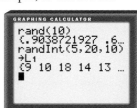

- Press $\boxed{\text{MATH}}$ $\boxed{\text{ ◄}}$ **5** to select **RandInt(**.
- Press **5** $\boxed{\text{ ,}}$ **2 0** $\boxed{\text{ ,}}$ **1 0** $\boxed{\text{)}}$ to perform ten trials in the range 5 to 20.
- Press $\boxed{\text{STO►}}$, then $\boxed{\text{2nd}}$ **1** for [$\mathbf{L_1}$], then $\boxed{\text{ENTER}}$ to place the ten random numbers directly into list $\mathbf{L_1}$.

Calculate Single-Variable Statistics

To obtain statistics for a data set, for example the values 9, 10, 18, 14, 13, 10, 14, 10, 17, 5:

- Enter the data set in a list, for example **L₁** (see Enter and Sort Data Using the STAT Editor, page 428).
- Press $\boxed{\text{STAT}}$ and cursor across to select the **CALC** menu.
- Press **1** or $\boxed{\text{ENTER}}$ to select **1-Var Stats**.
- Press $\boxed{\text{2nd}}$ **1** for [**L₁**], then $\boxed{\text{ENTER}}$.

A set of statistics are displayed.

- The first line, "$\bar{x}=$," gives the mean.
- The fourth line down, "Sx=," gives the standard deviation for a population, as calculated from a sample, using $\sqrt{\dfrac{\sum(x-\bar{x})^2}{n-1}}$ (see page 187).
- The fifth line down, "σx=," gives the standard deviation as calculated from a population, using $\sqrt{\dfrac{\sum(x-\bar{x})^2}{n}}$ (see page 185).
- The final line on the first screen, "n=," gives the number of data.

You can cursor down for more statistics.

- "minX=" gives the minimum value in the data set.
- "Med=" gives the median of the data set.
- "maxX=" gives the maximum value. Thus, for this data set, the range is 18 − 9 = 9.

There are two key statistics not given by the **1-Var Stats** function, the mode and the variance. To find the mode of the data set:

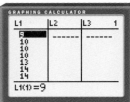

- Press $\boxed{\text{STAT}}$ **2** to select **SortA(**.
- Press $\boxed{\text{2nd}}$ **1** $\boxed{\text{)}}$ $\boxed{\text{ENTER}}$ to sort **L₁**.
- Press $\boxed{\text{STAT}}$ **1** to view **L₁** as sorted.

Repeated data items are now easy to spot. In this case, only the number 10 appears three times, so the mode is 10.

To find the variance:

- Press $\boxed{\text{2nd}}$ $\boxed{\text{STAT}}$ for [**LIST**] and cursor over to the **MATH** menu.
- Press **8** to select **variance(**.
- Press $\boxed{\text{2nd}}$ **1** $\boxed{\text{)}}$ $\boxed{\text{ENTER}}$ to calculate the variance of the data in **L₁**.

The **LIST MATH** menu also holds the individual functions **min(**, **max(**, **mean(**, **median(**, and **stdDev(**.

Create a Scatter Plot

To create a scatter plot of the data in the table:

Mark	Absences
35	6
42	5
63	3
70	1
71	1
70	2
60	2
50	4
87	0
68	2
60	3
88	1
87	0
55	4
63	3

- Enter the two data sets in lists L_1 and L_2.
- Press 2nd Y= for [**STAT PLOT**].
- Press **1** or ENTER for **Plot 1…**
- Press ENTER to select **On**.
- Cursor down, then ENTER to select the top left graphing option, a scatter plot.
- Cursor down and press 2nd **1** for [L_1].
- Cursor down and press 2nd **2** for [L_2].
- Cursor down and select a mark style by pressing ENTER.
- Press 2nd MODE for [**QUIT**] to exit the stat plot editor when the data are entered.

To display the scatter plot:
- Press Y= and use CLEAR to remove any graphed equations.
- Press 2nd MODE for [**QUIT**] to exit the **Y=** editor.
- Press WINDOW and enter appropriate settings. Use X and Y ranges from slightly below the minimum value to slightly above the maximum value for each data list.
- Press GRAPH to display the scatter plot.

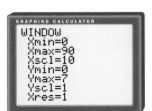

You can enter settings for up to three statistical plots simultaneously. You can also turn off all three, without losing the settings:

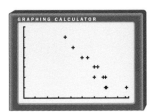

- Press 2nd Y= **4** to select **PlotsOff** from the **STAT PLOT** menu.

Perform Regression

To perform linear regression on the data sets in lists **L**$_1$ and **L**$_2$:
- Press ⟨ 2nd ⟩ **0** for [**CATALOG**], scroll to **DiagnosticOn**, and press ⟨ENTER⟩. This allows the correlation coefficient to be displayed.
- Press ⟨ STAT ⟩ and cursor to **CALC**.
- Press **4** to select **LinReg(ax+b)**. Press ⟨ENTER⟩.
- The equation of the line of best fit is given by the values "a=" and "b=."
- The line "r =" gives the correlation coefficient.

You can also use the **LinReg** function to graph the line of best fit with the raw data.
- Press ⟨ STAT ⟩ ⟨ ▶ ⟩ **4** ⟨ 2nd ⟩ **1** ⟨ , ⟩ ⟨ 2nd ⟩ **2** as above, but do not press ⟨ENTER⟩.
- Press ⟨ , ⟩ ⟨VARS⟩, cursor to **Y-VARS**, press **1** or ⟨ENTER⟩ for **Function Y**$_1$, and press ⟨ENTER⟩. This selects **Y**$_1$ as the best-fit line.
- Press ⟨ENTER⟩ again to complete the regression.
- Follow the steps described in "Create a Scatter Plot" (above) to set up a scatter plot of the data (L$_1$, L$_2$).
- Press ⟨GRAPH⟩ to view the scatter plot with the best-fit line.

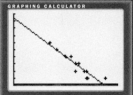

To perform quadratic regression on the data sets in lists **L**$_1$ and **L**$_2$:
- Press ⟨ STAT ⟩ ⟨ ▶ ⟩ **5** ⟨ 2nd ⟩ **1** ⟨ , ⟩ ⟨ 2nd ⟩ **2** ⟨ , ⟩ ⟨VARS⟩ ⟨ ▶ ⟩ **1** ⟨ENTER⟩ to key in **QuadReg L**$_1$**,L**$_2$**,Y**$_1$
- The equation of the best-fit quadratic is given by the values "a=," "b=," and "c=."

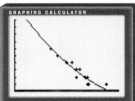

OTHER

CBR™ (Calculator Based Ranger)

To access the CBR™ through the TI-83 Plus:
- Connect CBR™ to the TI-83 Plus with the calculator-to-CBR cable. Make sure both ends of the cable are firmly in place.
- Press ⟨APPS⟩. Select **2:CBL/CBR.**
- When the CBL/CBR menu is presented, press ⟨ENTER⟩.
- To access the programs available, select **3:Ranger.**
- When the **Ranger** menu is presented, press ⟨ENTER⟩.

To set the units:
- Press **3:APPLICATIONS**. Then, select **1:METERS.**
  The **APPLICATIONS** menu will appear with a choice of programs.

MICROSOFT® EXCEL SPREADSHEETS
THE BASICS

Open

You can access Microsoft® Excel from the **Start/Programs** menu or directly from the desktop.

The Screen

The Microsoft® Excel spreadsheet screen is called a worksheet.

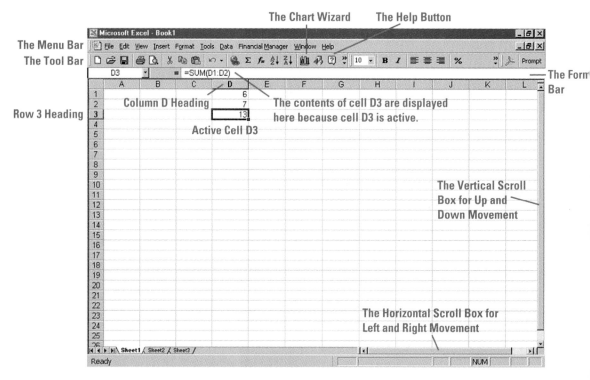

- Each cell has a **cell address**. For example, the address of the active cell above is D3 (column D, row 3).
- The **Tool Bar** contains the most commonly used functions also found in the **Menu Bar**, such as **Save**, **Print**, **Copy**, and **Paste**.
- The **Formula Bar** displays the address and the contents of the cell in which you are currently working.
- The **Chart Wizard** takes you through the steps of creating a chart (graph) that displays the data in the spreadsheet.

Enter Information

To enter information in a cell, click in the cell, and then, use the keyboard to type your heading, data, or formula. Finally, click on the checkmark or press ENTER on the keyboard.

Delete

Select the cell(s), column(s), or row(s) for which you want to delete the contents (see *Selecting Columns, Rows, and Cells*), and then, press **Delete** on the keyboard.

Save

To save a spreadsheet, go to **Save as** in the **File** menu or click on the floppy disk icon on the toolbar. Enter a name for your file, and then, save it in an appropriate folder.

Print

To print a spreadsheet, go to **Print** in the **File** menu. You have several printing options to choose from, such as **Number of copies** and **Print range**. The default settings are **1 copy** and **all pages**. Click **OK** when you are ready to print. (Note that clicking on the printer icon on the toolbar will automatically print using the default settings.)

SELECTING CELLS, COLUMNS, AND ROWS

In order to perform many of the spreadsheet functions, you must first select or highlight the appropriate cells, columns, or rows.

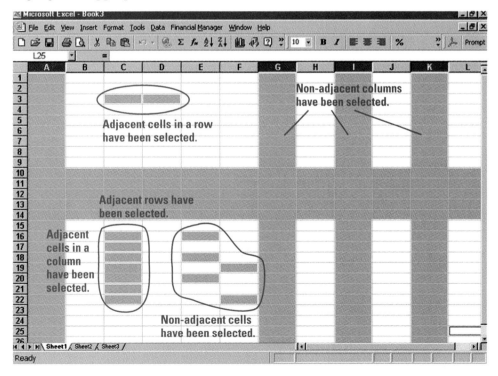

Single Cell

To select a single cell, click on the cell.

Single Column or Single Row

To select a single column or single row, click on the column heading or row heading.

Adjacent Columns or Adjacent Rows

To select a group of adjacent columns or adjacent rows, you can do either of the following:

- Click and hold on the middle of the first column or row heading in the group, and then, drag to highlight all the other columns or rows in the group.
- Click on the first column or row heading in the group, hold the **Shift** key, and then, click on the last column heading or row heading in the group.

Non-adjacent Columns or Non-adjacent Rows

To select a group of columns or rows that are not adjacent, click on the first column heading or row heading in the group, hold the **Ctrl** key down, and then, click on each of the other column headings or row headings in the group.

Adjacent Cells

To select a group of adjacent cells, you can do either of the following:

- Click and hold on the first cell in the group. Then, drag to highlight the other cells in the group.
- Click on the first cell in the group, hold the **Shift** key, and then, click on the last cell in the group.

Non-adjacent Cells

To select a group of cells that are not adjacent, click on the first cell in the group, hold the **Ctrl** key down, and then, click on each of the other cells in the group.

GRAPHS

Circle Graph (Pie Chart)

Select the cells that contain the data to be graphed. Then, click on the **Chart Wizard** button, or choose **Chart** from the **Insert** menu, and follow these steps:

	A	B	C
	Section	Amount	
	A	120	
	B	60	
	C	30	
	D	15	

Step 1: Select a **Pie Chart type** and **Chart sub-type**. Then, click **Next**.

Step 2: Make sure the correct cells (**Data range**) and **Series** have been selected.

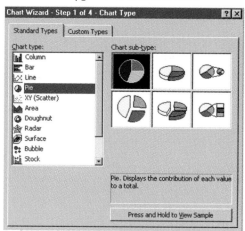

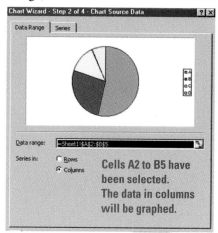

Cells A2 to B5 have been selected. The data in columns will be graphed.

Step 3: Click on the **Titles** tab and enter the **Chart title**. Click on the **Legend** tab and choose the **Placement** of the legend. Click on the **Data Labels** tab and choose the **Data labels** to display. Then, click **Next**:

Step 4: Choose to locate the chart **As an object in** the worksheet. Then, click **Finish**:

The pie chart will appear on the worksheet.

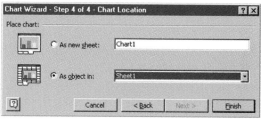

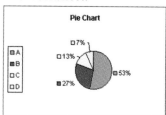

COREL® QUATTRO® PRO 8 SPREADSHEETS
THE BASICS
Open

You can access **Corel® WordPerfect® Suite 8** from the **Start/Programs** menu. Then, click on **Corel® Quattro® Pro 8.**

The Screen

The Quattro® Pro screen is called a notebook.

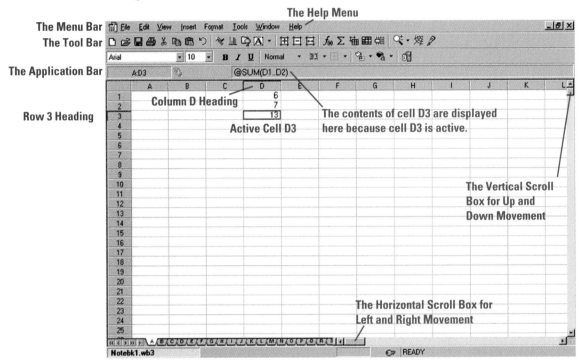

- Each cell has a **cell address**; for example, the address of the active cell above is D3 (column D, row 3).
- The **Tool Bar** contains the most commonly used functions also found in the **Menu Bar**, such as **Save**, **Print**, **Copy**, and **Paste**.
- The **Application Bar** displays the address and the contents of the cell in which you are working.

Enter Information, Save, Print, Delete

See Microsoft® Excel instructions on page 432.

Selecting Cells, Columns, and Rows

See Microsoft® Excel instructions on pages 433 to 434.

GRAPHS

Circle Graph (Pie Chart)

Select the cells that contain the data to be graphed. Then, choose **Chart** from the **Insert** menu and follow these steps:

	A	B	C
1	Section	Amount	
2	A	120	
3	B	60	
4	C	30	
5	D	15	
6			

Step 1: Make sure the correct cells (**Chart data**) are selected, and then, click **Next**.

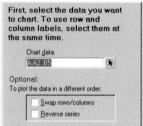

Steps 2 and 3: Choose **Pie** chart, click **Next**, and choose a **specific** pie **chart type**. Then, click **Next**:

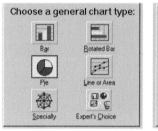
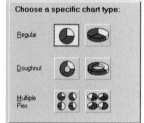

Step 4: Choose a color scheme and click **Next**:

Step 5: Enter the chart **Title** and **Subtitle**. Choose **Current Sheet** for **Destination**. Then, click **Finish**:

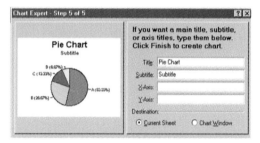

Step 6: Click on the notebook and the chart will appear.

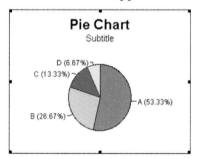

If you want to make changes, left click on one of the wedges of the pie, and then, right click. Choose **Pie Chart Properties** and the following window will appear:

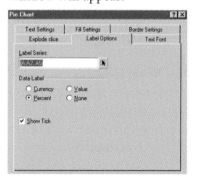

FATHOM™ STATISTICAL SOFTWARE
THE BASICS

Fathom™ is a dynamic statistics software package. It allows you to enter data, modify views of data, and filter and sort data. This tutorial gives a fast introduction, which will allow you to start working with Fathom™ quickly. A complete interactive guide to Fathom™ is available on the Fathom™ CD-ROM.

Start Fathom™

You can access Fathom™ from the **Start/Program** menu, or directly from the desktop.

The Screen

The Fathom™ screen is a working space in which you can place collections, case tables, graphs, and panels for statistical functions. Each panel can be activated by single-clicking on it.

- The **menu bar** contains all the standard Fathom™ functions.
- The **tool bar** contains icons for key Fathom™ objects. You can place new objects in the work space by clicking on any of these icons and dragging to the work space.

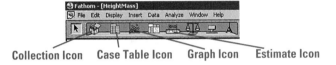

Collection Icon   Case Table Icon   Graph Icon   Estimate Icon

Create a Case Table

Start entering data into Fathom™ by creating a case table. For example, to enter height data into a case table:

- Select **Case Table** from the **Insert** menu, or click and drag the case table icon from the tool bar into the work space.
- Click on the space labelled **<new>**, which is the first attribute heading in the case table. Name this attribute by typing "Height_cm" and ENTER.
- A cell appears under the Height_cm attribute heading. You can enter the first piece of height data here.
- Press ENTER to enter each height and create a new cell for the next height.

Each column in the case table represents a separate "case." Notice that when you entered the attribute heading "Height_cm," a box marked "Collection 1" appeared. Then, when you entered the first piece of data, a group of gold balls appeared in the box.

Fathom™ organizes data into collections of cases. Each gold ball symbolizes an individual case, which can be a single data point, as here, or a set of several attributes.

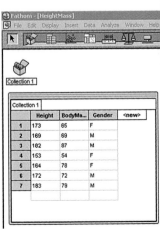

For example, add the attributes BodyMass_kg and Gender to the case table. This allows each case to store an individual's height, mass, and gender.

- Click on the space labelled <**new**>, to the right of the "Height_cm" heading in the case table. Name this attribute by typing "BodyMass_kg" and (ENTER).
- Another <**new**> attribute will appear. Click on this and type "Gender" and (ENTER).
- Enter body mass data in the second column and gender data in the third column.
- Double-click on the heading "Collection 1" and rename the collection by typing "Height + Mass Data" and (ENTER).

Save and Open Collections

To save the Fathom™ work space:

- Use the **Save** or **Save As...** command from the **File** menu.
- Supply an appropriate name for the collection.

Note that Fathom™ files are saved with the postscript ".ftm", and appear in folders with a gold ball icon.

Print, Cut, and Paste

The **Print** command in Fathom™, under the **File** menu, prints a copy of the whole work space.

Individual panes in the work space, such as case tables and graphs, can be copied, and pasted into other documents.

Import and Export

Fathom™ has powerful capabilities to import or export data from or to other programs, and to import from the Internet. For more about the commands **Import From File...**, **Import From Url...**, and **Export File...**, found under the **File** menu, see "Get Data into Fathom" in the Fathom™ reference manual.

CALCULATE SINGLE-VARIABLE STATISTICS

Measures

Fathom™ calculates statistics as "measures." For example, to find the mean height for the Height + Mass Data collection:

- Right-click on the Height + Mass Data collection.
- Select **Inspect Collection** from the menu that appears.
- A new pane appears, titled **Inspect Height + Mass Data**. This feature is called the inspector. Click on the **Measures** tab in the inspector to bring up the Measures pane.
- Click on <**new**> and type "meanHeight" and (ENTER).
- Double-click in the space under the **Formula** heading.

- A pane called **meanHeight formula** appears. This is the Fathom™ formula editor. Type "mean(". The word "mean" turns red to show Fathom™ recognizes this formula name.
- Now type "Height_cm." The word turns blue to show a recognized attribute name.
- Click on "OK" to exit the formula editor.

In the inspector, the mean of the heights now appears under the Value heading. The formula used appears under the Formula heading. If new data are added, the measure meanHeight is automatically recalculated. Other statistical measures, such as median, range, and standard deviation, can be found.

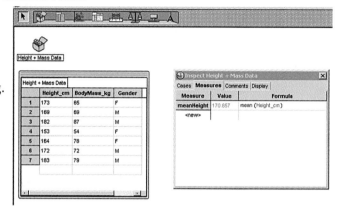

Sort Data

Cases in a collection can be sorted by any attribute, or by more than one attribute in sequence. For example, to sort the Height + Mass Data collection by gender and then by height:

- Right-click on the Height_cm attribute and select **Sort Ascending** from the menu that appears.
- The data are now sorted by height. Click on Gender and select **Sort Ascending**.
- The data are sorted by gender first, since that was the last sort performed. However, the sort by height is preserved within male and female genders.

GRAPHING

Create a Scatter Plot

To create a scatter plot of body mass versus height:

- Select **Graph** from the **Insert** menu, or click and drag the graph icon from the tool bar into the work space.
- Click on the **Height_cm** heading in the case table and drag it to the graph, over the words "Drop an attribute here." The height data will appear along the horizontal axis, with a scale.
- Click on the **BodyMass_kg** heading and drag it to the vertical axis of the graph. You will see a narrow bold rectangle appear when you are in the right place.

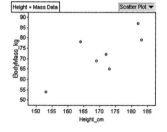

- A scale will appear on the vertical axis, and the data points will be plotted as a scatter graph. Notice that the tab in the top right corner of the graph panel now reads "Scatter Plot."

Create a Histogram

To create a histogram of the height data:
- Select **Graph** from the **Insert** menu, or click and drag the graph icon into the work space.
- Click on the **Height_cm** heading in the case table and drag it to the horizontal axis of the graph.
- Click on the tab that reads "Dot Plot." A menu of graph types appears. Select **Histogram** from the menu.
- Right-click on the graph and select **Show Graph Info** from the menu that appears.
- A pane appears below the graph, as shown. Highlight the bin width and type 5. Then, highlight the starting number and type 150. This sets the histogram intervals to start at 150 cm and to be 5 cm wide.

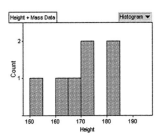

Create a Bar Graph

Fathom™ classifies attributes as continuous (numerical data) and categorical (descriptive data). A bar graph is used for descriptive data. For example, go to the case table, add an attribute called Eyes, and enter an eye colour for each individual.
- Select **Graph** from the **Insert** menu, or click and drag the graph icon into the work space.
- Click on the **Eyes** heading in the case table and drag it to the horizontal axis of the graph.

Fathom™ automatically selects a bar graph and counts the frequency of each colour.

PERFORM LINEAR REGRESSION

To perform linear regression on the height and body mass data:

With the graph pane active, pull down the **Graph** menu and select the **Least-Squares Line**.

The equation of the line of best fit is shown, in red, below the graph. You can write the equation in the more usual form as $y = 0.809x - 66$.

Notice that the value of r^2 is given. So, the linear correlation coefficient, r, is $\sqrt{0.61}$, or approximately 0.78.

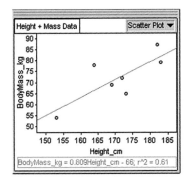

THE GEOMETER'S SKETCHPAD ® GEOMETRY SOFTWARE
THE BASICS
Menu Bar

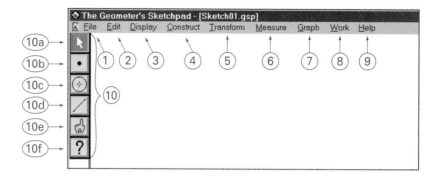

1. **File** menu—open/save/print sketches

2. **Edit** menu—undo/redo actions

3. **Display** menu—control appearance of objects in sketch

4. **Construct** menu—construct new geometric objects based on the objects in sketch

5. **Transform** menu—apply geometric transformations to selected objects

6. **Measure** menu—make various measurements on objects in sketch

7. **Graph** menu—create axes and plot measurements and points

8. **Work** menu—make script from sketch/manipulate windows/play scripts

9. **Help** menu—to use help system; this is an excellent reference guide

10. **Toolbox** consists of tools used for creating, marking, and transforming points, circles, and straight objects (segments, lines, and rays), also includes text and information tools

    10a) **Selection Arrow Tool** (Arrow)—to select and transform objects
    By holding down the mouse button, three options are revealed:

    Translate tool—to select and translate objects

    Rotate tool—to select and rotate objects

    Dilate tool—to select and dilate objects

    10b) **Point Tool** (Dot)—to draw points

    10c) **Compass Tool** (Circle)—to draw circles

    10d) **Straightedge Tool**—to draw line segments, rays, and lines
    By holding down the mouse button, three options are revealed:

    Segment tool—to draw line segments

    Ray tool—to draw rays

Line tool—to draw lines

10e) **Text Tool** (Pointing Finger)—to label points and to write text

10f) **Object Information Tool** (Question Mark)—to get information about objects

TUTORIAL

Create a Sketch or Open an Existing Sketch

• Under the **File** menu, select **New Sketch** to start with a new work area.

To open an existing sketch:

• Under the **File** menu, select **Open…**. The dialog box shown will appear.

• Select the sketch you wish to work on. Then, select **OK**.

OR

• Type in the name of the sketch in the **Filename:** entry box.

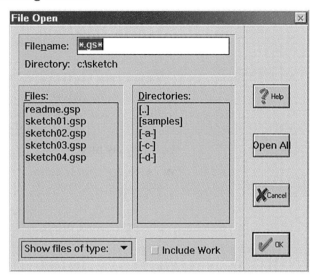

Save a Sketch

If you are saving for the first time in a new sketch:

• Under the **File** menu, select **Save As**. The dialog box shown will appear.

• You can save the sketch with the name assigned by *The Geometer's Sketchpad®*.

OR

• Press the backspace (or the delete key) to clear the name, and key in whatever you wish to name the sketch/file.

If you have already given your file a name, then select **Save** from under the **File** menu.

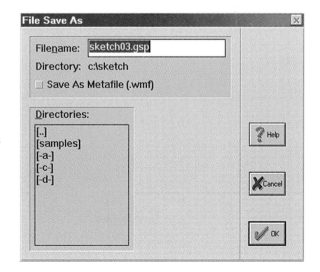

Close a Sketch With or Without Exiting

- Under the **File** menu, select **Close** to close the sketch, but to remain in *The Geometer's Sketchpad*®.
- Under the **File** menu, select **Exit** to close the sketch and exit *The Geometer's Sketchpad*®.

Select Points and Objects

- Select the **Selection Arrow Tool**. The mouse cursor appears as an arrow.

If you wish to select a single point:

- Select the point by moving the cursor to the point and clicking on the point. The selected point will now appear as a darker point, similar to a *bull's eye*.

If you wish to select a number of points:

- Hold the **Shift** key down, and select each of the points you require by moving the cursor to each of the points and clicking on the points.

If you wish to select a circle:

- Select a point on the circumference of the circle by moving the cursor to the circumference and selecting it. The whole circle will be selected.

Draw and Label a Polygon (Triangle, Rectangle, Square, Hexagon, etc.)

- Under the **Graph** menu, select **Snap to Grid**. Then, select **Hide Axes**, if axes are not required.
- Set the **Straightedge Tool** to the **Segment tool**.
- Hold the mouse button down and draw a line segment. Repeat as many times as is required.
- Label the vertices. Select the **Finger Tool** of the Toolbox. Label each of the vertices by moving the finger cursor to each point and clicking.

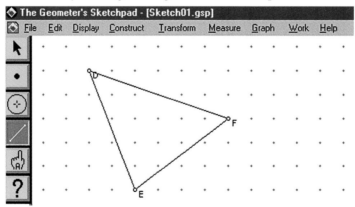

Calculate Ratios

- Select the **Selection Arrow Tool**. Holding the **Shift** key down, select two lengths to be compared.
- Select **Ratio** from under the **Measure** menu. The ratio of the lengths will appear in the upper left-hand corner of the screen.

Calculate an Angle

In order to measure an angle, the angle must be identified. Three points must be selected, with the second point being the vertex of the angle to be measured.

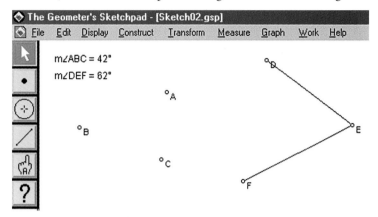

- Select the **Selection Arrow Tool.** Holding the **Shift** key down, select each point of the angle.
- Select **Angle** from under the **Measure** menu.

The angle measure will appear in the upper left-hand corner of the sketch.

In the example shown, two angles have been constructed and measured. One is made up of three points; the second is made of two line segments.

Calculate the Area of a Polygon

Polygon (Triangle, Rectangle, Square, Hexagon, etc.)

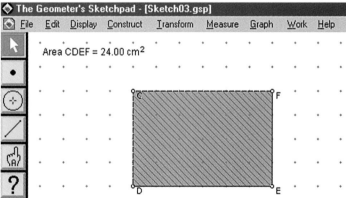

- Select all vertices of the polygon using the **Selection Arrow Tool.**
- Under the **Construct** menu, select **Polygon Interior**. The interior of the polygon will become shaded.

To measure the area:

- Select **Area** from the **Measure** menu. The area will be displayed in the upper left-hand corner of the sketch.

CORELDRAW® INTERACTIVE DRAWING SOFTWARE
THE BASICS
Menu Bar

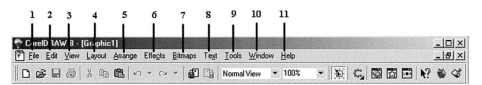

1. **File** menu—Open, Save, Print, and other file operations
2. **Edit** menu—Cut, Copy, Paste, and other editing operations
3. **View** menu—display choices, colour palettes, ruler, grid, toolbars, and other view options
4. **Layout** menu—Page Setup, graphic and text styles, snap settings, and other page options
5. **Arrange** menu—Transformations, Alignment, Lock, Join, and other options
6. **Effects** menu—Perspective, Extrude, Lens, colour adjustment, and other effects
7. **Bitmaps** menu—Conversion, Mask, 2D, 3D, Blur, Sharpness, and other options
8. **Text** menu—Format text, writing tools, Extrude text, Align text, and other text options
9. **Tools** menu—general options, palette editor, pattern creation, and other tools
10. **Window** menu—select drawing window, arrange windows, and other options
11. **Help** menu—Help Topics, Corel Tutor, technical support, and version information

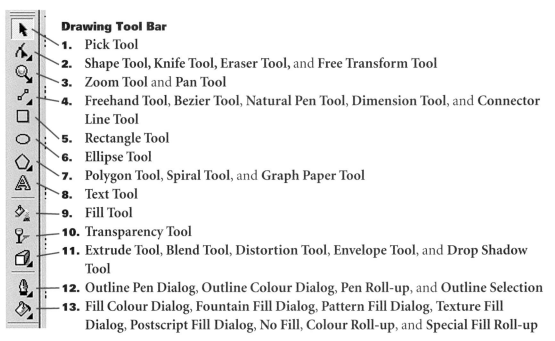

Drawing Tool Bar

1. Pick Tool
2. **Shape Tool, Knife Tool, Eraser Tool,** and **Free Transform Tool**
3. Zoom Tool and Pan Tool
4. **Freehand Tool, Bezier Tool, Natural Pen Tool, Dimension Tool,** and **Connector Line Tool**
5. Rectangle Tool
6. Ellipse Tool
7. **Polygon Tool, Spiral Tool,** and **Graph Paper Tool**
8. Text Tool
9. Fill Tool
10. Transparency Tool
11. **Extrude Tool, Blend Tool, Distortion Tool, Envelope Tool,** and **Drop Shadow Tool**
12. **Outline Pen Dialog, Outline Colour Dialog, Pen Roll-up,** and **Outline Selection**
13. **Fill Colour Dialog, Fountain Fill Dialog, Pattern Fill Dialog, Texture Fill Dialog, Postscript Fill Dialog, No Fill, Colour Roll-up,** and **Special Fill Roll-up**

TUTORIAL

Create a Drawing or Open an Existing Drawing

When you first run *CorelDRAW®*, you will see the Welcome Screen:

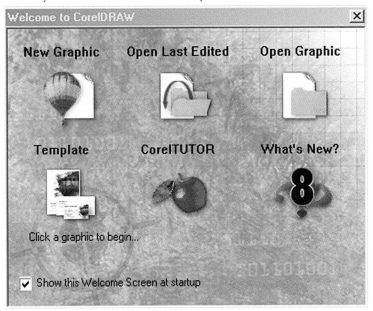

Select **New Graphic** to create a new drawing. Select **Open Last Edited** if you want to work on the drawing you were working on the last time you used the program. Select **Open Graphic** and select a file from the file dialog box if you want to work on another drawing. You can also choose to work from one of *CorelDRAW®*'s built-in **Templates**.

If you are already running *CorelDRAW®*, you can open a new drawing by using the **File** menu, or by clicking on the **New Drawing** icon.

If you want to open an existing file, use the **File** menu or click on the **Open Existing Drawing** icon.

Save a Drawing

You can save the drawing you are working on with the same name by choosing **Save** from the **File** menu, or by clicking on the **Save** icon. You can save a drawing under a different name by choosing **Save As** from the **File** menu. This will open the **File** dialog box, and allow you to select a name and directory for the drawing.

Close a Drawing With or Without Exiting

To close a drawing, you can select **Close** from the **File** menu.

To close a drawing and exit from *CorelDRAW®*, select **Exit** from the **File** menu.

Draw an Object

CorelDRAW® offers a wide range of drawing tools. This tutorial introduces the most important of these tools. To draw freehand, open a new drawing in *CorelDRAW*® ™ and then select the **Freehand Tool** from the drawing tool bar. Move the cursor near the upper left corner of the drawing, and click and release the left mouse button. Move the cursor diagonally to the right and click again. Note that you have drawn a straight line. Move the cursor near the centre of the page, and hold down the left mouse button while moving the cursor. Note that you are drawing as if holding a pencil or pen.

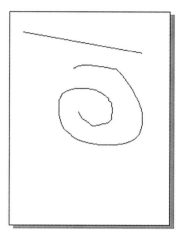

Open a new drawing. Select the **Rectangle Tool**. Move the cursor near the upper left corner of the drawing. Hold down the left mouse button, and move the cursor downward and to the right. Notice that a rectangle is drawn as you move the mouse. Release the mouse button to "freeze" the rectangle. Now select the **Ellipse Tool**. Start below your rectangle near the left-hand side of the page. Hold down the left mouse button, and move the cursor downward and to the right. Release the mouse button. Your drawing may look something like this:

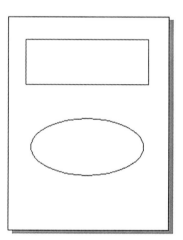

Experiment with other tools. If you want some information on what a tool does, select **Help Topics** from the **Help** menu, and take some lessons from **Corel Tutor**, or browse through **Corel Hints**.

Select Points and Objects

If you have an object already drawn, and you want to select it, click on the **Pick Tool** in the drawing tool bar, and then click on the object. To see how this works, open a new drawing and draw a rectangle with the **Rectangle Tool**. Click on the **Pick Tool** and then on the rectangle. Notice that the rectangle is outlined. If you click in the middle of the rectangle and drag, you can move the rectangle around the screen. Still using the **Pick Tool**, move the cursor to the lower left corner of the rectangle, click and hold the left mouse button, and drag the corner toward the right. This stretches the rectangle.

Modify an Object

Create a new drawing, and draw a rectangle. Select the **Pick Tool**, and click on the rectangle with the right mouse button to open a menu box. Select **Convert to Curves**. Keeping the **Pick Tool**, move the cursor to the lower left-hand corner of the rectangle. Click and hold the left mouse button, and drag the corner downward. Notice that you are now stretching the rectangle into a trapezoid, as shown.

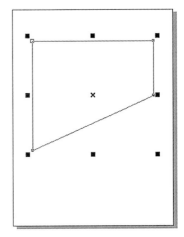

The points on your rectangle that you can move to stretch it are called **nodes**. You can add more nodes using the **Shape Tool**. Select the **Shape Tool**, and move it to the midpoint of the top edge of the trapezoid. Right-click the mouse to open a menu box. Select **Add**. This will put another node on the top edge that you can drag. Left-click on this node and drag it downward. Your final shape might look something like the one shown.

Note: Before you can stretch a shape like this, you must select it and convert it to curves, as explained above; otherwise, dragging the **nodes** will not have this effect.

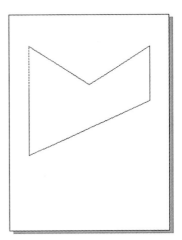

Use the Extrude Tool

You can add three-dimensional perspective to your drawings by using the **Extrude Tool**. To see how this works, open a new drawing, and draw a small rectangle near the middle of the drawing. Select the **Extrude Tool**, and left-click on the rectangle. Move the cursor to the lower right corner of the rectangle. Click and hold the left mouse button. Drag the cursor to the upper right corner of the drawing, and release the mouse button. You will end up with a perspective drawing of a box, as shown.

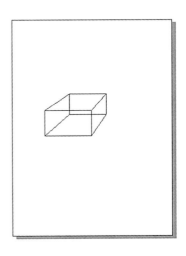

AUTOCAD LT® INTERACTIVE DRAWING SOFTWARE
THE BASICS

Introduction

The most widely used drawing software in industrial design is *AutoCAD®*, a program that allows drawing and manipulation of objects in a fully three-dimensional environment. *AutoCAD LT®* is a "lighter" version of *AutoCAD®* that works in only two dimensions. Two numbers, an *x*-coordinate and a *y*-coordinate, are used to represent all locations on the screen.

For example, a drinking glass could be drawn as four lines on a screen that connect specific points. The lines connect the points (2, 0) to (3, 0) to (4, 4) to (1, 4), then back to (2, 0).

AutoCAD LT® accepts commands from either the keyboard or from the menu and tool bars. It is a good idea to keep your eyes on the **Command Line** section in the bottom part of the screen. The **Command Line** section will tell you what current command *AutoCAD LT®* is running, and may prompt you to enter some information that the computer needs to complete the current command.

Screen and Toolbars

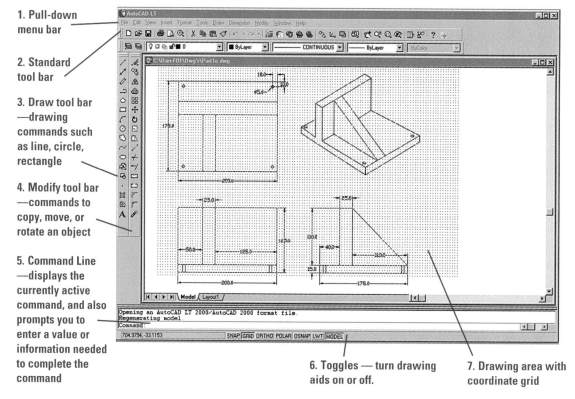

1. **Pull-down menu bar**

2. **Standard tool bar**

3. **Draw tool bar** —drawing commands such as line, circle, rectangle

4. **Modify tool bar** —commands to copy, move, or rotate an object

5. **Command Line** —displays the currently active command, and also prompts you to enter a value or information needed to complete the command

6. **Toggles** — turn drawing aids on or off.

7. **Drawing area with coordinate grid**

TUTORIAL

Create a Drawing

To create a new drawing:

- Under **File** menu, select **New**, then **Start from Scratch**.
- When you draw from scratch you also need to select the measurement system.
- Alternatively, use the Wizards or Templates.

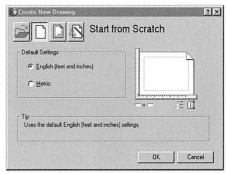

Lines, circles, and arcs are all drawn on the screen using x- and y-coordinates. To create a line on the screen, place a point for the start of the line and another point for the end of the line. The start- and end-points can be placed either by a mouse click on the screen or by entering the coordinates using the keyboard. Press (ENTER) to complete the line.

Scale with *AutoCAD LT*®

You can draw straight onto the screen and not worry about the scale of the objects. However, if you are designing something that you want to build or produce from a printed copy of the plan, then you must consider scale. In *AutoCAD LT*®, all objects are always drawn using actual sizes. When scaling for a printed copy, it is the text that must be scaled to correspond with the object size.

If you are drawing a large object and need to include text on the drawing, you will need to adjust the text size. As a guideline, if you are going to print your drawing at 1:10 scale, set the text size to be ten times the default setting. For example, if the default text size is 2.5, you could multiply by 10 and set the text size to 25.0. This will make your text visible on the screen and on the printed page.

The Draw Tool Bar

The **Draw** tool bar can be made visible by using the pull-down menu item **View** and then **Toolbars**. Placing a check in the checkbox beside each toolbar name makes it appear on the screen. A range of drawing commands is available: **Line**, **Construction Lines**, **Double Line** and **Polyline**, **Polygon**, **Rectangle**, **Arc**, and **Circles**, **Revcloud** and **Spline**, **Ellipse** and **Insert Block** and **Make Block**, **Point**, and lastly **Hatch**, **Region**, and **Text**.

The Modify Tool Bar

The **Modify** tool bar allows you to adjust a drawing object. **Modify** commands are **Erase**, then **Copy Object**, **Mirror**, **Offset**, **Array**, **Move Rotate** and **Scale**, then **Stretch**, **Lengthen**, **Trim**, and **Extend**, then **Break 1 point**, then **Break 2 point**, and lastly **Chamfer, Fillet**, and **Explode**.

Erase an Object

Erase is one of the **Modify** tools, but it is also available as **Clear** on the **Edit** menu. As a shortcut:

- Press (ENTER).
- Click on the object that you want to erase.
- While in the erase command, each time you click an object, the object is selected and forms part of a selection group.
- When the selection is complete, press (ENTER) to end, and then, press (ENTER) again to erase the selection group.

Redraw Command

After an erase command, cross marks called "blips" are usually left on the screen by the erased objects. To remove these, you can make *AutoCAD LT®* redraw the screen:

- Select **Redraw** from the **View** menu.

Copy or Move an Object

To copy or move an object:

- Click on **Copy Object** or **Move** on the **Modify** tool bar.
- Once the **Copy** or **Move** tool is selected, click on the objects that you want to copy or move.
- While in these commands, each time you click an object, the object is selected and forms part of a selection group.
- When you have all the items selected, press (ENTER) to end the selection process.
- The **Command Line** section of the screen asks for a base point. Click on any convenient point to form the "from" point of the command.
- You are now prompted for a "to" point (the destination). Click on your chosen destination point. The object or selection group will now be moved or copied.

Cancel a Command

To cancel any command, press the Escape key.

Drafting Settings

Select the **Drafting Settings** on the **Tools** menu. This selection opens the dialog box that is shown and has many useful features.

Grid on/off: The grid appears as dots on the screen, which can be used to guide drawing, like the lines printed on graph paper. The grid can be turned on or off by clicking in the **Grid** check box on the **Drafting Settings** dialog box, by pressing (F7), or by clicking

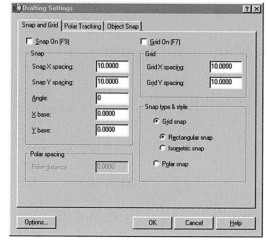

the **Grid** toggle along the bottom of the screen. The spacing of the grid can be set with the **Drafting Setting** dialog box. It is a good idea to turn on the grid, as it shows you the size of paper that you are working on and the direction of the vertical and horizontal.

Snap on/off: In *AutoCAD LT*® you can set the software to "snap" the points you draw to a preset alignment. This helps to make accurate drawings. There are two types of snap feature available in *AutoCAD LT*®: "snap to grid" and "object snap." The snap to grid feature can be turned on and off by pressing ⌐F9⌐, or by clicking the **Snap** toggle.

Ortho on/off: If all the lines in a drawing are to be vertical or horizontal, you may prefer to turn on the **Ortho** feature. This feature "snaps" all of the lines you draw either to the vertical or the horizontal, whichever is closer. The **Ortho** feature can be turned on and off through the **Drawing Assist** dialog box, by pressing ⌐F8⌐, or by clicking the **Ortho** toggle.

Object Snaps: The **Object Snap** tab on the **Drafting Settings** dialog box opens a pane with many "snap to object" settings. The **Osnap** (object snap) feature can be set to snap to points on objects that have already been drawn on the screen. For example, **Osnap** could be set to snap to the endpoint of a line, to the midpoint of a line, or to the intersection of two or more lines, depending on the **Osnap** setting. **Osnap** can be turned on and off by pressing ⌐F3⌐.

Example

To draw a cone:

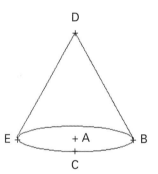

- Create a new drawing, starting from scratch.
- Set **Osnap** to "endpoints" and "tangent."
- Select **Ellipse** from the **Draw** tool bar (or use the **Draw** menu) to draw an ellipse.
- Click to place the centre of the ellipse, at point A.
- Move horizontally and click to place the endpoint of the ellipse's horizontal axis, at B.
- Move to below the first click point (the centre), and click to place the endpoint of the ellipse's vertical axis, at C. The ellipse is now created on the screen.
- Select **Line** from the **Draw** tool bar (or use the **Draw** menu), and click to start a line from directly above the ellipse centre, at point D.
- Move to the far right edge of the ellipse, click, and then, press ⌐ENTER⌐ to complete the line command.
- Use the **Line** tool again to draw a line from the tip of the cone to the far left edge of the ellipse, at E. The completed cone is now drawn on the screen.
- When you have completed this diagram, select **Save As…** from the **File** menu, and save the file to the directory your teacher has assigned.

Canadian Population Projections by Age Group (thousands of persons)

		Total	Under 5 years	5–9 years	10–14 years	15–24 years	25–34 years	35–44 years	45–54 years	55–64 years	65–74 years	75+ years
2011	Male	16 512	855	880	958	2261	2312	2343	2668	2129	1240	866
	Female	16 850	811	836	906	2155	2245	2305	2643	2200	1369	1372
2021	Male	19 165	1080	1102	1111	2317	2625	2656	2559	2664	1905	1147
	Female	19 534	1022	1042	1050	2194	2531	2594	2541	2719	2112	1729
2031	Male	20 362	1082	1131	1173	2410	2590	2834	2699	2467	2323	1653
	Female	20 855	1024	1070	1108	2280	2488	2753	2656	2515	2536	2424
2041	Male	21 110	1108	1137	1175	2499	2680	2799	2870	2598	2154	2090
	Female	21 742	1049	1075	1109	2364	2573	2711	2811	2625	2349	3076

Source: Statistics Canada's Internet Site, http://www.statcan.ca/english/Pgdb/People/Population/demo23b.htm, Dec 21, 2001

Life Expectancy for Newborns by Province for 1996

	Both Genders	Male	Female
Canada	**78.4**	**75.4**	**81.2**
Newfoundland	77.2	74.4	80.2
Prince Edward Island	77.9	74.5	81.5
Nova Scotia	77.7	74.8	80.6
New Brunswick	78.0	74.8	81.2
Québec	77.9	74.6	81.0
Ontario	78.6	75.9	81.3
Manitoba	77.9	75.1	80.5
Saskatchewan	78.3	75.4	81.4
Alberta	78.6	75.9	81.3
British Columbia	79.0	76.2	81.8
Yukon Territory	75.2	72.3	79.2
Northwest Territories excluding Nunavut	76.6	74.4	79.1
Nunavut	70.1	68.3	71.3

Source: Statistics Canada's Internet Site, http://www.statcan.ca/cgi-win/CNSMCGI.EXE, December 21, 2001

Secondary School Graduates

	1995	1996	1997	1998	1999
Canada	**301 739**	**304 451**	**307 772**	**310 626**	**316 810**
Newfoundland	7 318	7 476	6 999	7 073	6 715
Prince Edward Island	1 628	1 605	1 624	1 741	1 643
Nova Scotia	9 574	9 341	10 180	10 387	10 161
New Brunswick	9 281	9 064	8 960	8 754	8 778
Québec	86 174	89 397	85 354	82 216	82 200
Ontario	107 323	105 493	109 734	112 103	116 913
Manitoba	11 845	11 837	11 994	11 970	11 829
Saskatchewan	11 382	11 099	11 840	11 584	11 838
Alberta	24 858	24 980	25 689	26 433	27 386
British Columbia	31 892	33 720	34 943	37 739	39 331
Yukon	163	180	179	245	295
Northwest Territories	301	259	276	381	441

Source: Statistics Canada's Internet Site, http://www.statcan.ca/english/Pgdb/People/Education/educ18.htm, December 21, 2001.

Average Annual Income for Full-Time Workers in Canada by Gender

Year	Women's Income ($)	Men's Income ($)
1989	28 219	42 767
1990	29 050	42 913
1991	29 654	42 575
1992	30 903	42 984
1993	30 466	42 161
1994	30 274	43 362
1995	30 959	42 338
1996	30 606	41 897
1997	30 484	43 804
1998	32 553	45 070
1999	32 026	45 800

Source: Statistics Canada's Internet Site, http://www.statcan.ca/english/Pgdb/People/Labour/labor01b.htm, December 21, 2001

Average Number of Hours per Week of Television Viewing, by Province and Age/Gender, in 2000

	Canada	NF	PE	NS	NB	PQ			ON	MB	SK	AB	BC
						English	French	Total					
Total Population	**21.5**	**24.1**	**21.1**	**23.1**	**22.7**	**20.5**	**24.5**	**24.0**	**20.5**	**20.5**	**20.9**	**19.7**	**20.6**
Men													
18+	20.9	22.8	21.1	22.6	21.9	19.7	23.2	22.6	20.3	20.1	20.7	18.9	20.6
18–24	13.2	17.1	11.5	15.6	16.3	11.4	13.0	12.8	13.9	10.0	11.6	10.8	13.7
25–34	16.9	19.8	18.0	17.4	16.9	14.3	17.2	16.8	17.6	15.7	16.1	16.3	15.3
35–49	18.5	22.3	19.5	21.9	19.1	17.6	21.3	20.7	16.9	17.9	19.2	17.8	18.8
50–59	22.2	22.0	22.2	22.0	24.1	23.6	25.7	25.0	20.5	23.3	22.6	21.7	20.8
60+	32.3	30.3	30.8	32.0	31.8	28.9	37.3	35.5	31.7	31.0	30.5	28.5	31.8
Women													
18+	25.5	28.0	24.7	27.0	26.5	23.8	30.1	29.1	24.1	24.4	25.2	23.6	23.5
18–24	16.5	23.4	16.7	17.2	15.2	14.8	17.0	16.6	15.3	16.3	20.7	17.5	16.4
25–34	21.2	28.4	20.6	24.0	23.1	19.5	24.0	23.6	20.0	20.5	20.6	20.2	19.6
35–49	22.0	25.5	22.6	25.0	25.3	20.3	26.6	25.6	20.2	20.5	21.5	21.0	20.4
50–59	27.9	29.4	27.2	27.7	29.8	24.9	33.4	32.0	26.0	26.5	25.6	26.4	25.8
60+	35.8	32.7	32.1	34.9	32.4	33.6	42.9	41.2	34.8	34.0	33.9	33.4	32.4
Teens													
12–17	14.1	17.3	12.5	15.0	13.7	13.4	15.6	15.4	13.2	13.4	13.5	14.0	14.2
Children													
2–11	15.5	20.0	16.6	17.2	18.7	15.4	16.3	16.4	14.8	15.1	15.0	14.7	15.2

Source: Statistics Canada's Internet Site, http://www.statcan.ca/Daily/English/011023/d011023a.htm, December 21, 2001

Action Movies—Length by Date

Movie Title	Release Date	Length (min)
The Three Musketeers	1948	125
King Solomon's Mines	1950	103
Ivanhoe	1952	107
Battle Cry	1955	150
Around the World in Eighty Days	1956	167
The Guns of Navarone	1961	157
Goldfinger	1964	108
Thunderball	1965	130
The Dirty Dozen	1967	151
Bullitt	1968	113
Diamonds Are Forever	1971	119
The French Connection	1971	102
The Poseidon Adventure	1972	117
Papillon	1973	150
Earthquake	1974	129
Jaws	1975	124
King Kong	1976	135
Superman	1978	143
Rocky II	1979	120
Raiders of the Lost Ark	1981	116
Rocky III	1982	100
Superman III	1983	125
Indiana Jones and the Temple of Doom	1984	118
Rocky IV	1985	91
Top Gun	1986	110
Beverley Hills Cop II	1987	103
Batman	1989	126
Indiana Jones and the Last Crusade	1989	126
Robin Hood: Prince of Thieves	1991	144
Batman Returns	1992	126
Jurassic Park	1993	126
True Lies	1994	141
Die Hard: With a Vengeance	1995	128
Batman Forever	1995	123
Eraser	1996	115
Batman & Robin	1997	125
The Fifth Element	1997	127
Armageddon	1998	144
The Thin Red Line	1998	170
End of Days	1999	109
Three Kings	1999	114
The Sixth Day	2000	123
Rush Hour 2	2001	92

Carbon Dioxide Emissions in Canada

Year	Carbon Dioxide (megatonnes)	Carbon Dioxide (tonnes per capita)
1960	189	10.54
1961	194	10.61
1962	206	11.07
1963	220	11.59
1964	235	12.14
1965	254	12.92
1966	265	13.23
1967	281	13.78
1968	302	14.55
1969	314	14.93
1970	334	15.67
1971	342	15.51
1972	363	16.28
1973	376	16.66
1974	379	16.56
1975	381	16.40
1976	396	16.82
1977	390	16.38
1978	394	16.39
1979	412	16.96
1980	414	16.82
1981	402	15.79
1982	387	14.91
1983	380	14.31
1984	393	14.70
1985	385	15.20
1986	376	14.80
1987	404	15.21
1988	432	16.14
1989	452	16.55
1990	432	15.54
1991	422	14.99
1992	436	15.33
1993	443	15.38
1994	456	15.65
1995	452	14.25
1996	475	15.32
1997	481	15.47
1998	486	15.05
1999	497	14.35
2000	505	15.14

Source: Statscan Catalogue no. 11-509-XPE

Waste Management in Ontario

Residential Recycling in Ontario
Composition of Materials Recycled (kilograms per household)

Material	1996	1997	1998	1999
Paper	105	115	120	123
Glass	30	27	26	25
Plastic	4.5	5.4	5.7	5.6
Metal	13	11.8	12.3	10.9
Total	153	159	164	165

Household Special Waste Quantities Diverted in 1999

Material Type	Thousands of Kilograms
Oil	2033
Paints	1379
Flammables	1027
Car Batteries	814
Propane Tanks	268
Other	843

Changes in Composting Activity in Ontario

Centralized Composting

	1996	1997	1998	1999
Number of Municipalities Providing Service	150	241	235	217
Households With Access to Service (thousands)	3323	3560	3746	3874
Tonnes Processed (thousands)	219	280	290	281

Backyard Composting

	1996	1997	1998	1999
Number of Municipalities Providing Compost Bins	481	427	419	393
Households Provided With Compost Bins (thousands)	1000	1050	1110	1120
Approximate Tonnes Diverted (thousands)	100	105	111	112

Source of data: WDO Fact Sheet 1999, Municipal 3Rs in Ontario, http://www.rco.on.ca/factsheet/3Rs1999.pdf

Summer Olympics—Winning Times in Selected Track and Field Events

100-metre Run

Year	Men	Time (s)	Women	Time (s)
1948	Harrison Dillard, USA	10.30	Fanny Blankers-Koen, NED	11.90
1952	Lindy Remigino, USA	10.40	Marjorie Jackson, AUS	11.50
1956	Bobby Morrow, USA	10.50	Betty Cuthbert, AUS	11.50
1960	Armin Hary, GER	10.20	Wilma Rudolph, USA	11.00
1964	Bob Hayes, USA	10.00	Wyomia Tyus, USA	11.40
1968	Jim Hines, USA	9.95	Wyomia Tyus, USA	11.08
1972	Valery Borzov, USSR	10.14	Renate Stecher, E. GER	11.07
1976	Hasely Crawford, TRI	10.06	Annegret Richter, W. GER	11.08
1980	Allan Wells, GBR	10.25	Lyudmila Kondratyeva, USSR	11.06
1984	Carl Lewis, USA	9.99	Evelyn Ashford, USA	10.97
1988	Carl Lewis, USA	9.92	Florence Griffith Joyner, USA	10.54
1992	Linford Christie, GBR	9.96	Gail Devers, USA	10.82
1996	Donovan Bailey, CAN	9.84	Gail Devers, USA	10.94
2000	Maurice Greene, USA	9.87	Marion Jones, USA	10.75

200-metre Run

Year	Men	Time (s)	Women	Time (s)
1948	Mel Patton, USA	21.10	Fanny Blankers-Koen, NED	24.40
1952	Andy Stanfield, USA	20.70	Marjorie Jackson, AUS	23.70
1956	Bobby Morrow, USA	20.60	Betty Cuthbert, AUS	23.40
1960	Livio Berruti, ITA	20.50	Wilma Rudolph, USA	24.00
1964	Henry Carr, USA	20.30	Edith McGuire, USA	23.00
1968	Tommie Smith, USA	19.83	Irena Szewinska, POL	22.50
1972	Valery Borzov, USSR	20.00	Renate Stecher, E. GER	22.40
1976	Donald Quarrie, JAM	20.23	Bärbel Eckert, E. GER	22.37
1980	Pietro Mennea, ITA	20.19	Bärbel Eckert Wockel, E. GER	22.03
1984	Carl Lewis, USA	19.80	Valerie Brisco-Hooks, USA	21.81
1988	Joe DeLoach, USA	19.75	Florence Griffith Joyner, USA	21.34
1992	Mike Marsh, USA	20.01	Gwen Torrence, USA	21.81
1996	Michael Johnson, USA	19.32	Marie-Jose Perec, FRA	22.12
2000	Konstantinos Kenteris, GRE	20.09	Marion Jones, USA	21.84

400-metre Run

Year	Men	Time (s)	Women	Time (s)
1964	Mike Larrabee, USA	45.10	Betty Cuthbert, AUS	52.00
1968	Lee Evans, USA	43.86	Colette Besson, FRA	52.03
1972	Vince Matthews, USA	44.66	Monika Zehrt, E. GER	51.08
1976	Alberto Juantorena, CUB	44.26	Irena Szewinska, POL	49.29
1980	Viktor Markin, USSR	44.60	Marita Koch, E. GER	48.88
1984	Alonzo Babers, USA	44.27	Valerie Brisco-Hooks, USA	48.83
1988	Steve Lewis, USA	43.87	Olga Bryzgina, USSR	48.65
1992	Quincy Watts, USA	43.50	Marie-Jose Perec, FRA	48.83
1996	Michael Johnson, USA	43.49	Marie-Jose Perec, FRA	48.25
2000	Michael Johnson, USA	43.84	Cathy Freeman, AUS	49.11

Summer Olympics—Winning Times in Selected Track and Field Events

800-metre Run

Year	Men	Time (min:s)	Women	Time (min:s)
1960	Peter Snell, NZE	1:46.30	Lyudmila Shevtsova, USSR	2:04.30
1964	Peter Snell, NZE	1:45.10	Ann Packer, GBR	2:01.10
1968	Ralph Doubell, AUS	1:44.30	Madeline Manning, USA	2:00.90
1972	Dave Wottle, USA	1:45.90	Hildegard Falck, W. GER	1:58.55
1976	Alberto Juantorena, CUB	1:43.50	Tatyana Kazankina, USSR	1:54.94
1980	Steve Ovett, GBR	1:45.40	Nadezhda Olizarenko, USSR	1:53.42
1984	Joaquim Cruz, BRA	1:43.00	Doina Melinte, ROM	1:57.60
1988	Paul Ereng, KEN	1:43.45	Sigrun Wodars, E. GER	1:56.10
1992	William Tanui, KEN	1:43.66	Ellen van Langen, NED	1:55.54
1996	Vebjoern Rodal, NOR	1:42.58	Svetlana Masterkova, RUS	1:57.73
2000	Nils Schumann, GER	1:45.08	Maria Mutola, MOZ	1:56.15

4 × 100-metre Relay

Year	Men	Time (s)	Women	Time (s)
1948	United States	40.60	Holland	47.50
1952	United States	40.10	United States	45.90
1956	United States	39.50	Australia	44.50
1960	Germany	39.50	United States	44.50
1964	United States	39.00	Poland	43.60
1968	United States	38.23	United States	42.87
1972	United States	38.19	West Germany	42.81
1976	United States	38.33	East Germany	42.55
1980	Soviet Union	38.26	East Germany	41.60
1984	United States	37.83	United States	41.65
1988	Soviet Union	38.19	United States	41.98
1992	United States	37.40	United States	42.11
1996	Canada	37.69	United States	41.95
2000	United States	37.61	Bahamas	42.20

Men's Triple Jump

Year	Gold Medalist	Distance (m)	Year	Gold Medalist	Distance (m)
1896	James Connolly, USA	13.71	1956	Ademar da Silva, Brazil	16.35
1900	Meyer Prinstein, USA	14.47	1960	Jozef Schmidt, Poland	16.81
1904	Meyer Prinstein, USA	14.35	1964	Jozef Schmidt, Poland	16.85
1908	Timothy Ahearne, GBR	14.92	1968	Viktor Saneyev, USSR	17.39
1912	Gustaf Lindblom, Sweden	14.76	1972	Viktor Saneyev, USSR	17.35
1920	Viho Tuulos, Finland	14.50	1976	Viktor Saneyev, USSR	17.29
1924	Anthony Winter, AUS	15.53	1980	Jaak Uudmae, USSR	17.35
1928	Mikio Oda, Japan	15.21	1984	Afrederick Joyner, USA	17.26
1932	Chuhei Nambu, Japan	15.72	1988	Khristo Markov, Bulgaria	17.61
1936	Naoto Tajima, Japan	16.00	1992	Michael Conley, USA	18.17
1948	Arne Ahman, Sweden	15.40	1996	Kenny Harrison, USA	18.09
1952	Ademar da Silva, Brazil	16.22	2000	Jonathon Edwards, GBR	17.71

Passenger Aircraft Operating Statistics

Aircraft	Number of Seats	Mean Speed Airborne (km/h)	Mean Flight Length (km)	Mean Fuel Consumption (L/h)	Mass (t)
B747-100	410	834	4638	13 752	341
B747-400	400	867	8148	13 040	397
B747-200/300	369	851	5344	14 229	378
L-1011-100/200	305	801	2193	9 081	212
B-777	291	826	3944	7 711	243
DC-10-10	286	801	2403	8 453	200
DC-10-40	284	811	3159	10 020	260
DC-10-30	272	830	3829	9 937	260
A300-600	266	752	1812	6 325	172
MD-11	260	843	5235	9 085	281
L-1011-500	222	842	4820	9 289	229
B767-300ER	216	797	3751	6 019	187
B757-200	187	747	1878	3 967	116
B767-200ER	181	782	3436	5 421	179
MD-90	154	710	1258	3 093	63
B727-200	148	708	1194	4 876	87
A320-100/200	148	737	1772	3 089	77
B737-400	144	666	1130	2 998	63
MD-80	141	695	1284	3 498	64
B737-300	131	669	969	3 165	57
DC-9-50	121	602	555	3 399	55
B737-100/200	112	624	711	3 146	53
B737-500	110	663	917	2 813	52
DC-9-40	109	623	784	3 168	52
DC-9-30	100	626	753	3 096	49
F-100	97	618	805	2 669	44
DC-9-10	71	612	665	2 790	41

Passenger Aircraft Operating Statistics

Aircraft	Takeoff Runway Length (m)	Landing Runway Length (m)	Wingspan (m)	Length (m)
B747-100	2668.80	1875.80	59.70	70.70
B747-400	3355.00	2074.00	64.50	70.70
B747-200/300	3324.00	2119.80	59.70	70.70
L-1011-100/200	3428.20	1967.30	47.40	54.20
B-777	2562.00	1586.00	61.00	63.80
DC-10-10	2992.10	1775.10	47.40	55.60
DC-10-40	3126.30	1781.20	50.40	55.10
DC-10-30	3153.70	1820.90	50.40	55.40
A300-600	2318.00	1433.50	44.90	54.10
MD-11	3202.50	2119.80	51.70	61.20
L-1011-500	2867.00	2074.00	50.10	50.10
B767-300ER	3111.00	1586.00	47.60	55.00
B757-200	2348.50	1546.40	38.10	47.40
B767-200ER	2836.50	1494.50	47.60	48.60
MD-90	1982.50	1392.30	32.90	39.80
B727-200	3050.00	1616.50	32.90	46.70
A320-100/200	2074.00	1464.00	43.10	37.60
B737-400	2357.70	1488.40	28.90	36.50
MD-80	2211.30	1451.80	32.90	45.10
B737-300	2031.30	1396.90	28.90	33.40
DC-9-50	2531.50	1290.20	28.50	40.40
B737-100/200	2092.30	1299.30	28.40	30.60
B737-500	1860.50	1357.30	28.90	31.00
DC-9-40	2260.10	2260.10	28.50	38.30
DC-9-30	2260.10	2260.10	28.50	36.40
F-100	1721.10	1348.10	28.10	35.60
DC-9-10	1982.50	1363.40	27.30	31.80

Ontario Provincial Parks

Name	Area (ha)	Number of Campsites	Name	Area (ha)	Number of Campsites
Northwestern Region			**Northeastern Region (excluding Lake Superior)**		
Aaron	117	98	Esker Lakes	3 237	100
Blue Lake	353	196	Fushimi Lake	5 294	44
Calliper Lake	147	83	Greenwater	5 350	90
Kakabeka Falls	500	169	Ivanhoe Lake	1 589	120
MacLeod	74	94	Kap-Kig-Iwan	428	64
Neys	3 445	144	Kettle Lakes	1 261	137
Ojibway	2 630	45	Missinaibi	99 090	35
Pakwash	3 993	69	Nagagamisis	8 131	86
Quetico	475 782	107	Obatanga	9 409	132
Rainbow Falls	575	133	Pancake Bay	490	325
Rushing River	340	215	Rene Brunelle	3 015	90
Sandbar Lake	5 083	75	The Shoals	10 644	44
Sioux Narrows	130	60	Tidewater	980	20
Sleeping Gaint	24 400	200	Wakami Lake	8 806	65
Near North Region			White Lake	1 726	187
Arrowhead	1 237	388	**Southeastern Region**		
Chutes	108	130	Bon Echo	6 643	530
Driftwood	422	82	Bonnechere	162	128
Fairbank	105	160	Charleston Lake	2 334	236
Finlayson Point	37	117	Ferris	198	163
Grundy Lake	2 554	486	Fitzroy	185	235
Halfway Lake	4 730	215	Lake St. Peter	478	65
Killarney	48 500	126	Murphys Point	1 239	167
Killbear	1 756	880	Presqu'ile	937	394
Marten River	400	190	Rideau River	187	184
Massassauga, The	13 105	135	Sandbanks	1 509	549
Mikisew	131	258	Sharbot Lake	69	167
Mississagi	4 900	90	Silent Lake	1 450	167
Oastler Lake	32	148	Silver Lake	43	148
Restoule	1 200	278	Voyageur	1 464	416
Samuel de Champlain	2 550	215	**Southwestern Region**		
Six Mile Lake	94	217	Awenda	2 915	333
Sturgeon Bay	14	81	Bronte Creek	640	142
Windy Lake	13 105	135	Craigleith	66	172
Central Region (excluding Algonquin)			Earl Rowe	312	365
Balsam Lake	448	505	Long Point	150	254
Bass Lake	65	182	MacGregor Point	1 204	360
Darlington	208	315	Pinery	2 532	1000
Emily	83	299	Point Farms	307	200
Mara	40	105	Port Burwell	231	232
McRae Point	138	202	Rock Point	187	178
Sibbald Point	225	512	Rondeau	3 254	262
			Sauble Falls	20	152
			Selkirk	73	142
			Turkey Point	316	235
			Wheatley	241	220

Source of data: http://www.OntarioPark.com, © Queen's Printer for Ontario, 2002

The Solar System

Planet/Star	Diameter (km)	Relative Mass (Earth = 1)	Period of Orbit About the Sun	Period of Rotation on Axis (days)	Distance From the Sun (millions of km)	Number of Moons
Sun	1 392 000	332 830	—	25.38	—	—
Mercury	4 878	0.06	88.0 days	58.60	58	0
Venus	12 104	0.8	224.7 days	243.00	108	0
Earth	12 756	1.0	365.3 days	0.99	150	1
Mars	6 787	0.1	1.88 years	1.02	228	2
Jupiter	142 800	317.8	11.86 years	0.41	779	16
Saturn	120 000	95.2	29.63 years	0.42	1425	18
Uranus	51 200	14.5	83.97 years	0.45	2870	15
Neptune	48 680	17.2	164.80 years	0.67	4497	8
Pluto	2 300	0.002	248.63 years	6.38	5866	1

Canadian Lakes

Lake	Surface Area (km²)	Maximum Depth (m)	Volume (km³)
Athabasca	7 900	120	204
Erie	25 821	64	458
Great Bear	31 153	446	2 240
Great Slave	28 568	614	2 090
Huron	59 570	228	3 540
Melville	3 069	256	313
Michikamau	2 030	80	65
Mistassini	2 115	183	2 115
Nipigon	4 510	165	248
Ontario	19 009	224	1 640
Superior	87 367	406	12 200
Willistor	1 779	166	70
Winnipeg	23 750	36	284

Chapter 1 – Trigonometry

Get Ready, pp. 2–3

1. a) 31.3 **b)** 0.25 **c)** 0.06 **d)** 6.71 **e)** 56 **f)** 95.52 **g)** 1.77
2. a) 86° **b)** 116° **c)** 45° **3. a)** 10 cm **b)** 18.6 m **c)** 13.7 m
4. a) 3 **b)** 2:3 **c)** $\frac{4}{3}$ **d)** 7:2:3 **e)** 16:18:37 **f)** 5:2:7:25
5. a) $\frac{5}{4}$ **b)** $\frac{12}{7}$ **c)** $\frac{1113}{940}$ **6. b)** C, D, F **8. a)** 0.5000 **b)** 0.3907
c) 0.2679 **d)** 0.5000 **e)** 0.1564 **f)** 0.9945 **g)** 0.1045 **h)** 0.7071
i) 0.7071 **j)** 1.0000 **9. a)** 65.4° **b)** 29.0° **c)** 30.0° **d)** 65.9° **e)** 30.0°
f) 45.0° **g)** 60.0° **h)** 68.7° **i)** 56.8°

1.1 Using Trigonometry to Find Lengths, pp. 8–9

1. a) Use sine ratio; 11.6 **b)** Use cosine ratio; 10.1 **c)** Use
tangent ratio; 6.9 **d)** Use tangent ratio; 10.6 **2. a)** 11.4 **b)** 8.5
c) 11.3 **3.** 4.8 m **4.** 11.1 m **5.** 147.7 m **6.** Answers may vary.
For example, the wire, pole, and ground form a right triangle.
The angle opposite the pole has a sine of 0.5, and so is 30°. The
length of the adjacent side can now be found using the cosine
ratio. **7. a)** 48.1 m **b)** 47.0 m **8. a) i)** 273 000 AU
ii) 373 000 AU **iii)** 1 676 000 AU **iv)** 27 258 000 AU **b)** Answers
may vary.

1.2 Using Trigonometry to Find Angles, pp. 13–14

1. a) tangent; 66.0° **b)** cosine; 45.9° **c)** sine; 36.9° **2.** a = 8.0 cm,
∠A = 53.1 °, ∠B = 90°, ∠C = 36.9° **3. b)** PR = 18.0 cm,
∠P = 56.3°, ∠R = 33.7° **4.** 4.6° **5.** 74.5° **6.** 4.6° **7. a)** 41.8°
b) The rope is taut. **8.** 7.6° **9.** 24.7° **10.** Yes; the angle of
elevation is only 10.8°. **11.** 31.0 m **12.** 45.2° **14. a)** 180.1 m
b) 12 m/min **15.** 2.4°

1.3 Trigonometric Ratios of Obtuse Angles, pp. 19–20

1. a) 0.5000 **b)** −0.0872 **c)** −0.1763 **d)** 0.9848 **e)** −0.9563
f) −1.4281 **2. a)** 179.2° **b)** 97.8° **c)** 124.2° **d)** 116.8° **e)** 139.6°
f) 141.0° **3. a)** sin A = $\frac{1}{\sqrt{26}}$, cos A = $\frac{5}{\sqrt{26}}$, tan A = $\frac{1}{5}$;

∠A = 11.3° **b)** sin A = $\frac{1}{\sqrt{2}}$, cos A = $-\frac{1}{\sqrt{2}}$, tan A = −1;

∠A = 135° **c)** sin A = $\frac{3}{\sqrt{10}}$, cos A = $-\frac{1}{\sqrt{10}}$, tan A = −3;

∠A = 108.4° **d)** sin A = $\frac{1}{\sqrt{5}}$, cos A = $-\frac{2}{\sqrt{5}}$, tan A = $-\frac{1}{2}$;

∠A = 153.4° **4. a)** 0.61 m **b)** 1.05 m **5. a) i)** 29.9 m **ii)** 27.2 m
b) i) 2.6 m **ii)** 12.7 m **6. a)** sine: negative, cosine: negative,
tangent: positive **b)** sine: negative, cosine: positive, tangent:
negative

1.4 The Sine Law, pp. 25–27

1. a) a = 11.6 cm, c = 9.0 cm **b)** p = 11.8 cm, q = 17.4 cm
c) d = 10.2 m, e = 11.8 m **2. a)** ∠B = 58.6°, ∠C = 46.4°
b) ∠P = 51.4°, ∠R = 53.6° **c)** ∠J = 33.1°, ∠K = 100°, ∠L = 46.9°
3. a) b = 8.3 cm **b)** e = 14.5 cm **c)** ∠H = 52.5° **d)** ∠L = 51.1°
4. ∠C = 51.6°, ∠D = 70.4°, e = 8.3 cm; ∠Q = 32.1°, ∠S = 32.9°,
q = 14.6 cm **5. a)** 66 m **b)** 61 m **6.** 55°, 60° **7.** Yes; the angle of
inclination is 26.9°. **8.** Answers may vary. **10.** 25.4°
11. a) 59° **b)**

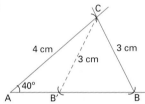

c) A calculator
gives ∠AB'C as 59°. **d)** Two possible triangles can be drawn for
the given measures. Since sin(180° − 59°) is the same as 59°, it is
hard to recognize by just applying the sine law.

1.5 The Cosine Law, pp. 31–33

1. a) 8.0 cm **b)** 22.5 cm **c)** 85.6° **d)** 53.3° **2. a)** 8.6 cm **b)** 20.6 cm
c) 31.1 m **3. a)** ∠A = 87.1°, ∠B = 44.4°, ∠C = 48.5°
b) ∠P = 41.4°, ∠Q = 82.8°, ∠R = 55.8° **c)** ∠J = 29.4°,
∠K = 117.5°, ∠L = 33.1° **4. a)** e = 8.9 cm, ∠C = 49.7°,
∠D = 72.3° **b)** r = 27.9 cm, ∠Q = 35.8°, ∠S = 29.2° **5.** 233 m
6. 189 nautical miles **7.** 8.6° **8. a)** 103° **b)** Answers may vary.
9. Answers may vary. **10.** cos A = $\dfrac{b^2 + c^2 - a^2}{2bc}$; This form
would be used to find the measure of ∠A when three side
lengths are given.

1.6 Problem Solving With Non-Right Triangles, pp. 38–41

1. 6.4 light years **2.** Sandra, by 7.7° **3.** 301.6 km **4.** 119.8°
5. 17.4° **6.** 2.6 m **7.** 330.2 m **8.** 2.3 km, 2.8 km **9.** 74.9 km,
99.9 km **10.** Questions and answers may vary.
11. 1 387 531.8 km **12.** 6.3 m **13. a)** 40.5° **b)** 23.7 m
14. ∠X = 35.5°, ∠Y = 45° **15.** 17.1 cm to 15.7 cm, 49.9° to
52.5° **17.** 10.5° **18.** Answers may vary. **19. a)** 10:02:15
b) 2530.4 km **c)** 34.3°

Review, pp. 44–47

1. a) $4\sqrt{13}$; $\sin A = \dfrac{3}{\sqrt{13}}$, $\cos A = \dfrac{2}{\sqrt{13}}$, $\tan A = \dfrac{3}{2}$

b) $2\sqrt{119}$; $\sin D = \dfrac{5}{12}$, $\cos D = \dfrac{\sqrt{119}}{12}$, $\tan D = \dfrac{5}{\sqrt{119}}$

c) $4\sqrt{11}$; $\sin G = \dfrac{4\sqrt{11}}{15}$, $\cos G = \dfrac{7}{15}$, $\tan G = \dfrac{4\sqrt{11}}{7}$

2. 10.8 m **3.** 8.7 m **4. a)** sine **b)** tangent **5. a)** 34.8° **b)** 37.6°
c) 66.4° **6.** 17.5° **7.** Yes; the angle of inclination is 6.9°.
8. Yes; the shadow cast will be less than 12.6 m during these hours of the day. **9.** Spotlight B is closer by 1.3 m.
10. a) 0.8660 **b)** 0.2588 **c)** −0.8391 **d)** 0.9272 **e)** −0.9205
f) −0.2679 **11.** 36.9° **12.** The measure of ∠B, the measure of another angle, and the length of the side opposite that angle are known. **13. a)** ∠B = 74.4°, ∠C = 43.6°, c = 8.6 cm
b) ∠S = 23.2°, ∠T = 36.8°, t = 30.4 cm **14.** 5 m **15.** Use the form $p^2 = q^2 + r^2 - 2qr\cos P$ and solve for ∠P after substituting the side lengths. **16. a)** 9.5 cm **b)** 15.0 m **17. a)** 51.0° **b)** 55.0°
18. 9.1 m **19.** 22.6 m **20.** 4.7 km, 3.3 km **21.** 86.9°

Practice Test, pp. 48–49

1. $\sin P = \dfrac{7}{12}$, $\cos P = \dfrac{\sqrt{95}}{12}$, $\tan P = \dfrac{7}{\sqrt{95}}$; 35.7°

2. 137.4 m **3.** 59.0° **4. a)** 170.5° **b)** 124.3° **c)** 111.4°
5. a) 18.5 m **b)** 102.2° **6. a)** sin A increases from 0 to 1.
b) cos A decreases from 0 to −1. **7.** No; the ladder should be moved at least 0.75 m toward the wall. **8.** 7056.9 m
9. a)

Ship 68°, 131 m, 72°, Diver, Treasure

b) 324.2 m **c)** 25.7 m

Chapter 2 – Problem Solving With Measurement

Get Ready, pp. 52–53

1. a) 0.065 **b)** 420 000 **c)** 0.5 **d)** 0.056 78 **e)** 0.08 **f)** 0.003 15
2. a) 0.56 m **b)** 34 L **c)** 78 kg **d)** 500 000 cm **e)** 50 000 mL **f)** 50 mg
3. a) $\dfrac{3}{8}$ **b)** $\dfrac{7}{16}$ **c)** $\dfrac{3}{4}$ **d)** $\dfrac{1}{2}$ **e)** $\dfrac{5}{8}$ **f)** $\dfrac{15}{16}$ **4. a)** $\dfrac{3}{8}, \dfrac{1}{2}, \dfrac{9}{16}, \dfrac{3}{4}, 1$
b) $\dfrac{1}{4}, \dfrac{5}{16}, \dfrac{11}{32}, 1\dfrac{1}{2}$ **c)** $\dfrac{1}{4}, \dfrac{3}{8}, \dfrac{1}{2}, \dfrac{5}{8}, \dfrac{3}{4}, \dfrac{7}{8}$ **5. a)** $\dfrac{7}{16}$ **b)** $\dfrac{9}{16}$ **c)** $\dfrac{3}{4}$
d) $14\dfrac{1}{2}$ **e)** $30\dfrac{3}{8}$ **f)** $13\dfrac{1}{16}$ **6. a)** 48 **b)** 5 **c)** 2.4 **7. a)** 86.01 **b)** 63
c) 1.02 **8. a)** 7.2 **b)** 100 **c)** 12.2 **d)** 15.5 **9. a)** 6.5×10^{-4}
b) 1.2×10^{7} **c)** 3×10^{-3} **d)** 3.4×10^{4} **e)** 1×10^{-7} **f)** 4.07×10^{-3}

2.1 Systems of Measure, pp. 61–66

1. a) 2 m **b)** 60 L **c)** 20 kg **d)** 300 mL **e)** 12 cm **f)** 1000 kg **g)** 20°C
2. a) 1950 mL **b)** 0.001 95 s **c)** 0.02 terabytes **d)** 1500 megahertz
e) 9000 millivolts **f)** 0.06 kilowatts **g)** 55 **3. a)** 100 cm
b) 10 000 cm² **c)** 1 m² = 10 000 cm² **d) i)** 1 000 000 m²
ii) 100 mm² **4. a)** 67 cm² **b)** 0.001 485 m² **c)** 0.045 268 1 km²
d) 15.45 cm² **5. a)** 100 cm each **b)** 1 000 000 cm³
c) 1 m³ = 1 000 000 cm³ **d) i)** 1000 mm³ **ii)** 1 000 000 000 mm³
iii) 1000 cm³ **e)** 1000 L **6. a)** A: $\dfrac{3}{4}''$, B: $1\dfrac{1}{2}''$, C: 2″, D: $3\dfrac{1}{4}''$,
E: $4\dfrac{3}{4}''$ **b)** F: $1\dfrac{3}{8}''$, G: $2\dfrac{5}{8}''$, H: $3\dfrac{1}{8}''$, I: $4\dfrac{1}{2}''$, J: $4\dfrac{3}{4}''$ **c)** K: $\dfrac{3}{16}''$,
L: $\dfrac{3}{8}''$, M: $\dfrac{11}{16}''$, N: $1\dfrac{7}{8}''$, P: $2\dfrac{1}{8}''$, Q: $2\dfrac{3}{4}''$, R: 3″, S: $4\dfrac{13}{16}''$,
T: $5\dfrac{1}{16}''$, U: $5\dfrac{1}{4}''$ **7. a)** 7 ft **b)** 15 gal **c)** 45 lb **d)** 10 fl oz **e)** 5 in.
f) 1 ton **g)** 70°F **8.** Answers may vary. **9. a)** 874.8 cm,
25 615.8 cm² **b)** 180 m, 1327 m² **c)** 41.8 m, 114.6 m²
10. a) $4\dfrac{3}{8}''$ **b)** $3\dfrac{1}{2}''$ **c)** $3\dfrac{5}{16}''$ **11.** $\dfrac{69}{128}$ in.²
12. a) 84.8 in.³ **b)** 10″ round **c)** 9.6 in.³ **d)** 8″ round: 16,
10″ round: 25, half slab: 40, quarter slab: 20
13. a) 530.9 in.², 1150.3 in.³ **b)** 5085 in.³, 8100 in.³
c) 3311.2 in.², 6427.7 in.³
14. a)

	Floor Area (ft²)	Ceiling Area (ft²)	Wall Area (ft²)
Bathroom	38.5	38.5	approx. 190
Bedroom	115.8	115.8	approx. 300
Hallway	32.3	32.3	approx. 70
Laundry	85.5	85.5	approx. 280
Rec Room	452.8	452.8	approx. 700

b) Answers may vary. For example, carpet and underpadding for bedroom and rec room approximately 3700 ft², total cost $8144.11; tiles for hallway, bathroom, and laundry approximately 160 tiles, cost $278.46 (for 14 dozen).
15. a) 41.7 in.² **b)** 374.7 ft² **c)** 13.4 in.²
16. a) 53.3 in.³ **b)** 1.96 kg **c)** 1960 g **d)** $\dfrac{5}{12}''$ **e)** 612

2.2 Converting Between Metric and Imperial Systems, pp. 70–73

1. a) 25.4 cm **b)** 54.48 kg **c)** 9.46 L **d)** 510.3 g **e)** 1.83 m
f) 236.8 mL **g)** 2413.5 km **h)** 2.56 L **i)** 681 kg **2. a)** 22.8″
b) 16.9 fl oz **c)** 10.6 gal **d)** 17.6 oz **e)** 42.2 qt **f)** 55.5 lb **g)** 39.0 mi
h) 3.2″ **i)** 10.6 oz **3.** 15.24 cm and 30.48 cm **4.** 1.8 kg ground beef, 1.4 L tomato juice, 227 g tomato paste, 114 g chopped onion, 829 mL baked beans, 414 mL kidney beans, chili powder to taste **5.** turkey **6. a)** 29.0 m **b)** 46.3 m **c)** 64.5 m
7. 8 **8. a) i)** $93\dfrac{1}{2}$ in.² **ii)** 603.08 cm² **iii)** 60 308 mm² **b)** 0.1 mm
9. no **10.** 69.2 min **11. a)** 2.54 cm **b)** 16.39 cm³
c) 1 in.³ = 16.39 cm³ **d) i)** 1.32 **ii)** 35.31 **iii)** 2.59
12. a) 15 cups **b)** 5 cups **13.** 290.25 cm²

14. a)

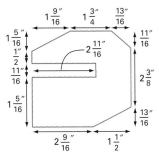

b) 11.82 in.^{2}
c) 10 697.8 in.^{3} (or 175.3 mL), assuming that the plastic is poured and there is no waste

15. 40.9 kg, 34.1 kg to 45.4 kg, 72.6 kg, 63.6 kg, 1.09 T, 1.14 T
16. a) 115 m^3 or 115 000 L **b)** 6 h 23 min 20 s
17. a) Option A: 69.75 m^2, Option B: 750.75 ft^2 **b)** Multiply Option B by 0.0929. **c)** Divide Option A by 0.0929.
18. a) No; the area is 81 024 ft^2, which is less than 2 acres
b) $30 384.02 **19.** 2 of 22.5″ by 58.5″ at $32.95 each, 1 of 34.5″ by 34.5″ at $39.95, 2 of 46.5″ by 58.5″ at $47.95 each; total cost $201.75 **20. a) i)** 44 640 ft^2; 14 651 ft^2 **ii)** 1.03 acres; 0.34 acres
b) 4147 m^2; 1361 m^2 **c)** 8400 ft^2, $5.06/ft^2; 8700 ft^2, $4.60/ft^2; Answers for the lot may vary. For example, 9130 ft^2, $5.15/ft^2 or 8775 ft^2, $5.36/ft^2 **d)** Answers may vary.

2.3 Measurement for Your Health and Fitness, pp. 76–79

1. a) 7 lb 5 oz **b)** 6 lb 5 oz **c)** 8 lb 14 oz **2. a)** 3.3 kg, 50.8 cm
b) 2.9 kg, 48.3 cm **c)** 3.7 kg, 53.3 cm **d)** 4.1 kg, 55.2 cm
3. a) i) 177.8 cm, 81.7 kg, 37.3°C **ii)** 157.5 cm, 64.9 kg, 38.2°C
iii) 174.0 cm, 73.5 kg, 38.4°C **iv)** 182.9 cm, 95.3 kg, 37.1°C
b) i) 5′11″, 154 lb, 98.6°F **ii)** 5′4.5″, 121 lb, 100.8°F **iii)** 5′2″, 150 lb, 102.2°F **iv)** 5′7″, 132 lb, 104.2°F **4. a)** 1.2 fl oz/lb
b) 41.2 qt **c)** 0.5 to 0.7 L **d)** 7.6 to 11.4 kL **e)** 2.1 pt **f)** 40°C
5. 734 mg **7. a)** 4-1000 mL bags, 1-250 mL bag, 1-100 mL bag
b) 1-250 mL bag, 2-100 mL bags **c)** 6-1000 mL bags, 1-500 mL bag **8. a)** 15.6°C to 48.9°C **b)** 32°F, 212°F **9. a)** yes
b) 105 mg **c)** 15 **d)** 25 **e)** 0.45% **10. a)** Multiply the number of kilometres by 0.6. **b) i)** 6 mi **ii)** 13.2 mi **iii)** 2.4 mi
11. a) 112 mL **b)** 151.2 mL **12. a)** 23 171 cm^3 **b)** 92 **c)** 6 days
d) 200 **13. a)** Atrazine: 0.000 003 338 oz/gal, Dichloromethane: 0.000 006 675 oz/gal, Carbofuran: 0.000 012 016 oz/gal, Chloramines: 0.000 400 529 oz/gal, Triallate: 0.000 030 707 oz/gal **b)** 3.338 oz **c)** Answers will vary; 5 mg if the bathtub holds 100 L. **14. a)** 1.03 m^2 **b)** 1.33 m^2
c) 70.2 mg **15. a)** Humidifier B **b)** Humidifier A
c) Humidifier C

2.4 Measurement for Home Improvement and Construction, pp. 82–86

1. $2\frac{5}{8}$″ **2. a)** $53\frac{3}{4}$″ **b)** 15.3 ft **3. a)** 40 **b)** 84 **c)** 3 **d)** 6
4. a) $68.40 **b)** $l = 6h + 2(w_1 + w_2 + w_3)$ **c)** Answers will vary.
5. a) i) $1\frac{1}{2}$″ by $7\frac{1}{4}$″ **ii)** $\frac{3}{4}$″ by $4\frac{1}{2}$″ **b)** 2

6. a) 1.2 yd^3 **b)** $176 **7.** 6″ **8. a)** 85 **b)** 82 **9. a)** 39 gal **b)** 50
10. a) 1241 gal **b)** 1′5.2″ **11. b)** 28 yd^2, 24 yd^2 **c)** With the single piece, there are no seams, but there is 7 yd^2 of waste. With two pieces, there is only 3 yd^2 of waste, but there is a seam. **12.** 23 bags, $91.54 **13. a)** 472 ft^2 **b)** 4.4 L to 5.5 L
c) 2 cans for a single coat **d)** $121.85 **14.** no **15.** 21 rolls membrane, 10 rolls felt, 124 boxes shingles
16. a) 1938 ft^2 **b)** 6.5 ft^2 **17. a)** $2r + t = 28$
b)

Riser Height (in.)	Tread Length (in.)
4.0	20
4.5	19
5.0	18
5.5	17
6.0	16
6.5	15
7.0	14
7.5	13
8.0	12

18. a) circle **b)** the sum of the 8 lengths divided by 8
c) 34 ft^2 **d)** 21 ft **19.** Answers will vary.

2.5 Measurement for Technology and Design, pp. 93–97

1. a) 1.2708″, 1.2690″ **b)** 2.002″, 1.998″ **c)** 3.0675″, 3.0575″
d) 10.227 mm, 10.179 mm **e)** 289.012 mm, 288.998 mm
f) 64.89 mm, 64.83 mm **2. a)** $\frac{9}{10\,000}$″ **b)** $\frac{1}{500}$″ **c)** $\frac{1}{200}$″
d) $\frac{3}{125}$ mm **e)** $\frac{7}{1000}$ mm **f)** $\frac{3}{100}$ mm **3. a)** 2.005″, 2.001″
b) 3.6435″, 3.6423″ **c)** 1.0865″, 1.0775″ **d)** 24.16 mm, 24.10 mm
e) 117.72 mm, 117.65 mm **f)** 397.225 mm, 397.216 mm
4. 2.43 mm, 2.37 mm **5. b)** 4 **6.** No, the second hole is $\frac{1}{16}$″ too far from the first hole. **7. a)** 45.1 in.^{2} **b)** $A = \frac{\pi}{2}(R^2 - r^2)$
8. 7.1″ **9. a)** 1.67 m **b)** yes **c)** 145.5 ft **10. a)** $390.57 **b)** 64 mm by 55.4 mm by 74 mm **c)** 65 624 mm^3 **d)** 511.9 g **11.** 16.62 in.^{2}
12. a) 27.09 in.^{3} **b)** 240 cm^3 **c)** 66.7% **13. a)** 3.751 in.
b) 8.12 in.^{2} **c)** 11.426 in. **d) i)** $62.84 **ii)** $142.83 **14.** Answers will vary. **15. a)** 12.6 ft^2 **b)** yes **16.** Answers will vary. **17.** 9

Career Profile, p. 98

1. a) 5 cc **b)** 15 mg **2.** 100 units **3.** 30 mL; 7500 mg

Review, pp. 100–103

1. a) $4\frac{1}{6}$ ft **b)** 160 fl oz **c)** 490 000 lb **d)** 15.5 yd^3 **e)** 0.28 mi
f) $3\frac{1}{4}$ lb **2. a)** 7′8$\frac{1}{2}$″, 325.5 in.^{2} **b)** 105.1 cm, 519.9 cm^2
3. a) 37.7 in.^{2}, 7.9 in.^{3} **b)** 7752 mm^2, 32 190 mm^3 **4. a)** 217.6 ft^3
b) 8.1 yd^3 **5. a)** 13.2 L **b)** 113.5 kg **c)** 5.5 m **d)** 354.4 g **e)** 34.6 mpg **f)** 37.3 mi **g)** 0.67 in. **h)** 0.50 lb **6.** $412.50 **7. a)** 11.8 yd^3
b) 59 125 lb **8. a)** Detroit, Michigan, at $0.470/L **b)** Port Huron, Michigan, at $0.492/L **c)** Buffalo, New York, at $0.478/L **d)** Sault Ste. Marie, Michigan, at $0.496/L **9. a)** 68.9°F
b) 7.8°C **10. a)** 75.4 ft **b)** 192 laps **c)** 3.1 m/s, 7.0 mph

11. a) 1481.2 mg **b)** 3.0 mL **12. a)** 52.4 ft^3, 1.9 yd^3 **b)** 98
c) 32.7 ft **d)** 88 **13. a)** 720 ft^2 **b)** 114 696 L
c) 2 days 15 h 43 min 12 s **d)** 12.6 yd^3 **14. a)** 321.4 ft^2 **b)** 560 ft^3
15. a) i) 0.917″ **ii)** 0.700″ **b) i)** $\dfrac{1}{640}$″ **ii)** $\dfrac{1}{2000}$″
16. a) 1.247″, 1.243″ **b)** $3\dfrac{3}{128}$″, $3\dfrac{1}{128}$″ **c)** 34.725 mm,
34.720 mm **d)** 42.5934 cm, 42.5924 cm **17. a)** 36.6° **b)** Yes, it is
26° less. **c)** Answers may vary. **d)** Answers may vary.

Practice Test, pp. 104–105

1. a) 2400 m^3, including the base; 6000 m^3 **b)** 48 825.9 ft^2,
1 014 494.3 ft^3 **2. a)** 37.3 mpg **b)** 5 mm **c)** yes **d)** 71″ by 55″
e) $4.00/lb **3. a)** 14 190 mm^2 **b)** 22 in.^{2} **c)** 1 213 975 mm^3,
74.1 in.^{3} **d)** $148.20 **4. a)** Vernier calipers **b)** micrometer
5. a) 12.058″, 12.038″ **b)** 13.5863 mm, 13.5853 mm

Chapter 3 – Geometric Shapes

Get Ready, pp. 108–109

1. a) square, triangle, pentagon, hexagon, octagon **b)** 360°,
180°, 540°, 720°, 1080° **3. a)** triangle, square, triangle,
pentagon, triangle **b)** Tetrahedron: 4, 6, 4; Cube: 6, 12, 8;
Octahedron: 8, 12, 6; Dodecahedron: 12, 30, 20; Icosahedron:
20, 30, 12 **4.** Answers may vary. **5.** Answers may vary.
6. a) 17′6″ by 11′4″, less a corner of dimension 3′6″ by 4′; $184\dfrac{1}{3}$ ft^2
b) 57′8″ **c)** approximately 768 ft^2 **7.** Answers will vary.

3.1 Geometric Shapes in Design, pp. 116–119

1. Answers will vary. **2.** The faces of polyhedra are polygons,
but a polyhedron is a three-dimensional solid, whereas a
polygon is a plane figure. **3.–7.** Answers will vary. **8. a)** Each
rectangle is divided into a square and a smaller rectangle. The
same process is applied to the smaller rectangle. The width of
the rectangle and the width of the square are in the golden
ratio. A quarter circle of radius equal to the side length of the
square is contained in each square. **b)** 1.618 units, 2.618 units,
1.618:1 **c)** The width of each larger square is the sum of the
next two smaller widths; the Fibonacci sequence
9.–10. Answers will vary. **11. b)** triangles, square,
parallelogram **d)** human figure **12.** Answers may vary.
13. a) curved; flat, with five equilateral triangles, the sum of the
angles around the vertex is only 300°. **b)** It is not a true regular
polyhedron because it is not completely the same in every
aspect.

3.2 Representing Three-Dimensional Objects, pp. 129–130

1.–2. Answers will vary. **3.** 64 cm **4. a)** 1 cm represents 10 km
b) Answers will vary. **c)** 190 km; 1 cm represents 20 km

5. 1″ represents $3\dfrac{1}{3}$′ **6.–8.** Answers may vary.
10. c) $F + V = E + 2$

3.3 Creating Nets, Plans, and Patterns, pp. 135–137

1.–3. Answers will vary. **4. a)** icosahedron **b)** tetrahedron
c) cube **d)** dodecahedron **e)** octahedron **5. a)** No; the faces are
not comprised of only one polygon. **b)** Answers may vary; for
example, a truncated cube has either six hexagonal faces and
eight triangular faces or six square faces and eight triangular
faces. **6.** Answers may vary. **7. a)** both have volume 1000 cm^3
b) Net A: 700 cm^2, Net B: 600 cm^2; Net B **c)** Net A: 25 cm by
30 cm, 750 cm^2, Net B: 20 cm by 40 cm, 800 cm^2; Net A uses
less area of card. **d) i)** Net A—easier to store and to pour from
ii) Net A—can accommodate more variations in sizes of gifts
9. Answers will vary. **10 a)** 1 **d)** no **e)** You would obtain a
Klein bottle.

3.4 Designing and Constructing Physical Models, pp. 142–143

1.–6. Answers will vary. **7. a)** 133 cm **b)** no **c)** 31 cm
8. Answers may vary.

Career Profile, p. 144

1. a) 29.9% **b)** 37.5% **c)** 56.6%

Review, pp. 146–147

1.–2. Answers will vary. **3. a)** square, triangles **4.–7.** Answers
will vary. **8. a)** Answers will vary. For example, 2.5 cm by 8.0 cm
by 10.0 cm **b)** 1.55 **c)** Answers will vary. Using the example
given, 3.9 cm by 12.4 cm by 15.5 cm. **d)** because volume is a
three-dimensional measure **d)** Each of the three dimensions is
increased by this factor. **9.–14.** Answers may vary.

Practice Test, pp. 148–149

1. Answers may vary. For example, stop sign: octagon, yield
sign: triangle, no parking: rectangle **2.** circles, cylinders,
hemispheres **3. a)** 886 ft^2 including the garage **b)** 550 ft^2
4. Answers may vary. **5.** storage box; 40 cm by 30 cm by
25 cm; storing files or binders **6.** Answers may vary.

Cumulative Review, Chapters 1–3, pp. 150–153

1. a) 23.6 m **b)** 73.6° **2.** 10.3 m **3.** 27.8° **4. a)** $v = 34.9$ m,
$w = 41.9$ m **b)** $\angle A = 10.6°$, $x = 2.8$ m, $y = 15.3$ m, $z = 26.6$ m
c) 70.3 m **5. a)** 74.4 in. **b)** 256 oz **c)** 2800 lb **d)** 2640 yd
e) 6 lb 4 oz **f)** 45 in. **g)** 63 360 in. **h)** 3 pt 8 fl oz
6. a) 264 in.^{3} **b)** 476.3 cm^3 **c)** 11 796 ft^3 **7. a)** 164 tiles
b) 14 **c)** $383.69 **d)** There will be 4 full tiles left over, along with
9 pieces. **8. a)** 4′ wide; all windows may be covered with a full
piece (either vertically or horizontally) **b)** two 8-ft lengths,

three 6-ft lengths, and one 4-ft length leaves 26″ waste
9. a) 392.7 mg **b)** 473.8 mg **10. a)** 26.61″ **b)** $292.71 **c)** 47.1 in.³,
14.1 lb **d)** 18.75″ by 18″ **e)** $405 **12.–16.** Answers may vary.

Chapter 4 – Single-Variable Statistics

Get Ready, pp. 156–157
1. a) 9 **b)** 13 **c)** 6.708 **d)** 15.652 **e)** 10.724 **f)** 17.564 **2. a)** 8
b) 0.48 **c)** 0.6 **3. a)** Friday **b)** 100 km **c)** 1050 km **d)** bar graph
4. a) 5 **b)** 8 **c)** 2 **d)** 75% **e)** circle graph **5. a)** 74.5 kg
b) June 5–June 12; 2 kg **c)** June 19–June 26; 1 kg **d)** 76 kg **e)** line
graph

4.1 Collecting Data: Sampling Techniques, pp. 161–162
1. Methods used to choose the samples will vary. **a)** population:
all teenagers living in the local municipality; for a large
municipality, it would not be feasible to survey the entire
population **b)** population: all people aged 18–24 in the city; it
likely would not be feasible to survey the entire population
c) population: all students attending your school; it would be
possible to survey the entire population **d)** population: all car
owners in your neighbourhood; the feasibility of surveying the
entire population would depend on the size of the
neighbourhood **2.** The population is the collection of all people
who shop at clothing stores. How the sample would be selected
depends on the budget for the survey. An inexpensive method is
a convenience sample, where interviewers would collect data
from shoppers at various shopping malls. Depending on the
desired information, shoppers could be asked to either name
their favourite clothing store or rank selected stores on a
questionnaire. **3. a)** The survey contains a measurement bias.
Musical groups that play music of the type normally played on
this radio station would be favoured by the listeners of this
station. The listeners of this station are not necessarily
representative of the entire population. A simple random sample
of the entire population may produce vastly different results.
b) The number of people obtained in this fashion is not going to
reflect the number of people needing the services on a weekly
basis. There is no way to determine the weekly requirements
from these data. **4.** If they know their audience is 60% males
and 40% females, they could choose a stratified sample in
proportion 6:4. If, based on past data collection, they also know
the percent audience by age, they could use stratified samples
based on this information. **5.** Determine the total number of
students in grades 1, 2, and 3. To determine the number of
students surveyed from grade 1 (for example), divide the
number of students enrolled in grade 1 by the total number of
grade 1, 2, and 3 students, and multiply by 30. This number of
students should be randomly selected from those attending

grade 1, and similarly for grades 2 and 3. **6.** From a numbered
list of the 650 students enrolled at the college, two groups of 20
could be selected by obtaining two sets of 20 random numbers.
There must be no duplicates of random numbers among the 40
chosen, and these numbers should be selected from the range 1
to 650. Students are then selected from the list according to these
numbers. **7. a)** Numbers generated will vary. **b)** Unless the
players can play all positions, this method of selection may not
yield the required number of players for each position. The skill
level of players will not be considered. **8.–9.** Answers will vary.

4.2 Methods of Collecting Data, pp. 166–167
1. a) 1 260 589 **b)** secondary; the newspaper obtained its
information from the polsters **2.** Collection methods/sources
will vary. **a)** Unless the performers are interviewed personally
by the data collector, the data source is secondary. **b)** secondary
c) secondary, if the information is gathered from the parents;
primary, if you ask the children yourself **3.–5.** Answers will
vary. **6. e)** Since you collected the data yourself, the source is
primary. **7.–10.** Answers will vary.

4.3 Representing Data, pp. 175–177
1. a) The spread in the data is 53 cm; to have a manageable
number of bars, between 5 and 1, for example, the interval
length should be between 5 cm and 10 cm. One possibility is
the intervals [120, 125), [125, 130), [130, 135), …, [175, 180)
b) [0, 3), [3, 6), [6, 9), …, [27, 30) **c)** [3:00, 3:30), [3:30, 4:00),
[4:00, 4:30), …, [6:30, 7:00)

2. a)

Class Interval	Frequency
[60, 65)	1
[65, 70)	3
[70, 75)	5
[75, 80)	7
[80, 85)	2
[85, 90)	2

3. a)

Class Interval	Frequency
[0, 4)	5
[4, 8)	7
[8, 12)	12
[12, 16)	3
[16, 20)	2
[20, 24)	1

b) 9 **4.** Answers will vary.

5.

Class Interval	Frequency
[100, 150)	1
[150, 200)	3
[200, 250)	11
[250, 300)	4
[300, 350)	1
[350, 400)	1
[400, 450)	1
[450, 500)	2

6. a) Soviet Union **c)** Answers will vary.

7. a)

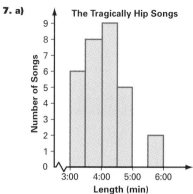

The Tragically Hip Songs

b) between 4:00 and 4:30 min is the most frequent, with 9 songs **8.** Answers will vary. **9. a)** Quetico
b) Sauble Falls; 1315.8 m² per campsite

c)

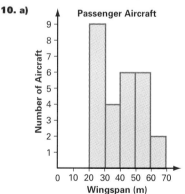

Ontario Provincial Parks

10. a)

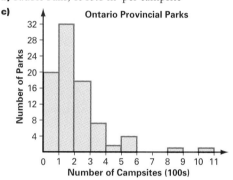

Passenger Aircraft

b)

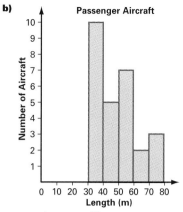

Passenger Aircraft

11.–12. Answers will vary.

4.4 Measures of Central Tendency, pp. 182–183

1. a) 50.7 **b)** 5.4, 5, 5 **c)** 121.3, 116.5, no mode **2.** $84 000, $89 000, no mode **3. a)** 33, 34.5, 4 modes: 34, 35, 36, 37
b) mean or median **4.** 72.4, 69.5, 60 **5. a)** 116, 117, 115
b) Answers may vary. **6. a)** 64.4, 61, 60 **b)** median; Values such as 80 and 85 have a high impact on the mean, when they may not be representative of typical commute times. These values raise the average artificially. **7. a)** 12.05%, 11.53%, 3 modes: 9.1%, 24.95%, 30.76% **b)** median; inordinately high (or low) values such as 52.48% skew the mean **8. a)** 383 **b)** $1.44, $1.50, $1.25 **9. a)** 124.7 min, 125 min, 126 min **b)** 84.1 min, 81 min, 75 min **c)** Family movies are approximately 40 min shorter on average. This is due to the characteristics of the viewing audience; the intended age group is much younger for family movies, and there may be a correspondingly shorter attention span. **10. a) i)** 2581.8 m, 2531.5 m, modes: 1982.5 m and 2260.1 m **ii)** 1675.7 m, 1586 m, modes: 1586.0 m, 2074.0 m, and 2260.1 m **b)** takeoff; the takeoff runway length is about 1.5 times the landing runway length to allow room for stopping if a takeoff needs to be aborted.

4.5 Measures of Dispersion, pp. 188–189

1. 302, 838, 7546.25, 86.87 **2.** 41, 71, 10.9
3. a) range: $132 900, variance: 1 321 853 000, standard deviation: $36 357.30 **b)** The data are fairly widely spread.
4. $154.94; $50.01 **b)** $207.50; $60.44 **c)** closing prices
5. a) 16.209, 1.2985 **b)** 16.13, 1.23 **c)** Answers may vary.
6. a) 2 390 915, 11 846 200, 3 418 188

4.6 Properties of Common Distributions, pp. 195–197

1. Normal: symmetric bell-shaped distribution; mean, median, and mode are all equal. Positively skewed: distribution of data has a long right-hand tail; the mean is greater than the median. Negatively skewed: distribution of data has a long left-hand

tail; the mean is less than the median. Bimodal: the distribution of data exhibits two distinct clusterings or peaks. Exponential: a skewed distribution with data distribution approximating an exponential curve. **2. a)** Mathematics: mean less than the median, single mode; Language Arts: mean and median close to equal, two modes or clusters; Chemistry: mean greater than the median; single mode **b)** Mathematics: negatively skewed; Language Arts: bimodal; Chemistry: positively skewed **3.** A: normally distributed; B: bimodal; C: positively skewed **5. a)** Sketches may vary. **b)** This is the most frequently purchased size. **c)** No, sizing has changed.
7. b) negatively skewed
8. a) bimodal

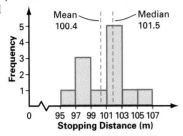

b) 100.4, 101.5 **c)** 2.7 **d)** 97.7–103.1, 68% **e)** 10, 83.3%, not normal
9. a) positively skewed
b) 124.7 min, 125 min
c) 18.3 min
d) 69.7%; 31 or 70.7%
e) 95%; 41 or 95.3%
f) positively skewed

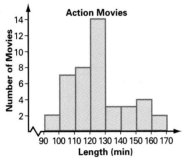

Career Profile, p. 198

1. a) approximately normal **b)** 150 mg/kg for both median and mean **c)** 50 mg **2. a)** positively skewed **b)** 100 min

Review, pp. 200–201

1. a) A random sample is a method of selecting respondents whereby each member of the population has the same chance of being selected. The selection of particular respondents does not affect the changes of selection of any other member of the population. **b)** Stratified sampling divides a population according to certain well-defined characteristics. Respondents are then selected from within a division in numbers according to the quotient of the number in the population with the number in the division. Stratified sampling ensures representation from each of the various strata, whereas simple random sampling does not. **2.** stratified sample **3. a)** This will result in a sample of size 10%, but is not a random sample.
b) Select distinct random numbers from the range 1 to the

number of bottles in a batch. The number of numbers selected should equal 10% of the number of bottles in a batch. Bottles should be selected from the arrangement according to these numbers. **4. a)** The population is the collection of all people or objects from which certain information is desired. **b)** A sample is a subset of the population from which data will be gathered. **c)** A primary source of data is data that you collect.
5. Depending on the degree of bias, the results of a biased survey may range from slightly skewed from the actual data to entirely distinct from the actual data. **6.** Histograms show frequencies of either discrete or continuous data in equally-sized contiguous intervals. Bar graphs present data grouped according to various categories. **8. a)** 6.9 days, 6.5 days, modes: 2, 4, 5, 7, and 11 days **b)** median; the outlier, 18, raises the average so that it is no longer representative **9. a)** 7945.5, 7940, 7650 **b)** mean or median; the mode is too low **10.** 3700, 1169.1 **11. a)** 18 **b)** 4.1, 4.4 **12. a)** A: positively skewed, B: bimodal, C: exponential **b)** A: average age is greater than median age, B: age groups cluster in groups of people less than 20 years old and between 50 and 70 years old, C: vast majority of population is quite young **13. a)** Distribution is approximately normal.
b) $25.80, $24.50, $35 **c)** $12.68 **d)** between $0 and $51.82, using sample standard deviation

Practice Test, pp. 202–203

1. simple random sample; methods will vary. **2. a)** bar graph **b)** 1st and 3rd quarters **c)** 75 **d)** 17 **3. a)** 31.6°C, 34°C, 34°C **b)** 18.8°C, 19°C, modes: 11°C, 19°C, 21°C, and 24°C **c)** range of high temperatures **4. a)** 44.9 days **b)** 22.4 days

c)

Class Interval	Frequency
[0, 5)	1
[5, 10)	5
[10, 15)	4
[15, 20)	5
[20, 25)	1
[25, 30)	6
[30, 35)	2
[35, 40)	4
[40, 45)	1
[45, 50)	1

e) 134.7, 11.6 days **f)** Answers will vary.

Chapter 5 – Two-Variable Statistics

Get Ready, pp. 206–207

1. a) positive **b)** negative **c)** zero
2. a)

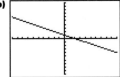

b)

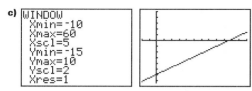

c)

3. Determine the highest and lowest values for the *x*- and *y*-variables. **4.** Answers will vary. **a)** This can occur if, while making a scatter plot of L1 and L2, there are no entries in the **Stat List Editor** for L1, L2, or both. To fix this, press 1 to quit, then press (STAT) 1, then add some entries under L1 and L2. **b)** This can occur when the minus sign is used instead of the (−) sign. To fix it, press 2 and press (−). **c)** This can occur when the wrong window settings are used. To fix it, change Xmin, Xmax, Ymin, and Ymax so that the graph is within the appropriate span. **d)** This can occur if, while making a scatter plot of L1 and L2, there are not the same number of entries in L1 as there are in L2. To fix it, press 1 to quit, then press (STAT) 1, and add or delete some entries so that there are the same number of entries in L1 and L2. **6. a)** A sample is cheaper to obtain and, if chosen appropriately, produces accurate results. **b)** experimental observations, data bases, surveys **7. a)** In a random sample, a fraction of the population is chosen, using a method that gives each member of the population the same chance of being selected. In a stratified sample, the population is divided by certain characteristics, and a sample is chosen from within each subgroup. **b)** Biased sampling refers to a sampling technique that has a tendency to deviate in some direction from a true value. **8. a)** positive **b)** tightly scattered **c)** Answers may vary. **d)** 3°C; 130°C

5.2 Methods for Collecting, Storing, and Retrieving Data, pp. 217–219

1. a) primary **b)** primary **c)** secondary **d)** primary **e)** secondary
2. a) dependent: level of hunger, independent: time of day
b) dependent: number of whales, independent: year
c) dependent: consumption of electricity, independent: population **d)** dependent: net worth, independent: age
3. a) secondary **b)** primary **c)** primary **d)** secondary
4. a) descriptive **b)** inferential **c)** inferential **5.** Answers may vary. For example, when determining the population of a province or country. **6.** This information is relevant when the insurance company is determining insurance premiums.
7. a) descriptive **b)** inferential **c)** descriptive **d)** descriptive
e) descriptive **8.** Answers may vary. **9.** Answers may vary.
10. a) Sample data: a record of the viewing habits of the members of 5000 households; Population: the residents of the United States **b)** inferential **c)** Answers may vary. **11.** This procedure contains response/non-response bias. **12.** Answers may vary. This can provide a good sample, as long as the homes surveyed in each city are selected at random.
13. a) Examples may vary. **b)** Obtaining data from the entire

population is costly and time-consuming. An appropriately chosen sample should produce similar results. **14.** Results will vary. **16. a)** descriptive; age of drivers, number of accidents per age group **b)** Members of age groups that have higher accident rates pay higher insurance premiums.

5.3 Scatter Plots, pp. 226–228

1. a) independent: area, dependent: cost **b)** independent: time, dependent: population **c)** independent: launch number, dependent: height **d)** independent: age, dependent: heart rate
2. a) i) linear **ii)** strong **iii)** negative **iv)** no **b) i)** linear **ii)** strong **iii)** positive **iv)** no **c) i)** linear **ii)** weak **iii)** positive **iv)** no
d) i) none **ii)** not applicable **iii)** not applicable **iv)** not applicable
3. a) This will occur when there are more entries in L1 than there are in L2 in the **Stat List Editor**. **b)** Change the window.

4. a)

b) linear **c)** Answers may vary. A student may compensate for days absent by asking for extra help. **d)** The more days absent, the lower the final mark. **5. a)** A linear relationship exists; the younger the employee, the more sick days they take.
b) Graph the data:

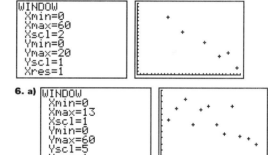

6. a)

b) There is a moderate correlation. **c)** yes; two employees
7. a) The graphs use different axes to represent the independent and dependent variables. **b)** Graph A is better since Year is the independent variable.

8. a)

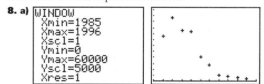

b) The number of cases of chicken pox has decreased exponentially with time.

9.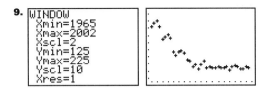

5.4 Line of Best Fit, pp. 235–238

1. a) 0.94 **b)** −0.85 **c)** 0.01

3. a)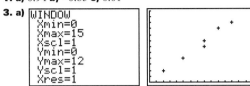

b) 0.9878 **c)** strong, positive linear relation

4. a)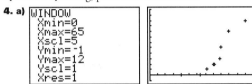

b) 0.9013 **c)** Answers will vary. There is a strong linear correlation; working longer hours results in more injuries and accidents in the workplace.

5. a)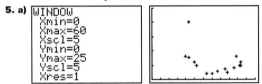

b) −0.3987 **c)** weak, negative linear correlation **d)** Answers may vary. **e)** Answers may vary. **6.** As distance increases, prices increase approximately linearly.

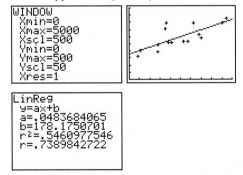

The correlation value is 0.7390. Reasons for outliers will vary.
7. a) The per capita amount gives a measure of personal contribution. While a total amount may appear low (or high) when compared to another country, the per capita amount may be much higher (or lower).

b)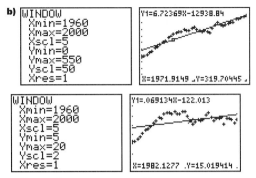

c) Comparisons may vary. **d)** Reports may vary.

8. a)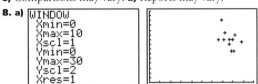

b) The point (6.94, 26) is an outlier. Reasons may vary.
c) This is probably not a good source of information for selecting next year's team. There is a low correlation.
9. a) linear relationship; strong, positive correlation; $y = 0.041x − 12.66$, $r = 0.9882$ **b)** An astronomer could identify the orbital period of a new planet through extrapolation or interpolation of the line of best fit.

5.5 Uses and Possible Misuses of Sample Data, pp. 241–245

1. a) An outlier has caused the line of best fit to be shifted.
b) The correlation coefficient is too weak to make predictions.
c) The correlation coefficient is too weak to make predictions.
d) The data set is too small to make predictions with confidence.
2. Answers may vary. **3. a)** A strong linear correlation exists. **b)** No; the data are sensitive to the political climate. **4.** A relatively strong correlation exists. **5. a)** $y = 0.01x + 62.53$, correlation: 0.001 429 **b)** The time for the sixth trial is an outlier and should be eliminated from the data set. **c)** yes

6. a)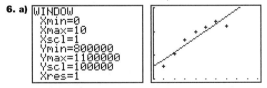

$y = 29\ 951x + 852\ 589$, where x is the number of years after 1990. **b)** No; the data is sensitive to various factors: environment, politics, and economics.
7. a) $y = −6453.8x + 495\ 133.9$, where x is the number of years after 1990. **b)** The model is moderate, with a correlation of −0.64. No; the data is sensitive to various factors: environment, politics, and economics. **c)** In 1997, the area reforested was only 45% of the area harvested.

8. a)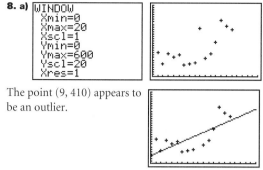

The point (9, 410) appears to be an outlier.

The correlation is moderately strong, 0.7287. It is not reasonable to extrapolate because as September nears, the weather is likely to become cooler, which will result in fewer forest fires.

b) 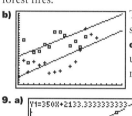 The correlation is moderately strong, 0.6948.
c) The number of fires in 2000 was unusually low. The outlier should not have been removed.

9. a) 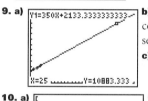 **b)** No; there is not enough data collected, and the data is sensitive to various factors.
c) Answers may vary.

10. a) $y = 1016.5x - 2115.7$, where x is the number of five-year periods after 1856. **b)** 27 362 800; 31 156 393 (Oct 2001) **c)** 29 395 800, 31 428 800; 33 362 000, 38 699 000; reasons will vary.

Career Profile, p. 246

2. a) crushed limestone, crushed granite, rounded gravel
b) rounded gravel; the line of best fit passes through or very close to all the data points. **c)** rounded gravel

Review, pp. 248–251

3. data published by others, data gathered from an experiment, data collected in an observational study **4. a)** inferential
b) inferential **c)** descriptive **5. a)** primary **b)** secondary
c) secondary **d)** primary **6.** Answers may vary. **7. a)** none
b) moderate; negative **c)** strong; positive **8. a)** independent: age; dependent: heart rate **b)** independent: amount of pollution; dependent: number of fish **c)** independent: age; dependent: number of words known **9. a)** If the variable exists on its own, it is independent. If the variable is influenced by other factors, it is dependent. **b)** In order to construct scatter plots and make reasonable predictions

10. a)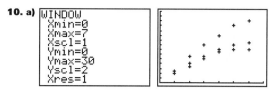

b) weak linear correlation

c) 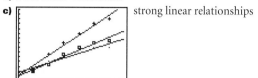 strong linear relationships

11. Answers may vary. **12. a)** very strong **b)** moderate
c) very weak **d)** strong **13.** Answers may vary.

14. a)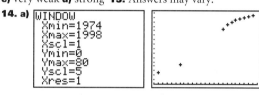

b) 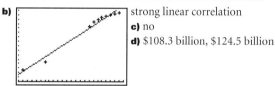 strong linear correlation
c) no
d) $108.3 billion, $124.5 billion

15. a) An outlier is a data point that deviates from the pattern formed by the other data points. **b)** Outliers can shift the line of best fit in their direction. **c)** An outlier can be deleted if it is identified as an error.

16. a) 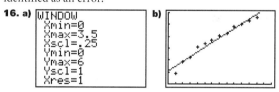 **b)**

c) strong linear correlation **d)** 17 kg **e)** no

Practice Test, pp. 252–253

1. Answers may vary. **2. a) i)** Data made available by others: secondary **ii)** Data resulting from an experiment: primary
iii) Data collected in an observational study: primary
b) Answers may vary. **3.** Steps: data collection, data organization, data summarizing, data analysis, data presentation **4.** Inferential statistics uses a sample of the population to draw conclusions about the entire population. For example, surveying 20 students at each grade level to determine the average number of study hours per week, and using this data to infer the number of study hours per week for a student of each grade. Descriptive statistics surveys the entire population. For example, surveying each member of a club to determine whether fees should increase. **5. a)** inferential

b) descriptive **6. a) i)** independent: dog years; dependent: human years **ii)** strong, positive linear relationship **iii)** none **b) i)** independent: number of floors; dependent: height of building **ii)** moderate, positive linear relationship **iii)** (14, 75)

7. a)

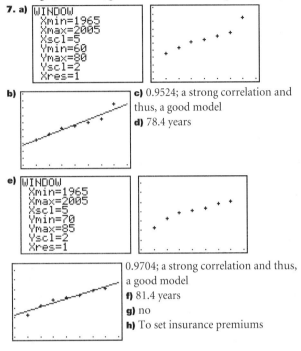

b) **c)** 0.9524; a strong correlation and thus, a good model

d) 78.4 years

e)

**0.9704; a strong correlation and thus, a good model

f) 81.4 years

g) no

h) To set insurance premiums

Chapter 6 – Interpreting and Analysing Data

Get Ready, pp. 256–257

1. Answers may vary. **a)** the value of a stock over time **b)** the proportion of market share of various companies that produce similar products **c)** the number of days with precipitation in a certain city each month **2.** Answers may vary. **3.** independent: time, dependent: thousands of dollars **4. a)** broken-line graph; explanations will vary.

5. a)

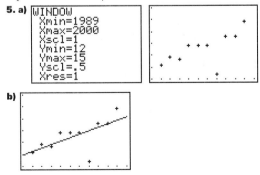

b)

$r = 0.6713$, so the correlation is not very strong **c)** 14.7, 15.5

d) Answers may vary. **6. a)** 73 min **b)** 5 min **c)** 50 min **d)** 14 min **7.** 32% **8. a)** The *x*-axis is broken, making the rate of growth appear greater. **b)** Time, the independent variable, should be on the horizontal axis.

6.1 Retrieving Information, pp. 265–270

1. a) secondary, static **b)** secondary, dynamic **c)** primary, static **d)** secondary, static **2. a)** increased **b)** by extrapolation **3. a)** Answers may vary. **b)** The number of youths by year. **4. a)** The information is displayed creatively, using relevant imagery. **b)** The average price of gas in the Toronto area on the previous day, and the weekly average since the beginning of the year. **c)** The vertical axis is broken to enable the graph to focus on the relevant data. The units of measure are cents on the vertical axis and time (months) on the horizontal axis. **5.** Descriptions may vary. The first graph uses smoke wafting upward to attract attention to the graph. The second graph shows relative proportions in a pie chart. The third graph shows differences in magnitude using bars. **6. a)** no **b)** Answers may vary slightly. Titanic: drama; Star Wars: science fiction; Star Wars: The Phantom Menace: science fiction; E.T.: science fiction; Jurassic Park: science fiction; Forrest Gump: drama; The Lion King: family; Return of the Jedi: science fiction; Independence Day: science fiction; The Sixth Sense: horror **c)** science fiction **d)** Not necessarily; the price of admission in the various years is required. **7. a)** 15:46 **b)** 5 h 56 min **c)** 18:15 **d)** no **e)** 55 km **f)** Answers may vary. **8.** Answers may vary. **9. a)** Sault Ste. Marie **b)** no **10. a)** Newfoundland, Nova Scotia, New Brunswick, Prince Edward Island, Quebec, Manitoba **b)** Saskatchewan, Alberta, British Columbia **c)** Divide the sum of the values for Total Registered by the sum of the values for Objective, then, multiply by 100. **11. a)** exponential **b)** Answers may vary. For example, an increase in the number of snowboarders. **c)** Answers may vary.

6.3 Representing Data in a Variety of Ways, pp. 279–283

1. – 7. Answers will vary.

8. a)

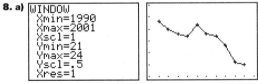

b) Viewing time is decreasing. Possible factors may vary. **c)** Answers will vary. **d)** Women watch more television than men; people over 50 years of age watch more television than the national average; Newfoundland has the greatest weekly average; Alberta has the least weekly average; teens in Ontario watch less television than the Canadian average for their age group; males 18–24 are below the national average.

9. a)   **Year**   **Population (millions)**

1871	3.463
1881	4.325
1891	4.833
1901	5.371
1911	7.207
1921	8.788
1931	10.377
1941	11.507
1951	14.009
1961	18.238
1971	21.568
1981	24.343
1991	26.994
2001	31.156

b) scatter plot or bar graph **c)** exponential **11.** Depending on the scale used on the axes, data can appear more highly correlated than it really is. The correlation coefficient gives the actual value. **12.** Graphs may vary. **a)** Graphs may vary. If the trend seen from 1997 to 2000 continues, the Northern Right Whale may become extinct. **13.** Answers may vary. **14. b)** The graph for the principal is a slowly increasing curve: this is exponential growth. The graph for the interest is a line: this is linear growth. The graph for the balance is a slowly decreasing curve: this is exponential decrease. **c)** The interest line would be the same. The principal and balance curves would be a little more curved since more principal would be paid each year and the balance would decrease more quickly.

6.4 Validity and Bias, pp. 289–291

1. Answers may vary. **2.** People attending the Jazz Festival are more likely to prefer jazz music. To validate such a statement, a random sample of Canadians should be asked for their favourite type of music. **3.** Question A is the most misleading. It solicits the desired response. **4.** The statement is misleading. Patrons of a health food store do not constitute a random sample of the population. **5.** The first graph; part of the vertical axis is missing in the second graph. This gives the visual impression that the number 6, for example, was drawn less than half the number of times the other numbers were drawn. **6.** The shortened vertical axis in the first graph gives a misleading impression of the number of deer; the bar graph gives the impression that the number of deer fluctuate over time. The shortened vertical axis in the second graph gives a misleading impression of the percent for Hipswing Inc. Additionally, since the data are reported as percents, the actual number for Hipswing Inc. may be larger in 1991 than it was in 1986, whereas the graph shows a decrease. **7.** Answers may vary. **8. a)** 28th percentile **b)** 63rd percentile **9. a)** Not necessarily for their own class; yes, if the results are grouped, their percentile rank would be the same as each other but not necessarily the same as it is for their own class. **b)** Katie's class: 55th percentile; Jenna's class: 48th percentile; grouped results: 51st percentile for both **10.** The survey is biased. Attending a Toronto Maple Leafs hockey game is expensive and so, the people who can afford to attend a game are not necessarily a representative sample. **11. a)** The graph decreases and might mislead some

people to believe that this is a decrease in injuries. **b)** Answers may vary. The graph shows the number of injury-free players *per team*. This is an average and the number of teams has increased. Overall, the total number of injuries has increased.

Career Profile, p. 292

1. dependent: wear rate; independent: hardness
2. composition and heat treatment **3.** exponential

Review, pp. 294–297

1. a) Static data is taken from a source that will not be updated, whereas dynamic data is taken from a source that is updated regularly. **b)** static: textbooks, almanacs; dynamic: electronic databases, Internet **2. a)** Bar segments of different colours represent the proportion of people with each opinion on the level of distraction. **b)** 19% **c)** combing hair, shaving, or applying makeup; reading a magazine **d)** Answers may vary. **3. a)** Bank of Montreal, Bank of Nova Scotia, CIBC, Citibank Canada, Ing Direct, Royal Bank, TD Canada Trust **b)** $5000 **c)** Certificates/Debentures at Ing Direct **4. a)** Consumer Price Index **b)** The CPI gives a measure of inflation by allowing a comparison of relative price levels. **c)** Answers may vary. For example, Industrial Product Price Index and New Housing Price Index. **d)** 1992 = 100 means that the price of an item in 1992 is considered to be 100, and so may be used as a base for comparison. **5.** Answers may vary. **6.** Answers may vary. Line graph: to track values over time; Histogram: to display frequency of occurrence; Circle graph: to display relative proportions; Table: to display data and associated computations. **7.** Answers may vary. **8.** Yes; $1341.93; $6441.26 **9. a)** The percent is increasing. **b)** Yes **c)** Canada has a higher percent than the United States, but a lower percent than Denmark and Sweden. **10. b)** Answers will vary. About 2 175 000. **c)** Answers will vary. About 2 535 000. **d)** Answers may vary. For example, the interpolated value in part b) may be more accurate. **11.** Answers may vary. **12.** The scale on the *y*-axis is reversed. **13. a)** A reader poll is biased and does not necessarily reflect the opinions of the population. There is no indication of the type of survey, the questions asked, or the number of respondents. **b)** The population is not defined, nor is the sampling method. There is no way to ascertain the validity of the statement. **c)** The sample size is not indicated, nor is there a definition of "relief." **14. a)** There are only two different activities compared. There is not enough evidence to support this claim. **b)** This is not necessarily the case. There are only three different countries on which to base the statement. Also, the data do not mention regular use for banking and downloading music, but rather users who used these services at least once. **c)** This statement is very likely false; although, in percent terms, it is true. The graph shows percent. A smaller percent of a large group can represent more people than a larger percent of a small group. **15.** 75th percentile

Practice Test, pp. 298–299

1. The CPI compares prices relative to prices in a particular year. The prices of items in that year are considered to be 100, and this is used as a basis for comparison of prices in other years. **2.** A percent is a fraction of 100, whereas a percentile is a ranking or position of an individual in a group.
3. 91st percentile **4. a)** The validity is difficult or impossible to determine. There is no mention of the population or the sample size or the method of taking the sample. The phrase "felt better" is subjective. **b)** The validity is difficult or impossible to determine. The popularity is not defined, nor is the sampling mechanism. **5. b)** The costs are decreasing. With competition, air time has become increasingly less expensive.
6. b) Enrolment is increasing over time. **c)** Extrapolations may vary, and may be verified using data from Statistics Canada.
7. a) 08:48 **b)** Depart from Ottawa at 06:45 and arrive in Montreal at 08:48. Depart from Montreal at 18:00 and arrive in Ottawa at 20:12. Journey time: 4 h 15 min **c)** Depart from Ottawa at 07:30 and arrive in Montreal at 09:35. Depart from Montreal at 18:00 and arrive in Ottawa at 20:12. Journey time: 4 h 17 min

Cumulative Review, Chapters 4–6, pp. 300–303

1. a) inferential, sample **b)** descriptive, population **c)** descriptive, population **2.** mean: the arithmetic average of the values; median: the midpoint of the distribution of the values; mode: the most frequently occurring value **4.** Sketches will vary. Bimodal distribution should have two humps; exponential distribution should have the majority of the data bunched to one side of the data. **5.** Answers may vary.
6. a) mean: 3.481; median: 3.935; mode: 4.00

b)

Class Interval	Frequency
[1.50, 2.00)	11
[2.00, 2.50)	4
[2.50, 3.00)	1
[3.00, 3.50)	4
[3.50, 4.00)	9
[4.00, 4.50)	13
[4.50, 5.00)	8

c)

d) not normal; close to bimodal **e)** range: 3.26; standard deviation: 1.08 **7. a)** dependent: number of applicants, independent: year **b)** dependent: number of collisions, independent: time **c)** dependent: total precipitation, independent: month

8. a)

b) The point (162 900, 16 000) is an outlier. Assume that it should be (162 900, 160 000). **c)** $y = 1.022x - 6187$
d) $r \doteq 0.997$; strong relationship **e)** \$180 328 **f)** The selling price is usually a little less than the asking price. **9. a)** The standards at all plants were inferred from the standards at one plant.
b) To make the claim valid, products from all plants would have to be tested, with similar results. **10. a)** 30 min
b) 11:15 **c)** You can call the bus company and arrange a pick-up at your home, if you live in the bus zone. **11. a)** December and January **b)** The consumption in June/July 2001 is much higher than in June/July 2000. It is possible that an air conditioner was used in June/July 2001. **12.** 73rd percentile
13. a) A primary source of data is gathered by yourself or your organization. A secondary source of data is gathered from that which is published by other organizations. Examples will vary.
b) Answers will vary. **14. a)** The question is subjective; two people who are equally healthy may rate themselves differently.
b) The initial preamble to the question casts the team as a poor one, and so introduces a bias. **c)** Patrons of a soccer game are not a representative sample of the entire population, and so, the results will be biased.

Chapter 7 – Exploring Quadratic Models

Get Ready, pp. 306–307

1. a) 1, 36; 2, 18; 3, 12; 4, 9; 6, 6; −1, −36; −2, −18; −3, −12; −4, −9; −6, −6 **b)** 1, 105; 3, 35; 5, 21; 7, 15; −1, −105; −3, −35; −5, −21; −7, −15 **c)** 1, 2; −1, −2 **d)** 1, 81; 3, 27; 9, 9; −1, −81; −3, −27; −9, −9 **e)** 1, 74; 2, 37; −1, −74; −2, −37 **f)** 1, 17; −1, −17
2. a) 2, 5 **b)** −4, 7 **c)** −9, 4 **d)** −7, −6 **3. a)** 3 **b)** 2 **c)** 2 **d)** 2 **e)** 4
f) 3 **4. a)** 1 **b)** −2.5 **5. a)** −11 **b)** −4 **c)** −0.91 **d)** −1, 4
e) 0.25, −1.75 **f)** −7, 2 **g)** −2, 0.67 **6. a)** 16 cm² **b)** 10 m²
c) 225 m² **d)** 49π cm² **7. a)** $6x^2 + 7x - 20$ **b)** $-12x^2 - 39x + 105$
c) $75x^2 + 30x + 3$ **d)** $-2x^2 + 12x - 13$ **8. a)** 47 **b)** −56 **c)** −3.5
d) 125.5 **9. a)** $5(2y - 3)$ **b)** $2(2x^2 + 5x - 6)$ **c)** $3a(2t + w + 3z)$
d) $6k(k - 8)$ **e)** $4m(4 + 2m + 3m^2)$ **f)** $5n(7 - 3n)$

10. a) 4 **b)** 274 **c)** 7 **d)** $-\dfrac{3}{16}$ **11. a)** 5, 5, 5, 5, 5 **b)** 1, 2, 4, 8, 16

c) −48, −24, 0, 24, 48 **d)** 95, 145, 195, 245 **12.** The graph in question 11a) is linear; the first differences for a linear relation are constant.

7.1 Recognizing and Understanding Quadratic Relationships, pp. 312–315

1. a) linear **b)** quadratic **c)** linear **d)** quadratic **e)** neither **f)** quadratic **2. a)** neither **b)** quadratic **c)** neither **d)** linear **e)** quadratic **f)** neither **3. a)** quadratic **b)** linear **c)** quadratic **d)** quadratic **e)** quadratic **f)** linear **g)** linear **h)** neither **i)** quadratic **4. a)** $y = 2x^2 - 4x + 5$ **b)** $y = 4x^2 + 8x - 3$ **c)** $y = x^2 - 4x + 1$ **d)** $y = -x^2 + 10x - 17$ **e)** $y = x^2 + 4x + 1$ **f)** $y = x^2 - 2x + 1$ **g)** $y = -x^2 - 10x - 17$ **h)** $y = x^2 + 4x + 7$ **5. a)** $y = (x - 1)^2 + 4$ **b)** $y = 2(x + 2)^2 + 5$ **c)** $y = 3(x - 4)^2 - 3$ **d)** $y = -\dfrac{1}{5}(x - 1)^2 + 5$ **6. a)** $y = (x - 4)^2 - 4$ **b)** $y = x^2 + 2x - 24$ **c)** $y = x^2 + 77x - 240$ **d)** $y = -2(x - 3)^2 + 1$ **e)** $y = (x + 3)^2 - 1$ **f)** $y = 2x^2 - 2x - 4$ **7. b)** quadratic **c)** Answers may vary. **d)** Answers may vary; for example, close the parlour for the months of December, January, and February, since the monthly expenses exceed sales. **8. a)** Answers will vary. **b)** No; the data fits a linear model. **c)** $200 - 10x$ **d)** $90 + 5x$ **e)** $y = (200 - 10x)(90 + 5x)$ **9.** $y = -\dfrac{26}{441}x^2 + 26$

7.2 Factoring to Find the Zeros of a Quadratic, pp. 321–324

1. a) 0 **b)** none **c)** 3, 6 **d)** none **e)** −7, −2 **f)** −5 **2. a)** −2, 8 **b)** −5, −2 **c)** −13, 2 **d)** −3, 20 **e)** 8, 13 **f)** −15, −4 **3. a)** $y = x^2 + 3x - 4$ **b)** $y = 3x^2 + 18x + 15$ **c)** $y = 6x^2 + 5x - 4$ **d)** $y = -4x^2 + 17x - 4$ **e)** $y = 12x^2 + 22x + 8$ **f)** $y = -3x^2 + 18x - 15$ **4. a)** $3a(a - 3)$ **b)** $2(2x^2 + 8x - 9)$ **c)** $-3(z - 7)(z - 1)$ **d)** $(2x - 1)(4x - 3)$ **e)** $(5t + 7)(3 - 2t)$ **f)** $4(1 - 3d)(2d + 1)$ **5. a)** $(x + 2)(2x - 3)$ **b)** $(2t - 5)(2t - 3)$ **c)** $(2x - 1)(5x + 3)$ **d)** $(a + 2)(2a + 1)$ **e)** $(n - 3)(5n + 2)$ **f)** $(7x + 2)(8x - 1)$ **6. a)** −2, 5 **b)** −7, $\dfrac{1}{3}$ **c)** $-\dfrac{1}{2}$, $\dfrac{5}{4}$ **d)** 0, 8 **e)** $-1, \dfrac{3}{2}$ **f)** $-6, \dfrac{8}{5}$ **7. a)** $(x + 1)(x + 12)$ **b)** $(x - 10)(x + 3)$ **c)** $(x - 3)(x - 5)$ **d)** $(x - 5)(x - 10)$ **e)** $(2x - 3)(2x + 3)$ **f)** $(x - 6)^2$ **8. a)** $(x + 3)(2x + 1)$ **b)** $(x + 2)(2x + 5)$ **c)** $(x + 4)(3x - 5)$ **d)** $(t - 2)(4t - 5)$ **e)** $(2x + 3)(4x - 3)$ **f)** $(2x + 1)(3x - 1)$ **9. a)** $2(x - 1)(2x - 3)$ **b)** $3(x + 2)(3x + 5)$ **c)** $-2(2x - 1)(3x - 1)$ **10. a)** 1, 4 **b)** −2, 5 **c)** $-5, \dfrac{2}{3}$ **d)** $3, -\dfrac{1}{4}$ **e)** −9, 3 **f)** $3, -\dfrac{5}{3}$ **11. a)** −7, 5 **b)** $\dfrac{1}{4}$ **c)** −1 **d)** −5, 4 **e)** $-\dfrac{4}{3}, \dfrac{2}{5}$ **f)** $2, \dfrac{4}{5}$ **12. a)** −8, −5 **b)** −4, −3 **c)** $\dfrac{3}{2}$ **d)** $-\dfrac{1}{6}, \dfrac{1}{3}$ **13. a)** $-\dfrac{3}{2}, \dfrac{5}{2}$ **b)** $-\dfrac{5}{2}, \dfrac{7}{2}$ **14. a)** 100 cm² **b)** $10 + 2x$, $5 + 2x$ **c)** $(10 + 2x)(5 + 2x) - 50$ **d)** $(10 + 2x)(5 + 2x) - 50 = 100$; 2.5 cm **15. a)** 12 000 **b)** 6000; $7200 **c)** 7000 **16. a)** $20 + 2x$, $10 + 2x$ **b)** $(20 + 2x)(10 + 2x) - 200$ **c)** 400 cm² **d)** 5 m **17. a)** 130 at $120 each **b)** $100; 140 **18. a)** $58 500 **b)** 3000 or 11 000 **c)** 7000 **20. a)** 14 **b)** To find the number of pairs of teams, there are n choices for the first team and $n - 1$ choices for the second team. So, there are $n(n - 1)$ ways to choose a first team and then, a second team. However, this method counts each pair of teams twice. For example, the game A vs. B is counted once as A vs. B, and then again as B vs. A. Thus, the total number of games is $\dfrac{n(n - 1)}{2}$. **21. a)** 2, 6 **b)** The x-coordinate of the vertex is 4, which is the average of the zeros. So, the coordinates of the vertex are (4, −4). **c)** $\dfrac{x_1 + x_2}{2}$

7.3 Solving Quadratic Equations Without Factoring, pp. 331–332

1. a) $3x^2 - 4x - 5 = 0$; $a = 3$, $b = -4$, $c = -5$ **b)** $8x^2 - 2x - 7 = 0$; $a = 8$, $b = -2$, $c = -7$ **c)** $x^2 + 17x + 3 = 0$; $a = 1$, $b = 17$, $c = 3$ **d)** $9x^2 - 7x = 0$; $a = 9$, $b = -7$, $c = 0$ **e)** $3x^2 - 6x + 8 = 0$; $a = 3$, $b = -6$, $c = 8$ **f)** $x^2 + 5x - 19 = 0$; $a = 1$, $b = 5$, $c = -19$ **2. a)** −1.82, −0.18 **b)** −2.37, −0.63 **c)** −5, 0.5 **d)** −0.69, 0.29 **3.** The equation in part c) **4. a)** $-\dfrac{4}{3}, \dfrac{3}{2}$ **b)** 0, 4 **c)** $-\dfrac{1}{4}, 3$ **d)** −1.35, 3.35 **e)** $-\dfrac{7}{4}, \dfrac{1}{3}$ **f)** −0.47, 0.61 **5. a)** 0, 60 **b)** 30 m; 900 m² **c)** 47.32 m by 12.68 m **6. a)** 307.9 m² **b)** $\pi(7 + x)^2$ **c)** 9.9 m **d)** 12.1 m **7. a)** 0, 40 **b)** 10 m **c)** no **d)** 3.3, 36.7 **e)** 36.7 m **8. a)** 82.8 km/h **b)** yes **9. a)** 2500 **b)** $P(x) = x^2 - 19x - 150$ **c)** −6, 25 **d)** Answers may vary; for example, they should keep the production level below 2500 plants to achieve a profit. **e)** 2329

7.4 Modelling With Quadratics, pp. 336–339

1. a) 11, 16, 22 **b)** quadratic; second differences are constant **c)** $y = \dfrac{1}{2}x^2 + \dfrac{1}{2}x + 1$ **d)** 121 **e)** 44 **2. b)** 360 m **c)** 110 m; 0 m **d)** $y = -15.36x^2 + 121.07x + 125.71$; x represents the time, in seconds, and y represents the height, in metres. **e)** $y = 0$ at $x = 8.8$ **f)** 364.3 m **3. a)** $C(t) = t^2 + 10t + 5000$ **b)** 770 days **4. a)** $y = -4.9t^2 + 50t$ **b)** 127.6 m **c)** $y = -7.1t^2 + 49.7t$ **d)** 460 m **5. a)** 9 ft by 9 ft **b)** 16 **6. a)** 150 m **b)** $y = 75x - \dfrac{1}{2}x^2$

c)
 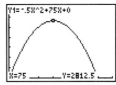

d) length: 75 m, width: 37.5 m **7. a)** 224 ft **b)** down **c)** 3.5 s **d)** 7.5 s **e)** 9.7 s **8.** 24 cm by 24 cm **10. b)** 10.5 mg **c)** No; the concentration continues to diminish. **d)** Answers may vary.

Career Profile, p. 340

1. a) $x[x(7x + 1) + 4] - 2$ **b)** $x[x(4x - 2) + 8] - 5$ **2. a)** $x^3 + 2x^2 + 3x - 6$ **b)** $4x^3 - x^2 - 5x + 3$

Chapter Problem Wrap-Up, p. 341

1. a) $190; $18 050 **b)** $150; 115 parkas
2. a) $R(x) = (200 + 10x)(90 - 5x)$

Review, pp. 342–345

1. a) linear **b)** neither **c)** quadratic **d)** neither **e)** quadratic
f) neither **g)** neither **h)** neither
2. a) $y = -2x^2 - 4x - 5$ **b)** $y = 0.4x^2 - 1.2x + 13.6$
c) $y = -x^2 + 16x - 64$ **3. a)** $y = -x^2 - 4x + 1$; $y = -(x + 2)^2 + 5$
b) $y = \frac{1}{4}x^2 - 2x + 1$; $y = \frac{1}{4}(x - 4)^2 - 3$
4. a) i) $y = 2x^2 - 4x + 7$ **ii)** $y = 8.56x^2 + 20.77x - 268$
b) For the data in part i), the equation is exact.
5. a) $(x - 4)(x + 1)$ **b)** $(2x + 5)(3x - 1)$ **c)** $-(x + 4)(5x - 3)$
d) $(x - 9)(x + 7)$ **e)** $(x + 8)^2$ **f)** $-4(x - 4)(x + 9)$
6. a) $-4, -3$ **b)** $-10, \frac{5}{2}$ **c)** $-3, 5$ **d)** $-5, -\frac{1}{2}$ **e)** $-3, 5$ **f)** $\frac{1}{2}, 2$ **7.** 1 in.
8. a) $y = 16x^2 + 56x + 52$; $a = 16, b = 6, c = 52$
b) $y = -\frac{1}{2}x^2 + 3x - 5$; $a = -\frac{1}{2}, b = 3, c = -5$
c) $y = -2x^2 - 12x - 22$; $a = -2, b = -12, c = -22$
d) $y = \frac{3}{4}x^2 + 18x - \frac{213}{2}$; $a = \frac{3}{4}, b = 18, c = \frac{213}{2}$
9. a) $-2.77, 1.27$ **b)** $0.38, 2.62$ **c)** no solutions **d)** $-0.61, 6.61$
e) $-2.64, 15.14$ **f)** $-0.5, 1.67$ **10. a)** 883 **b)** $26 217.80; 423
c) Between 0 and 883 **11. a)** 1, 10; 2, 5; 3, 2; 4, 1; 5, 2; 6, 5; 7,
10; 8, 17; 9, 26 **b)** The relationship appears to be quadratic.
c) 4 **d)** $y = x^2 - 8x + 17$ **12. b)** $y = -0.000 023 5x^2$ **c)** 574 m
d) $y = 0.000 051 4x^2 + 2$ **13. a)** yes **b)** 4 s

Practice Test, pp. 346–347

1. a) linear **b)** quadratic **c)** neither **d)** neither
2. a) $y = 3x^2 - 2x + 1$ **b)** $y = -\frac{1}{4}(x - 3)^2 + 4$ **3. a)** $(x + 5)(x - 3)$
b) $(2x + 5)(3x - 1)$ **c)** $-6(x - 7)(2x - 1)$
4. a) 1, 4 **b)** $-\frac{1}{2}, 3$ **c)** $-\frac{5}{2}, -\frac{3}{2}$ **d)** $-6, 3$ **e)** $-\frac{3}{2}, -\frac{1}{2}$
f) no solutions **5.** 70 m by 15 m or 35 m by 30 m
6. a), b)

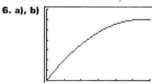

c) 54 m

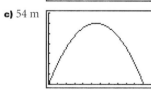

d) The curve is symmetric about a line through the maximum
point. **e)** $y = -0.0395x^2 + 2.1758x + 0.0022$ **f)** 0, 55.1
g) 29.9 m, 27.5 m **h)** 54.1 m

Chapter 8 – Interpreting Graphical and Algebraic Models

Get Ready, pp. 350–351

1. a) 16, 19, 22; arithmetic **b)** 24, 48, 96; geometric
c) 25, 36, 49; power **d)** 10 000, 1000, 0.1, 0.01; geometric
e) 10, 15, 20, 35, 40; arithmetic **f)** 0.36, 0.64, 1; power
2. a) 22 **b)** 16 **c)** -40 **d)** -127 **e)** 8 **f)** 70 **g)** -2700 **h)** 8
3. a) 0.5 **b)** 0.7071 **c)** 0.2679 **d)** 0.3682 **e)** 0.1389 **f)** 0.3333
g) $-247.6259°$ **h)** 25.9287° **i)** 0.0345° **4. a)** 8 **b)** 18 **c)** 36 **d)** 30
e) 11 **f)** 9 **g)** -7 **h)** -8 **5. a)** 3 **b)** 5 **c)** 3 **d)** $-\frac{1}{3}$ **e)** -8 **f)** -54 **g)** 14
h) 3.5 **i)** -14.4 **j)** 5 **k)** -9 **l)** -38 **6. a)** $6x + 9y$ **b)** $8k$ **c)** x **d)** $2e + 6f$
7. a) ii) x-intercept 2.5; y-intercept -5 **b) ii)** x-intercept 6;
y-intercept 3 **c) ii)** x-intercept 3; y-intercept 2 **d) ii)** x-intercept
6; y-intercept -5

8.1 Interpreting Graphical Models, pp. 357–361

1. The graph in part b); while pedalling up the steep hill, the
speed of the cyclist is low and so, the distance from the starting
point does not increase rapidly. After reaching the top of the
hill, the speed increases and the distance from the starting
point increases rapidly. **2.** The graph in part a); the distance
above the ground is zero at the start, and then, varies between
the high and low points of the Ferris wheel. **3. a)** Answers will
vary. **b)** approximately 148 beats/min **c)** approximately
162 beats/min **4.** Answers may vary. **5. a)** It is faster and easier
to look up values. **b)** 32°F **c)** 32°C **d)** $-40°$ **6. a)** 11.5 L/100 km
b) 60 km/h; 10.0 L/km **c)** It would be steeper, and translated
upward. **d)** It would be flatter, and translated downward.
7. a) 0.5 s **b)** 25 cm **c)** 100 cm **d)** shorten the length
8. a) 7.5 **b)** After too many viewings, consumer resistance
builds. **c)** Yes; if people see the advertisement too often, they
may be "turned off" the product. **9. a)** parabolic
b) 2 m; the height of the ball when it was released
c) 4 m **d)** 28 m **10. a)** Answers may vary. **b)** 2063 **c)** Answers
may vary. **11. a)** exponential **b)** 9.85 s **c)** 9.82 s **d)** Answers will
vary. **12. a) i)** 10 m; 15 m **ii)** 60 m; 90 m **iii)** 85 m; 130 m
b) 73 km/h **c)** Not in theory; in practice, the stopping distances
would be essentially identical for very low speeds.
13. a)

b) approximately $5000 loss

8.2 Evaluating Formulas for Any Variable, pp. 369–372

1. a) 2 **b)** 0.06 **c)** 480 km **d)** 20 m **e)** 12 mm **f)** 36.9°
2. a) $x = \frac{y - b}{m}$ **b)** $r = \frac{I}{Pt}$ **c)** $d = St$ **d)** $w = \frac{P}{2} - l$

e) $x = \pm\sqrt{r^2 - y^2}$ f) $\angle C = \sin^{-1}\left(\dfrac{a}{b}\right)$ 3. a) $n = \dfrac{C - 4}{0.28}$

b) 24 exposures c) up to 39 exposures 4. a) $A = \dfrac{230 - M}{1.2}$

b) i) 30 years ii) 80 years 5. a) $n = \dfrac{C - S}{4}$ b) 1018 c) 2125

6. a) $P = \dfrac{A}{1 + rt}$ b) $1428.57

c)

Rate, r	Principal, P
0.030	7692.31
0.035	7407.41
0.040	7142.86
0.045	6896.55
0.050	6666.67
0.055	6451.61
0.060	6250.00
0.065	6060.61
0.070	5882.35
0.075	5714.29
0.080	5555.56
0.085	5405.41
0.090	5263.16
0.095	5128.21
0.100	5000.00

7. a) $762 b) $r = \dfrac{C}{F}$ c) $0.0347/Jamaican dollar

d) $0.000 698/lire e) 1433.33 lire/$; This form is more useful when you are buying lire with Canadian dollars.
8. a) 1.1 s b) $l = 25p^2$ c) 25 cm d) 100 cm e) The length and period of a pendulum is not a linear relationship.

9. a) 1.01 s b) $t = \sqrt{\dfrac{k - h}{9.8}}$ d) after 2.26 s

e)

Height, h	Time, t
50	0.00
45	0.71
40	1.01
35	1.24
30	1.43
25	1.60
20	1.75
15	1.89
10	2.02
5	2.14
0	2.26

10. a) i) 45.3 N ii) 49.2 N b) $A = \cos^{-1}\left(\dfrac{F_H}{F}\right)$ c) i) 36.9° ii) 75.5°

11. a) $t = \dfrac{d}{S}$ b) 5 h c) 5 min 12. a) 0.012 b) $D = d + TL$; 4.1 cm

c) $d = D - TL$; 2.72 cm d) $L = \dfrac{D - d}{T}$; 60 cm

13. a) $A = \dfrac{g}{2}\sin^{-1}\left(\dfrac{d}{771.6}\right)$ b) i) 54.9° ii) 20.7° iii) not possible

iv) 3.4° 15. a) $h = \dfrac{S}{2\pi r} - r$ b) 7.0 m c) 1.4 m d) $n = \dfrac{S}{10}$

e) 4.4 m 16. a) 1.33 b) $I = \sin^{-1}(n \sin R)$ c) 55.4°

d) $R = \sin^{-1}\left(\dfrac{\sin I}{n}\right)$ e) 21.0° f) Answers will vary.

8.3 Graphing and Interpreting Linear Models, pp. 378–379

1. a) $\dfrac{1}{2}$, -4 b) 2, -1 c) $-\dfrac{1}{2}$, $\dfrac{3}{2}$ 2. a) 8 b) $\dfrac{1}{2}$ c) 3

3. a) 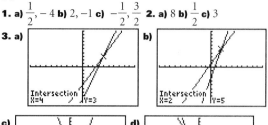 b)

c) 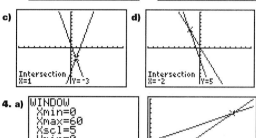 d)

4. a)

b) $2.50 per swim is the better deal, if you plan to swim fewer than 48 times in the 6-month period. 5. b) 9 years; $10 800

6. a) $y = 90x$ c) $60; $30

8.4 Solving Systems of Equations: Substitution Method, pp. 384–386

1. a) $x = -4$, $y = -11$ b) $x = 6.5$, $y = 36.5$ c) $x = 0$, $y = 1$

d) $x = -16$, $y = -29$ 2. a) $x = 3$, $y = 3$ b) $x = -1$, $y = -2$

c) $x = \dfrac{7}{6}$, $y = \dfrac{1}{12}$ d) $x = 1$, $y = -1$ e) $x = 6$, $y = 1$ f) $x = 2$, $y = -1$

g) $x = 5$, $y = 2$ h) $x = 6$, $y = -8$ i) $x = -1$, $y = 2$ 3. 52

4. a) Model (1) has $1500 in fixed costs plus $2.50 per student. Model (2) has $2060 in fixed costs plus $1.80 per student.
b) 800; $3500 5. a) Cantech: Base salary of $25 000 plus $2000 per year of service. Wordcom: Base salary of $30 000 plus $1500 per year of service. b) 10 years; $45 000 6. a) no solution
b) The lines are parallel and do not intersect.

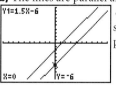 c) A system of equations will have no solution when the equations define parallel lines. 7. a) 23.3 years; 20.5%

b)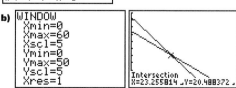

c) Answers may vary. 8. b) $x = 0$, $y = 1.8$ c) 1.8 m d) 2.4 m

9. a) Piece-a-Pizza: $y = 7.95 + 0.75x$, Flying Pizza:
$y = 8.40 + 0.60x$ **b)** 3 toppings; $10.20 **10. a)** sedan:
$y = 350 + 0.35x$, hatchback: $y = 285 + 0.41x$ **b)** 1083.3 km;
$729.17 **11. a)** $P = 100x^2 - 10\ 000$ **b)** 10 days **c)** 34 days
d) $192 500 **12.** Answers may vary. **a)** $2x + 3y = 19$,
$3x - 4y = -14$ **13. a)** parabola; straight line **b)** 2 **c)** $x = -1$, $y = 6$;
$x = 2.5$, $y = 11.25$ **d)**

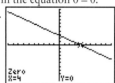

8.5 Solving Systems of Equations: Elimination Method, pp. 391–393

1. a) $x = 2$, $y = 1$ **b)** $x = 2$, $y = 7$ **c)** $x = 4$, $y = 4$ **d)** $x = 3$, $y = 4$
2. a) i) $2 \times (1)$, $3 \times (2)$ **ii)** $5 \times (2)$ **b) i)** $7 \times (2)$
ii) $8 \times (1)$, $3 \times (2)$ **c) i)** $3 \times (1)$, $5 \times (2)$ **ii)** $11 \times (1)$, $7 \times (2)$
d) i) $2 \times (1)$, $9 \times (2)$ **ii)** $4 \times (1)$, $3 \times (2)$ **3. a)** $x = -4$, $y = 2$
b) $x = 3$, $y = -1$ **c)** $x = -2$, $y = 4$ **d)** $x = -1$, $y = -3$
4. a) $x = 2$, $y = 3$ **b)** $x = \dfrac{90}{11}$, $y = \dfrac{32}{11}$ **c)** $x = -1$, $y = 2$
d) $x = -3$, $y = 5$ **e)** $x = -3$, $y = -7$ **f)** $x = 4$, $y = 0.5$
5. a) $x = 16$, $y = 34$ **b)** $x = 40$, $y = 5$ **c)** $x = -2$, $y = 4$
d) $x = 2000$, $y = 3000$ **6.** $45; $25 **7.** 50 g; 5 g **8.** 17 m³; 9 m³
9. a) Elimination results in the equation $0 = 0$.
b) The lines are identical.

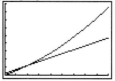

c) Some systems of equations have infinitely many solutions;
this occurs when each equation defines the same line.
10. a) $x + 2y = 7.5$, $x + 5y = 12$ **b)** $4.50; $1.50
11. a) $x + y = 220$, $\dfrac{x}{90} + \dfrac{y}{70} = 3$

b) 45 km at 90 km/h, 175 km at 70 km/h **12.** 5 **13.** Answers
may vary. **a)** $3x - 2y = 16$, $5x + 3y = -5$

Career Profile, p. 394

1. $80 080 **2.** full page, 6th advertisement

Chapter Problem Wrap-Up, p. 395

1. a)
```
WINDOW
 Xmin=0
 Xmax=45
 Xscl=5
 Ymin=0
 Ymax=450000
 Yscl=50000
 Xres=1
```

b) approximately $5000 loss **2.** $P(x) = 100x^2 - 10\ 000$
3. a) 10 days **b)** 33 days **c)** $192 500 **4.** The start-up costs are

$20 000, instead of $10 000. This would produce a lower profit.
5. The curve would be steeper. This would produce a higher
profit. **6.** Answers will vary.

Review, pp. 396–399

1. The person starts from home, accelerates for a while, then
walks at a constant speed for a time. The person turns around
and walks back toward home at a constant speed. For the last
part, the person has stopped some distance from home.
2. a) approximately 0.7 m **b)** 0.76 s **c)** No; there would not
be enough time for a full rotation. **3. a)** exponential
b) 14 years; 9 years; 7 years **c)** $4500; $10 000; $17 500 **d)** The
rate of increase would lessen. **e)** vertical translation by 1000
upward, and vertical stretch by a factor of 2 **4. a)** 10 cm
b) 5 cm; the length of the spring with no masses hanging from it
c) 27 kg **d)** 36 cm **5. a)** 3.3 cm **b)** 5 ohms **c)** 30°

d) 5 years **6. a)** $h = \dfrac{C - I}{r}$

b)

Initial Cost, I ($)	Hourly Rate, r ($)	Cost of 4-h Rental ($)
0.00	7.50	30.00
5.00	6.25	30.00
10.00	5.00	30.00
15.00	3.75	30.00
20.00	2.50	30.00
25.00	1.25	30.00

7. a) $r = \sqrt{\dfrac{V}{\pi h}}$ **b)** 1.6 m **8. a)** $\angle C = \sin^{-1}\left(\dfrac{2A}{ab}\right)$ **b)** 30°

9. 25 **10.** Answers may vary. **11. a)** $x = 5$, $y = 13$
b) $x = 4$, $y = -1$ **c)** $x = -1$, $y = -3$ **d)** $x = 3.36$, $y = 2.10$
e) $x = 10.94$, $y = -0.63$ **f)** $x = 10$, $y = 1$ **12.** 1440 km
13. 4400 stickers; $644 **14.** Answers may vary.
15. a) $x = -1$, $y = 2$ **b)** $x = 1$, $y = 1$ **c)** $x = -3$, $y = 2$
d) $x = -2$, $y = -5$ **e)** $x = 4$, $y = -5$ **f)** $x = 15$, $y = 9$ **16.** 300; 150

Practice Test, pp. 400–401

1. a) linear **b)** 0.4; For each 1-m increase in height, the shadow
increases by 0.4 m. **c)** 0.64 m **d)** 1.38 m **e)** 1.2 m **f)** The slope
would be less steep. **2. a)** exponential **b)** 1000 g; the initial
amount of sodium **c)** The slope is approaching zero, so the rate
of decay decreases. **d)** 250 g **e)** 109 g **f)** The decay would be
more rapid and so, the graph would decrease more quickly.
3. a) 25 **b)** 3 **c)** 39 **4. a)** $x = 7$, $y = 27$ **b)** $x = -2$, $y = 3$
5. a) $x = 1$, $y = -1$ **b)** $x = -1$, $y = 2$ **6.** 57 quarters, 83 loonies

7. a) 1413.7 cm³ **b)** $b = \sqrt{\dfrac{4D}{\pi ns}}$ **c)** 8.2 cm

Cumulative Review, Chapters 1–8, pp. 402–407

1. a) $\angle A = 54°$, $a = 6.9$ m, $c = 5.0$ m **b)** $r = 17.7$ cm,
$\angle P = 102.1°$, $\angle Q = 36.9°$ **c)** $\angle X = 58°$, $\angle Y = 64°$, $y = 0.85$,
$z = 0.8$ m **2.** 10.2 m **3.** $1539.72 (for 28 yd²) **4.** 224.25 mm to

225.75 mm **5.** Answers may vary. **6. b)** normal distribution
7. a) City; Weather Outlook; High/Low Temperature (°C)
b) Cornwall, Ottawa, and Timmins; Dryden and Thunder Bay
c) July **d)** 18.9°, 19°, 20° **e)** 7:48 p.m.
8. Answers may vary.
9. a)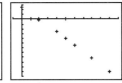

b) as altitude increases, temperature decreases
c) 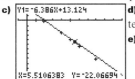 **d)** −6.5; the rate of change of temperature with respect to altitude **e)** $t = -6.5a + 11.25$ **f)** −66.8°C

10. a)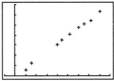

b) $y = 0.0798x - 0.0649$ **c)** 0.998 12; yes **d)** 0.82 mm

11. a)

Age	Blood Pressure
25	116.4
26	117.9
27	119.4
28	120.9
29	122.4
30	123.9
31	125.4
32	126.9
33	128.4
34	129.9
35	131.4
36	132.9
37	134.4
38	135.9
39	137.4
40	138.9
41	140.4
42	141.9
43	143.4
44	144.9
45	146.4
46	147.9
47	149.4
48	150.9
49	152.4
50	153.9

b) 141.9

12. a) −3, −2 **b)** $-\dfrac{3}{2}, \dfrac{3}{2}$ **c)** $-3, \dfrac{5}{2}$ **d)** 2, 4 **e)** $-1, \dfrac{2}{3}$ **f)** −7, 3
13. a) $y = -x^2 + 2x + 2$, $y = -(x-1)^2 + 3$
b) $y = \dfrac{1}{2}x^2 - 1$, $y = \dfrac{1}{2}(x-0)^2 - 1$

14. a) $y = -\dfrac{4}{3}x + 2$; slope: $-\dfrac{4}{3}$, y-intercept: 2
b) $y = 3x - 2$; slope: 3, y-intercept: −2 **c)** $y = 2x - 4$; slope: 2, y-intercept: −4 **15. a)** 5.6 m vs. 26 m **b)** 1.8 and 0, the initial height of the projectile **c)** 2.2 s, 10.2 s **d)** a vertical translation of 1.2 m upward
16. a)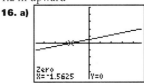

b) slope: 0.32, x-intercept −1.5625, y-intercept 0.5
c) about 1.6 months **17. a)** (−1, 2) **b)** (−4, −5) **c)** (4, 1)
18. a) A: $y = 8 + 0.55x$; B: $y = 1.25x$
b) 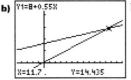 between 11 and 12 transactions

c) Option B is better with few transactions or with a balance of less than $1000.
19. Answers may vary.

Appendix A, Review of Prerequisite Skills, pp. 408–421

Adding or subtracting polynomials, p. 408
1. a) $10x + 5y$ **b)** $13x$ **c)** $-11y$ **2. a)** $7x - 3y$ **b)** $2x$ **c)** $7x$

Angles of a triangle, p. 408
1. a) 20° **b)** 60° **c)** 22°

Degree of polynomials, p. 409
1. a) 5 **b)** 1 **c)** 2 **d)** 2 **e)** 4 **f)** 0

Evaluating expressions, p. 409
1. a) −31 **b)** 2 **c)** 271 **d)** −422 **e)** 48 **f)** −227 **g)** −42 **h)** −699 **i)** −36

Exponential graphs, p. 410
1. a) 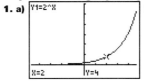 **b)**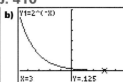

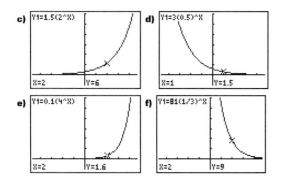

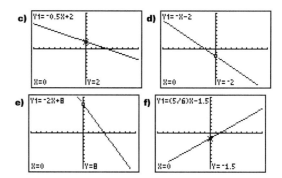

Factors of numbers, p. 410

1. a) 2; 1, 2, 4, 8, 16, 32 **b)** 2, 5; 1, 2, 5, 10 **c)** 2, 5; 1, 2, 5, 10, 25, 50 **d)** 3, 7; 1, 3, 7, 21 **e)** 5, 23; 1, 5, 23, 115 **f)** 13; 1, 13 **2. a)** 3, 9 **b)** 2, 8 **c)** 3, 5 **d)** −9, −3 **e)** no solution **f)** −5, −3 **g)** −3, 9 **h)** −9, 3 **i)** −2, 8 **j)** 2, −8 **k)** −3, 5

Factoring polynomials, p. 411

1. a) $3(m - 2n + 3)$ **b)** $9c(2c^2 + 4c - 1)$ **c)** $2j(7k + 8m - 5n)$
d) $10x^2y^2(20xy^3 + 3x^2 + 5x)$ **e)** $6a(1 - 2a^2)$ **f)** $cd(5 + 12c - 7d)$

First differences, p. 411

1. a) −3, −3, −3 **b)** 6, 6, 6 **c)** 2, 4, 8 **d)** −12, −4, 4, 12

Geometric shapes and solids, p. 411

1. equilateral triangle, square, pentagon, hexagon, octagon
3. square-based pyramid

Graphing and interpreting data, p. 412

1. b) Graphs that do not exhibit changes over time would not be appropriate; for example, a circle graph. **2. a)** 176 **c)** 0: 20%, 1: 14%, 2: 32%, 3: 15%, 4: 10%, 5: 7%, 6: 2% **e)** line graph

Graphing skills, p. 413

2. b) The points form a linear pattern; they satisfy the equation $y = x + 3$. **c)** (0, 3), (−3, 0), (1, 4)

Interpreting graphs, p. 414

1. a) increasing then decreasing then leveling off **b)** 4th, 1st **c)** 2nd and 3rd **d)** 4th and 5th **e)** A line graph shows a trend over time. **2. a)** 29 **b)** 7 **c)** 23 **d)** 17% **e)** 1 **f)** 6.6 **g)** Frequencies of marks are easily compared.

Linear graphs, p. 414

1. a) 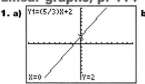 **b)**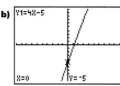

Metric and imperial measures, p. 415

1. a) 6.54 yd **b)** 3.64 m **c)** 72.45 km **d)** 62 mi **e)** 12.32 gal **f)** 149.6 lb **g)** 4′8″

Metric measurement, p. 415

1. a) 5.25 m **b)** 58.09 kg **c)** 240 000 L **d)** 240 000 cm **e)** 0.056 s **f)** 450 mm **2. a)** 52 mm, 0.052 m **b)** 18 mm, 0.018 m

Number skills, p. 416

1. a) 33 **b)** 195 **c)** 108 **d)** $3\frac{4}{15}$ **e)** $-\frac{1}{4}$ **f)** $1\frac{1}{2}$ **g)** $-\frac{1}{5}$ **h)** 0.5 **i)** 64.4

2. a) $\frac{1}{2}, \frac{7}{12}, \frac{3}{5}, \frac{5}{8}, \frac{2}{3}$ **b)** $3\frac{5}{9}, 3\frac{3}{4}, 3\frac{6}{7}, 3\frac{7}{8}$ **3. a)** L.S. = 5, R.S. = 7

b) L.S. = $x^2 + 2xy + y^2$, R.S. = $x^2 + y^2$ **c)** L.S. = $\frac{9}{6}$, R.S. = $\frac{7}{9}$

Patterns, p. 416

1. a) subtract 3; 6, 3, 0 **b)** multiply by 3; 108, 324, 972
c) perfect squares; 16, 25, 36 **d)** add 2; 8, 10, 12
e) divide by −3; $-\frac{1}{3}, \frac{1}{9}, -\frac{1}{27}$ **f)** divide by −2; −10, 5, $-\frac{5}{2}$
g) append an additional q: pqqqqq, pqqqqqq, pqqqqqqq
h) append twice the previous number of q's: p followed by 8 q's, p followed by 16 q's, p followed by 32 q's
i) alternate p's and q's, p's increase by 1, q's double each time; ppppp, qqqqqqqq, ppppp

Percents, p. 417

1. a) 75%, 0.75 **b)** 50%, 0.5 **c)** 840%, 8.4 **d)** $\frac{17}{50}$, 0.34
e) $\frac{3}{10\ 000}$, 0.0003 **f)** $\frac{7}{125}$, 0.056 **g)** $\frac{9}{20}$, 45% **h)** $\frac{3}{100}$, 3%
i) $2\frac{17}{25}$, 268%

Perimeter and area, p. 417

1. a) 24 cm, 23 cm² **b)** 9 cm, 4.86 cm²

Proportion, p. 418

1. a) 8.4 **b)** $3\frac{1}{3}$ **c)** 7.5 **d)** 10 **e)** 16.8 **f)** 1.575

Pythagorean theorem, p. 419

1. a) 13 **b)** 15.3 **c)** 43.3

Quadratic graphs, p. 419

1. a) 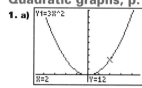 y-intercept 0, x-intercept 0

b) 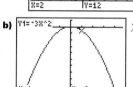 y-intercept 0, x-intercept 0

c) 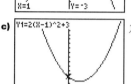 y-intercept 5, x-intercepts none

d) 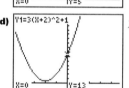 y-intercept 13, x-intercepts none

e) 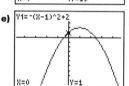 y-intercept 1, x-intercepts −0.4, 2.4

f) 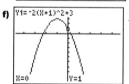 y-intercept 1, x-intercepts −2.2, 0.2

g) 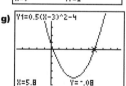 y-intercept 0.5, x-intercepts 0.2, 5.8

h) 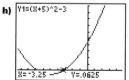 y-intercept 22, x-intercepts −6.7, −3.3

i) 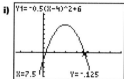 y-intercept −2, x-intercepts 0.5, 7.5

Reducing ratios, p. 420

1. a) 7:9 **b)** 1:3 **c)** 7:2:5 **d)** 5:3:7:20 **e)** 3:7 **f)** 7:4:12 **g)** 12:57:34:83 **h)** 20:10:45:5

Scientific notation, p. 420

1. a) 6.531×10^{13} **b)** 2.57×10 **c)** 8.143×10^9 **d)** 2.37×10^{-7} **e)** 4.1×10^{-5} **f)** 2.15×10^{-3} **g)** $7.892\ 34 \times 10^5$ **h)** 7.65×10^{-22}

2. a) 340 000 **b)** 0.009 34 **c)** 506.6 **d)** 0.000 283 **e)** 7 540 000 **f)** 8333.3

Simplifying expressions, p. 420

1. a) $x^2 - 4y^2$ **b)** $4x^2 - 20x + 25$ **c)** $x^4 + 2x^2y + y^2$ **d)** $25x^2 - 20xy + 4y^2$ **e)** $4x^2 + 7x + 1$ **f)** $22y^2 - 20y + 4$ **g)** $k^2 - 6k + 13$ **h)** $25m^2 - 5mn - 2n^2$ **i)** $50y^2 + 80xy + 32x^2$

Solving equations, p. 421

1. a) −3 **b)** −6 **c)** 1 **d)** 18 **e)** −2.42 **f)** −11 **g)** −4 **h)** ±6 **i)** ±11

Trigonometry, p. 421

1. a) 0.4226 **b)** 0.9063 **c)** 0.3746 **d)** 0.9397 **e)** 0.5774 **f)** 28.6363

2. a) 34.6° **b)** 77.7° **c)** 49.4° **d)** 51.9° **e)** 29.4° **f)** 1.2° **3. a)** 2.7 **b)** 1.4 **c)** 1.4 **d)** 0.1 **e)** 0.03 **f)** 36.9

lossary

A

acute angle An angle that measures less than 90°.

adjacent In right △ABC, BC is the side adjacent to ∠C that is not the hypotenuse.

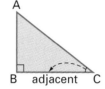

algebraic expression An expression that includes at least one variable.

$2t$, $3x^2 + 4x - 5$, and 2^x are algebraic expressions.

algebraic modelling The process of representing a relationship by an equation or a formula, or representing a pattern of numbers by an algebraic expression.

altitude The height of a geometric figure. In a triangle, an altitude is the perpendicular distance from a vertex to the opposite side.

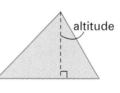

angle of depression The angle, measured downward, between the horizontal and the line of sight from an observer to an object.

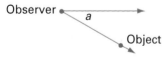

∠a is an angle of depression.

angle of elevation The angle, measured upward, between the horizontal and the line of sight from an observer to an object.

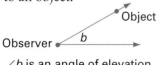

∠b is an angle of elevation.

application The use of mathematics to solve real-world problems.

area The number of square units contained in a region.

arithmetic sequence A sequence that has a common difference between consecutive terms.

1, 4, 7, 10, … is an arithmetic sequence.

astronomical unit (AU) The distance from Earth to the sun, approximately 150 000 000 km.

axis of symmetry The fold line of a symmetrical figure.

B

bar graph A graph that uses bars to represent data.

base (of a power) The number that is repeatedly multiplied.

In 3^4, the base is 3.

BEDMAS An acronym that lists the order of operations. BEDMAS stands for **B**rackets, **E**xponents, **D**ivision, **M**ultiplication, **A**ddition, **S**ubtraction.

bias In statistics, some factor influences a sample such that the data is not truly representative of the population.

binomial An algebraic expression with two terms.

$5x - 2$ is a binomial.

broken-line graph A graph that relates two variables as ordered pairs, with consecutive points joined by line segments.

C

capacity The greatest volume that a container can hold, usually measured in litres, millilitres, or kilolitres.

census A survey in which data are collected from every member of the population.

circle The set of all points in the plane that are equidistant from a fixed point called the centre.

circle graph A graph using sectors of a circle to represent data.

circumference The perimeter of a circle.

class interval A range of values for grouping data in a histogram.

The class interval [30, 35) includes 30, but not 35.

clustered sampling A method of choosing a sample in which a random sample is chosen from one particular group within the population.

coefficient The factor by which a variable is multiplied.

In the term $3x^2$, the coefficient is 3.

common difference The difference between consecutive terms of an arithmetic sequence.

For 2, 5, 8, 11, …, the common difference is 3.

common factor Any factor that two or more numbers, or two or more terms of a polynomial, share.

2 is a common factor of 4, 6, and 18.
$3x$ is a common factor of $3x^2 - 12x$.

common ratio The ratio of consecutive terms of a geometric sequence.

For 2, 6, 18, 54, …, the common ratio is 3.

composite shape A shape made up of two or more distinct shapes.

compound interest Interest that is calculated at regular compounding periods, and then, added to the principal for the next compounding period.

concurrent Two or more lines that have one point in common.

cone A three-dimensional object with a circular base and a curved lateral surface, which extends from the base to a point called the vertex.

congruence The property of being congruent. Two geometric figures are congruent if they are equal in all respects.

constant rate of change A linear relationship between two variables where equal intervals of the first variable correspond to equal intervals of the second variable.

The cost of gasoline at $0.80/L has a constant rate of change because 1 L costs $0.80, 2 L cost $1.60, and so on.

constant term A term that does not include a variable. A numerical term.

In $x^2 + 5x - 1$, the constant term is -1.

contained angle The angle that lies between two named sides of a polygon.

correlation coefficient A variable used to measure the strength of a relationship when regression is used to find an equation that approximates data.

cosine law The relationship between the lengths of the three sides and the cosine of an angle in any triangle.

$$a^2 = b^2 + c^2 - 2bc \cos A$$

cosine ratio In a right triangle, for $\angle A$, the ratio of the length of the side adjacent to $\angle A$ and the length of the hypotenuse.

$$\cos A = \frac{\text{adjacent}}{\text{hypotenuse}}$$

cross-multiply In an equation like $\frac{a}{b} = \frac{c}{d}$, multiplying both sides by bd gives $ad = bc$.

cube A three-dimensional shape with six congruent square faces.

curve of best fit A curve that approximates the distribution of points in a scatter plot.

cylinder A three-dimensional shape with two faces that are congruent circles, and a curved surface connecting the two circles.

decagon A polygon with ten sides.

decreasing function A function for which the first differences decrease.

degree of a polynomial in one variable The greatest exponent of the variable in any one term.

The degree of $x^3 + 6x^2 - 1$ is 3.

degree of a polynomial in two variables The greatest sum of the exponents in any one term.

The degree of $2a^3b^2 + a^2b - 7b$ is 5.

degree of precision The smallest measurable unit of a measurement tool.

dependent variable In a relation, the variable whose value depends on the value of the independent variable. On a coordinate grid, the values of the dependent variable are on the vertical axis.

In $d = 4.9t^2$, d is the dependent variable.

descriptive statistics The outcomes of data collection and analysis are used to describe the population.

diagonal A line segment joining two non-adjacent vertices of a polygon.

diameter of a circle A chord that passes through the centre of a circle.

distributive property $a(b + c) = ab + ac$

dodecagon A polygon with twelve sides.

dodecahedron A regular polyhedron with twelve faces.

domain of a relation The set of numbers for which a relation is defined. The set of all first coordinates of the ordered pairs in a relation.

double root The solution of a quadratic equation where both roots are the same.

dynamic data Facts presented in a medium that are often changed or updated regularly.

dynamic geometry software Computer software that allows the user to plot points on a coordinate system, measure line segments and angles, construct two-dimensional shapes, create two-dimensional representations of three-dimensional objects, and transform constructed figures by moving parts of them.

edge The line segment where two plane faces of a three-dimensional shape meet.

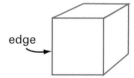

elements The individual members of a set.

elimination method A method of solving a system of equations by addition or subtraction of the equations to eliminate one variable.

equation A mathematical sentence formed by two equivalent expressions.

$5x - 3 = 2x + 6$ is an equation.

equilateral triangle A triangle with all sides equal.

equivalent algebraic expressions Expressions that are equal for all values of the variable.

$7t + 3t$ and $10t$ are equivalent algebraic expressions.

equivalent equations Equations that have the same solution.

equivalent fractions Fractions that represent the same part of the whole or group.

$\dfrac{1}{3}$, $\dfrac{2}{6}$, and $\dfrac{5}{15}$ are equivalent fractions.

equivalent ratios Ratios that are in proportion.

  2:5 and 6:15 are equivalent ratios.

evaluate To determine a value for.

expand To multiply, usually applied to polynomials.

  $4(n-3)$ expands to $4n-12$.

exponent A raised number in a power that indicates repeated multiplication of the base.

  In $(x+3)^2 = (x+3)(x+3)$, the exponent is 2.

exponential equation An equation that has a variable in an exponent.

  $3^x = 81$ is an exponential equation.

exponential form A shorthand method for writing repeated multiplication.

  4^3 is the exponential form for $4 \times 4 \times 4$.

exponential function A relation of the form $y = ab^x$, where $a \neq 0$, $b > 0$, and $b \neq 1$.

exponential growth Non-linear growth represented by an exponential equation and a graph with an upward curve. Geometric sequences and compound interest show exponential growth.

expression A mathematical phrase made up of numbers and/or variables.

  $x^2 + x - 5$, $2x$, and 3^n are expressions.

extrapolate Estimate values lying outside the range of the given data. To extrapolate from a graph means to estimate coordinates of a point beyond those that are plotted.

face A plane surface of a polyhedron.

factor To express a number as the product of two or more numbers, or an algebraic expression as the product of two or more algebraic expressions.

  $12 = 2 \times 2 \times 3$
  $4x^2 + 8xy = 4x(x+2y)$

Fibonacci numbers The sequence 1, 1, 2, 3, 5, 8, 13, …, formed by adding the preceding two terms.

first differences In a relation between two variables, the difference between successive values of the second variable for regular steps of the first variable.

$y = 2x + 1$		
x	y	First Differences
-2	-3	
-1	-1	$-1 - (-3) = 2$
0	1	$1 - (-1) = 2$
1	3	$3 - 1 = 2$
2	5	$5 - 3 = 2$

FOIL A method for multiplying two binomials. The sum of the product of **F**irst terms, **O**utside terms, **I**nside terms, and **L**ast terms.

$$(3x-1)(3x+5) = 9x^2 + 15x - 3x - 5$$
$$= 9x^2 + 12x - 5$$

frequency distribution table A table used to organize and count data.

function A relation in which each x-value (or first variable value) has only one corresponding y-value (or second variable value).

  $y = (3x-5)^2$ is a function.

geometric sequence A sequence that has a common ratio between consecutive terms.

  2, 6, 18, 54, … is a geometric sequence.

golden ratio The ratio 1:1.618 approximately.

golden rectangle A rectangle that is pleasing to the eye. The ratio of the length to the width is approximately 1.6:1.

greatest possible error (of a measuring tool) One half the precision of the tool.

heptagon A polygon with seven sides.

hexagon A polygon with six sides.

histogram A graph, like a bar graph, but with no space between the bars, used to show continuous data that can be organized into class intervals.

hypotenuse The longest side of a right triangle.

icosahedron A regular polyhedron with twenty faces.

increasing function A function for which the first differences increase.

independent variable In a relation, the variable whose value determines that of the dependent variable. On a coordinate grid, the values of the independent variable are on the horizontal axis.

In $d = 4.9t^2$, t is the independent variable.

infer from data To make a conclusion based on a relationship identified between variables in a set of data.

inferential statistics The outcomes of data analysis on a sample are used to make generalizations about the population.

integer A number in the sequence … −3, −2, −1, 0, 1, 2, 3, ….

intercept The distance from the origin of the Cartesian coordinate plane to the point at which a line or curve crosses a given axis.

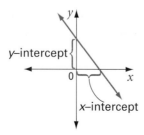

interest The amount earned on an investment or savings alternative, or the cost of borrowing money.

interest rate The rate, as a percent, at which an investment or savings alternative increases in value, or the cost of borrowing money, expressed as a percent.

interior angle An angle that is inside a polygon.

interpolate Estimate values for a relation that lie between given data points.

isometric drawing A two-dimensional representation, also called a corner view, of a three-dimensional object. It is often drawn on isometric dot paper; for example, a unit cube is shown.

isosceles triangle A triangle with exactly two equal sides.

like terms Terms that have exactly the same variables(s) raised to exactly the same exponent(s).
$3x^2$, $-x^2$, and $-7x^2$ are like terms.

line of best fit The line that best describes the distribution of points in a scatter plot. The line passes through, or close to, as many of the data points as possible.

line segment The part of a line that joins two points.

linear equation An equation that represents the relationship between two variables that have a linear relationship and form a straight line when graphed.

linear function A relation between two variables such that each value of the first variable has only one corresponding value. The relation can be expressed in the form $y = mx + b$, where m and b are constants.

linear growth Growth represented by a linear equation and a straight-line graph. Arithmetic sequences and simple interest show linear growth.

linear regression A method for determining the linear equation that best fits the distribution of points on a scatter plot.

linear relation A relationship between two variables that forms a straight line when graphed.

mathematical model A description of a real situation using a diagram, a graph, a table of values, an equation, a formula, a physical model, or a computer model.

mathematical modelling The process of describing a real situation in a mathematical form.

mean The sum of a set of values divided by the number of values.

measure of central tendency A value that can represent a set of data; for example, the mean, the median, or the mode.

measure of dispersion A value that quantifies the spread of a set of data; for example, the range, the variation, or the standard deviation.

median (data) The middle value when data are arranged in numerical order. If there is an even number of pieces of data, then the median is the average of the two middle values.

modal class interval In a histogram, the class interval in which most data fall.

mode The number (or attribute) that occurs most frequently in a set of data. A set of data can have more than one mode, or no mode.

net A pattern for constructing a three-dimensional shape.

non-linear relation A relationship between two variables that does not form a straight line when graphed.

normal curve A curve that is symmetrical and bell-shaped.

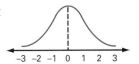

normal distribution A frequency distribution that can be represented by the normal curve. Data such as heights of men and life expectancy of light bulbs have a normal distribution.

oblique triangle A triangle that is not right-angled.

obtuse angle An angle that measures more than 90°, but less than 180°.

obtuse triangle A triangle containing one obtuse angle.

octagon A polygon with eight sides.

octahedron A regular polyhedron with eight faces.

opposite side In $\triangle ABC$, AB is the side opposite $\angle C$.

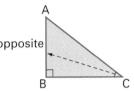

ordered pair A pair of numbers, such as (3, 8), used to locate a point on a graph.

origin The point of intersection of the x-axis and the y-axis on a coordinate grid. It is described by the ordered pair (0, 0).

orthographic drawings A two-dimensional representation of a three-dimensional object, showing how it appears from the top, front, and right side.

outlier A data point that does not conform to the pattern of the other data.

parabola A U-shaped curve that is symmetric about a line of symmetry.

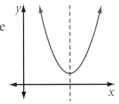

parallelogram A quadrilateral with two pairs of opposite sides that are parallel.

pattern A net that has been separated into individual shapes, for constructing a three-dimensional object.

pentagon A polygon with five sides.

percent A number that represents a fraction or ratio with a denominator of 100.

$\dfrac{34}{100}$ as a percent is 34%.

percentile rank If a number, X, is in the nth percentile of a set of numerical data, then n percent of the data lie below X.

perimeter The distance around a two-dimensional shape.

perspective drawing A two-dimensional representation of a three-dimensional object, using vanishing points to give the impression of distance and depth.

point of intersection The point that is common to two non-parallel lines.

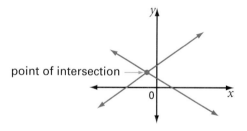

polygon A two-dimensional closed shape whose sides are line segments.

polyhedron A three-dimensional object with faces that are polygons.

population In statistics, the total number of individuals or items under consideration in a survey, from which a sample may be chosen.

power An abbreviation for repeated multiplication.

The power 6^3 means $6 \times 6 \times 6$.

primary source of data Use of a survey or an experiment to collect your own data.

prism A three-dimensional shape with two parallel, congruent polygonal faces. The prism is named according to the shape of these two faces; for example, triangular prism.

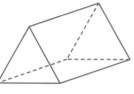

proportion An equation that states that two ratios are equal.

$\dfrac{3}{5} = \dfrac{x}{80}$ is a proportion.

pyramid A polyhedron with one base in the shape of a polygon and the same number of lateral triangular faces as there are sides in the base.

Pythagorean theorem In a right triangle, the square of the length of the hypotenuse is equal to the sum of the squares of the lengths of the other two sides.

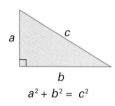

$a^2 + b^2 = c^2$

quadrant One the four regions formed by the intersection of the x-axis and the y-axis.

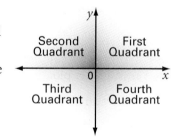

quadratic equation An equation that can be written in the form $ax^2 + bx + c = 0$, where a, b, and c are real numbers and $a \neq 0$.

quadratic formula The zeros, or solutions, of a quadratic equation of the form $ax^2 + bx + c = 0$, where $a \neq 0$, are given by

$$x = \dfrac{-b \pm \sqrt{b^2 - 4ac}}{2a}$$

quadratic function A relationship between two variables defined by an equation of the form $y = ax^2 + bx + c$, where a, b, and c are real numbers and $a \neq 0$. Its graph is a parabola.

quadrilateral A polygon with four sides.

radical sign The symbol $\sqrt{}$.

random number A number chosen from a set of numbers in such a way that each number has an equally-likely chance of being selected.

random sample A sample in which every member of a population has an equal chance of being selected.

range (data) The difference between the greatest and the least values in a set of data.

range of a relation The set of all second coordinates of the ordered pairs of a relation.

rate A special type of ratio that compares two quantities with different units.

   8.4 L/100 km is a rate.

ratio A comparison of quantities with the same unit.

   3 cans of water to 1 can of juice is 3:1.

regression A method for determining the equation of a curve or line that best fits the distribution of points on a scatter plot.

regular polygon A polygon in which all sides are equal and all interior angles are equal.

regular polyhedron A polyhedron in which all faces are congruent regular polygons and each vertex touches the same number of polygons. Also known as a Platonic solid.

relation A relationship between variables that can be represented by a table of values, a graph, or an equation.

response bias Bias that arises from the phrasing of a survey question so that it tends to provoke a particular type of answer.

rhomboid A polyhedron in which each face is a rhombus.

right angle An angle that measures 90°.

right triangle A triangle that contains a 90° angle.

rise The vertical distance between two points.

roots The solutions of an equation.

run The horizontal distance between two points.

sample A small group chosen from a population and examined in order to analyse the population.

sampling technique A process for collecting a sample of data.

scale diagram A drawing that is an enlargement of a small object, or a reduction of a large object.

scalene triangle A triangle with no sides equal.

scatter plot A graph showing two-variable data by means of points plotted on a coordinate grid.

scientific notation A method of writing large or small numbers that contain many zeros. The number is expressed in the form $a \times 10^n$, where a is greater than or equal to 1, but less than 10, and n is an integer.

secondary source of data Information that has been collected by someone else.

sector A part of a circle bounded by two radii and an arc of the circle.

selection bias Bias that occurs due to the type of sample selected.

sequence An ordered list of numbers or terms.

similar figures Figures having corresponding angles equal and corresponding lengths proportional.

simple interest Interest calculated only on the original principal using the simple interest formula $I = Prt$.

simplest form The form of a fraction or ratio that has no common factors.

simplest form of an algebraic expression An expression that has no like terms.

$5x + 3 - x + 2$ in simplest form is $4x + 5$.

sine law The relationship between the lengths of the sides and their opposite angles in any triangle.

$$\frac{a}{\sin A} = \frac{b}{\sin B} = \frac{c}{\sin C}$$

sine ratio In a right triangle, for $\angle A$, the ratio of the length of the side opposite $\angle A$ and the length of the hypotenuse.

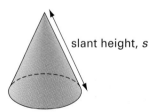

$$\sin A = \frac{\text{opposite}}{\text{hypotenuse}}$$

skewed distribution A distribution of two-variable data that is lopsided, or assymetrical.

slant height (of a cone) The distance from the vertex to a point on the edge of the base.

slant height, s

slant height (of a pyramid) The height of each triangular face.

slope A measure of the steepness of a line.

The slope, m, of a line containing points $P(x_1, y_1)$ and $Q(x_2, y_2)$ is

$$m = \frac{\text{vertical change}}{\text{horizontal change}} \text{ or } \frac{\text{rise}}{\text{run}}$$

$$= \frac{y_2 - y_1}{x_2 - x_1}, x_2 \neq x_1$$

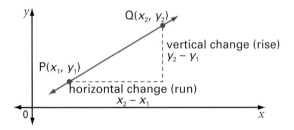

slope and y-intercept form of a linear equation A linear equation written in the form $y = mx + b$, where m is the slope and b is the y-intercept.

solve To find the value of a variable in an equation.

When $2x = 16$ is solved, $x = 4$.

standard deviation A measure of the spread of a set of data, found by taking the positive square root of the variance.

standard form of a linear equation A linear equation written in the form $Ax + By + C = 0$, where A, B, and C are integers, A and B are not both zero, and x and y are real numbers.

standard form of a quadratic function A quadratic function written in the form $y = ax^2 + bx + c$, where $a \neq 0$.

static data Facts presented in a medium that cannot be changed, such as an almanac.

stratified sampling A method of choosing a sample in which the population is divided into groups, or strata, from which random samples are selected.

substitution method A method of solving a system of equations by solving one equation for one variable, then substituting that value into the other equation.

surface area The number of square units needed to cover the surface of a three-dimensional shape.

table of values A table used to record and relate the coordinates of ordered pairs in a relation.

y = 2x + 3	
x	**y**
−1	1
0	3
1	5
2	7

tangent ratio In a right triangle, for ∠A, the ratio of the length of the side opposite ∠A and the length of the adjacent side.

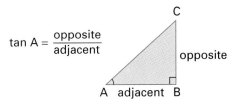

term A number or a variable, or the product or quotient of numbers and variables.

$3x^2 - 5$ has two terms: $3x^2$ and −5.

tetrahedron A regular polyhedron with four faces.

trapezoid A quadrilateral with one pair of parallel sides.

trinomial A polynomial with three terms.

unlike terms Terms that have different variables, or different powers of the same variable.

$2x^2$, −x, and 5y are unlike terms.

variance A measure of the spread of a set of data found by calculating the average squared difference of each value from the mean.

vertex A point at which two sides of a polygon meet.

vertex form of a quadratic function The equation is expressed in the form $y = a(x - h)^2 + k$, where a, h, and k are real numbers and $a \neq 0$. The vertex of the parabola is at (h, k).

vertex of a parabola The point where the axis of symmetry of the parabola intersects it.

volume The amount of space that an object occupies, measured in cubic units.

x-axis The horizontal number line in the Cartesian coordinate system.

x-coordinate The first number in the ordered pair describing a point.

For the point P(2, 8), 2 is the x-coordinate.

x-intercept The distance of the point from the origin where a line or curve crosses the x-axis.

y-axis The vertical number line in the Cartesian coordinate system.

y-coordinate The second number in the ordered pair describing a point.

For the point P(2, 8), 8 is the y-coordinate.

y-intercept The distance of the point from the origin where a line or curve crosses the y-axis.

zero of a function A value of x (or the independent variable) that results in a y-value of zero.

ndex

Credits

Photo Credits

iv both PhotoDisc; **v-vi** Eyewire; **vii** PhotoDisc;
viii-1 © Stephen Johnson/Getty Images/Stone; **4** © Mark
Wagner/Getty Images/Stone; **10** *www.comstock.com*; **15** NASA;
21 *www.comstock.com*; **28** Ivy Images; **33** Ian Crysler;
34 Artville; **40** © Will & Deni McIntyre/Getty Images/Stone;
42 Eyewire; **43** Ivy Images; **49** Courtesy of Krista Alexander;
50-51 CORBIS RF/MAGMA PHOTO; **53** Eyewire;
56 PhotoDisc; **58** Mitutoyo Canada Inc.; **67** Eyewire;
74 PhotoDisc; **87** © B. Busco/Getty Images/The Image Bank;
90 A-D Mitutoyo Canada Inc., **E** Dick Hemingway;
98 PhotoDisc; **99** Eyewire; **103 both** Mitutoyo Canada Inc.;
104 top © Stephen Johnson/Getty Images/Stone, **bottom**
Joseph Pearce/Valan Photos; **105 both** Mitutoyo Canada Inc.;
106-107 Copyright Art Resource, NY; **109** PhotoDisc;
110 *www.confederationbridge.com*; **111 top** CORBIS RF/
MAGMA PHOTO, **centre** Joseph Pearce/Valan Photos,
bottom Galleria del' Academia, Venice, Italy/Bridgeman Art
Library; **112 top** © Harald Sund/Getty Images/The Image
Bank, **bottom** Ian Crysler; **113** Copyright Art Resource, NY;
114 both Maya Real-Time Author (Maya RTA) Copyright
© 2000-2001 Alias|Wavefront, a division of Silicon Graphics
Limited. All rights reserved.; **115-116** PhotoDisc; **117** Artville;
123 © Jerry Nelson/Illustration Works, Inc; **131-132** Dick
Hemingway; **134** The Fine Art Society, London, UK/
Bridgeman Art Library; **138** Dakota Jackson Inc.;
144 Ian Crysler; **145** PhotoDisc; **146** The Granger Collection;
147 PhotoDisc; **154-155** Dick Hemingway; **157** PhotoDisc;
158 Eyewire; **163** CORBIS RF/MAGMA PHOTO; **168** Eyewire;
178 PhotoDisc; **185** Halifax Chronicle-Herald/Jim Krochak/CP
Picture Archive; **190** Eyewire; **198-199** PhotoDisc;
204-205 Krasemann/Photo Researchers Inc.; **207** Eyewire;
208 Andre Forget/CP Picture Archive; **212** Dick Hemingway;
220 CORBIS RF/MAGMA PHOTO; **229** PhotoDisc;
239 PhotoDisc; **242** Courtesy of Tess Miller; **243** PhotoDisc;
246 Gilles Delisle/Valan Photos; **247** Eyewire; **254-255** © Roger
Tully/Getty Images/Stone; **257** © Terje Rakke/Getty Images/
The Image Bank; **258** PhotoDisc; **267 All** Copyright © 2001,
Province of British Columbia, All rights Reserved:

left *http://www.hlth.gov.bc.ca/tobacrs/teen/index.html*,
centre *http://www.hlth.gov.bc.ca/tobacrs/teen/teen3.html*,
right *http://www.hlth.gov.bc.ca/tobacrs/teen/teen3.html#3.2*;
271 PhotoDisc; **274** Eyewire; **286** Artville; **292** PhotoDisc;
293 © Terje Rakke/Getty Images/The Image Bank;
304-305 Val Wilkinson/Valan Photos; **307** Michael Keller/
Firstlight.ca; **308** PhotoDisc; **315** PhotoDisc; **319** Eric Sanford/
Firstlight.ca; **323** Courtesy of Maria DeCambra; **328** © Jody
Dole/Getty Images/The Image Bank; **333** Courtesy of Maria
DeCambra; **340** © Kaz Mori/Getty Images/The Image Bank;
341 Michael Keller/Firstlight.ca; **343** Courtesy of Maria
DeCambra; **345** Grant Black/Firstlight.ca; **348-349** Dick
Hemingway; **351** *www.comstock.com*; **352** PhotoDisc;
362 top David Nunuk/Firstlight.ca; **bottom** Dick Hemingway;
374 © Farmhouse Productions/Getty Images/The Image Bank;
380 John Fowler/Valan Photos; **389** Eyewire; **393** CORBIS
RF/MAGMA PHOTO; **394** Eyewire; **395** *www.comstock.com*

Text Credits

208 Statistics Canada's Internet Site, *http://www.statcan.ca*,
November 7, 2001; **209-210** Statistics Canada's Internet Site,
http://www.statcan.ca/english/edu/index.htm, November 7,
2001; **228** Statistics Canada, #H510821, matrix 1089;
230 Statistics Canada's Internet Site, *http://estat.statcan.ca/
cgi~win/CNSMCGI.EXE*, table 326-0012, November 7, 2001;
237 Statistics Canada Catalogue No 11-509-XPE; **242** Statistics
Canada, Canada Year Book 2001, p. 64, #160200-XKE;
243 Statistics Canada, Canada Year Book 2001, p. 69;
245 Statistics Canada, Canada Year Book 2001, Population
data by year 1861-1996; **250** Statistics Canada, Canada Year
Book 2001, p. 130; **253** Statistics Canada, Canada Year Book
2001, p. 139; **256** Statistics Canada's Internet Site,
http://estat.statcan.ca/cgi~win/CNSMCGI.EXE, table 001-0008,
November 7, 2001; **258** Adapted from: Statistics Canada's
Internet Site, *http://estat.statcan.ca/cgi~win/CNSMCGI.EXE*,
table 202-0101, November 7, 2001; **260** Maclean's Guide to
Canadian Universities and Colleges 2001; **264** Statistics
Canada's Internet Site, *http://estat.statcan.ca/cgi~win/
CNSMCGI.EXE*, Population Percent Change 1991–1996, by
Provinces and Territories, November 7, 2001;

266 top Statistics Canada's Internet Site, *http://estat.statcan.ca/ cgi~win/CNSMCGI.EXE*, table 477-006, November 7, 2001, **centre** Statistics Canada's Internet Site, *http://estat.statcan.ca/ cgi~win/CNSMCGI.EXE*, table 252-0002, November 7, 2001, **bottom** Toronto Star Staff, Ministry of Energy, Science and Technology; **267** *www.movieweb.com/movie/alltime.html*; **268** Via Rail Canada; **269** Data from The Trans Canada Trail Progress Report, *http://www.tctrail.ca/regional.htm*; **271** Statistics Canada's Internet Site, *http://estat.statcan.ca/ cgi~win/CNSMCGI.EXE*, table 326-0002, November 7, 2001.; **272 top** Statistics Canada's Internet Site, *http://estat.statcan.ca/ cgi~win/CNSMCGI.EXE*, table 326-0002, November 7, 2001, **bottom** Adapted from: Statistics Canada's Internet Site, *http://estat.statcan.ca/cgi~win/CNSMCGI.EXE*, Statscan CPI for January by year 1980-2000, November 7, 2001; **276** Statistics Canada's Internet Site, *http://estat.statcan.ca/ cgi~win/CNSMCGI.EXE*, table 326-0009, November 7, 2001; **277** Statistics Canada's Internet Site, *http://estat.statcan.ca/ cgi~win/CNSMCGI.EXE*, table 326-0002, November 7, 2001; **279** Statistics Canada's Internet Site, *http://estat.statcan.ca/ cgi~win/CNSMCGI.EXE*, Instructions lead to E-STAT 051-0001, population data, November 7, 2001; **281** Statistics Canada's Internet Site, *http://estat.statcan.ca/cgi~win/ CNSMCGI.EXE*, Average weekly TV viewing time Canada, 1991-2000, from The Daily, October 23, 2001; **289** Data from The Western Canadian Lottery Corporation; **291 and 293** *http://proicehockey.about.com/library/weekly/aa111799.htm*; **294** CAA Member Opinion 2000; **296** This Week [Whitby], Tuesday Edition, June 26, 2001 and Maclean's Magazine, 3rd Annual Ranking of "Where We Get the Best Healthcare"; **297** Ipsos-Reid; **298** Adapted from: Statistics Canada's Internet Site, *http://estat.statcan.ca/cgi~win/CNSMCGI.EXE*, Full-time enrolment in post-secondary education in Ontario, by year 1987–1996, November 7, 2001; **299** Via Rail Canada; **454 top** Statistics Canada's Internet Site, *http://www.statcan.ca/ english/Pgdb/People/Population/demo23b.htm*, Dec 21, 2001, **bottom** Statistics Canada's Internet Site, *http://www.statcan.ca/ cgi-win/CNSMCGI.EXE*, December 21, 2001; **455 top** Statistics Canada's Internet Site, *http://www.statcan.ca/english/Pgdb/ People/Education/educ18.htm*, December 21, 2001, **bottom** Statistics Canada's Internet Site, *http://www.statcan.ca/ english/Pgdb/People/Labour/labor01b.htm*, December 21, 2001; **456** Statistics Canada's Internet Site, *http://www.statcan.ca/ Daily/English/011023/d011023a.htm*, December 21, 2001;

458 Statistics Canada, Catalogue no. 11-509-XPE; **459** WDO Fact Sheet 1999, Municipal 3Rs in Ontario, *http://www.rco.on.ca/factsheet/3Rs1999.pdf*; **464** *http://www.OntarioPark.com*, © Queen's Printer for Ontario, 2002; **465** *http://www.ilec.or.jp*

Statistics Canada information is used with the permission of the Minister of Industry, as Minister responsible for Statistics Canada. Information on the availability of the wide range of data from Statistics Canada can be obtained from Statistics Canada's Regional Offices, its world-wide web site at *http://www.statcan.ca*, and its toll-free access number 1-800-263-1136.

Technical Art and Illustration

Tom Dart, Greg Duhaney, Alana Lai, Claire Milne, and Kim Hutchison of First Folio Resource Group, Inc.